Biodiversity: Culture, Conservation, and Ecodevelopment

Biodiversity

Culture, Conservation, and Ecodevelopment

EDITED BY

Margery L. Oldfield
and Janis B. Alcorn

Foreword by E. O. Wilson

1991

Westview Press
BOULDER • SAN FRANCISCO • OXFORD

This Westview softcover edition is printed on acid-free paper and bound in library-quality, coated covers that carry the highest rating of the National Association of State Textbook Administrators, in consultation with the Association of American Publishers and the Book Manufacturers' Institute.

Published in 1991 in the United States of America by Westview Press, Inc., 5500 Central Avenue, Boulder, Colorado 80301-2847, and in the United Kingdom by Westview Press, 36 Lonsdale Road, Summertown, Oxford OX2 7EW

Library of Congress Cataloging-in-Publication Data
Biodiversity : culture, conservation, and ecodevelopment /edited by
Margery L. Oldfield and Janis B. Alcorn.
p. cm.
Includes bibliographical references.
ISBN 0-8133-7680-7
1. Germplasm resources. 2. Biological diversity conservation.
3. Agricultural conservation. 4. Agricultural ecology.
I. Oldfield, Margery L. II. Alcorn, Janis B.
S494.3.B56 1991
631.5'23—dc20 91-31102
CIP

Printed and bound in the United States of America

The paper used in this publication meets the requirements of the American National Standard for Permanence of Paper for Printed Library Materials Z39.48-1984.

10 9 8 7 6 5 4 3 2 1

Dedicated to the memory of

Deborah Rabinowitz,

for her energetic pursuit of
novel ideas that placed her on the
leading edge of conservation biology

Contents

Tables

Figures

About the Contributors

Janis B. Alcorn has been actively involved in environmental research and international development for 20 years, particularly in Asia and Latin America. At present, she is a Program Officer for the Biodiversity Support Program, an A.I.D. funded, joint venture of World Wildlife Fund, The Nature Conservancy, and World Resources Institute, housed at World Wildlife Fund in Washington, D.C. Her publications include *Huastec Mayan Ethnobotany* and a forthcoming book from Columbia University Press entitled *Ethnobiology: Culture & Biodiversity*.

Dorothea Bedigian has broad-based interests in biodiversity conservation and economic botany, although she specializes in origins of agriculture, plant domestication processes, and drylands agriculture. Her principal geographic areas of interest have been Africa and Asia, and she is currently a Research Associate in the Departments of Biology and Anthropology at Washington University in St. Louis.

J. Artie Browning participated in the release of 14 modern cultivars used to control oat crown rust and sought holistic management strategies to maximize natural control processes for oat diseases for 28 years at Iowa State University. In 1981, he became Head of the Department of Plant Pathology and Microbiology at Texas A&M University, and is a past president of the American Phytopathological Society. During 1990-91, he was a Fulbright Scholar in the "Institute for Cereal Crops Improvement Through the Utilization of Gene Pools of Their Wild Relatives" at Tel Aviv University in Israel.

Mac Chapin is Projects Director in the Arlington, Virginia office of Culture Survival, Inc. He has been active in Latin American rural development for over 25 years, including work as a social science analyst for U.S. Agency for International Development and as an Interamerican Foundation Country Representative. He has conducted anthropological research on Kuna medical systems, indigenous natural resource management, and the role of grassroots organizations in ecodevelopment. His writings frequently appear in *Cultural Survival Quarterly*.

Jason Clay is an anthropologist who has served as research director at Cultural Survival, Inc. (CSI) in Cambridge, Massachusetts and founding editor of *Cultural Survival Quarterly*. The author of two books on the Ethiopian famine, a book on indigenous peoples in tropical forests, and more than 100 articles on human rights, indigenous peoples, peasant agriculture, and "development" issues, he is currently the director of the innovative CSI marketing program aimed at demonstrating the employment and income-generating potential of healthy tropical rain forest ecosystems.

Raymond F. Dasmann contributed to the development of the UNESCO Man and the Biosphere Programme and has worked for UNESCO and international conservation agencies such as the IUCN and the World Wildlife Fund for more than two decades. The author of *Ecological Principles for Economic Development* and *African Game Ranching*, he is now Professor Emeritus of Environmental Studies at the University of California at Santa Cruz.

Hank Fitzhugh is an animal breeder and livestock production systems specialist and the author of *Hair Sheep–A Genetic Resource for the Tropics*. While at Winrock International from 1975-90, he was responsible for livestock development programs in Africa and Latin America. Currently, he is Deputy Director General (Research) at the International Livestock Centre for Africa and chair of the Animal Genetic Resources Committee for the NAS/NRC study on "Managing Global Genetic Resources." **Ann E. Wilhelm** is an animal scientist with a specialization in animal breeding and biometry. She has studied animal selection in a variety of production environments and is currently operating a small farm in Grandy, Connecticut.

Stephen R. Gliessman is Director of the Agroecology Program at the University of California at Santa Cruz, and from 1977-80 he was Chair of the Department of Ecology, Colegio Superior de Agricultural Tropical in Cardenas, Tabasco, Mexico. He has conducted research on tropical agroecosystems for over 20 years and has authored numerous publications on the ecology of agricultural systems.

William P. Gregg, Jr. initiated his career as an ecologist in the Washington D.C. office of the National Park Service (NPS) 20 years ago when he served as a consultant, preparing management plans and related environmental assessments for national parks. Since 1980, he has served as NPS Man and the Biosphere Program Coordinator and has been instrumental in developing the U.S. Biosphere Reserve network. Recently, he has specialized in regional and global influences affecting natural ecosystems, and has coordinated park-based research programs on acidic deposition and global change.

Raymond Hames has done extensive research among the Yanomamö and Ye'kwana Indians of the Venezuelan Amazon since 1975. His major interests are behavioral ecology and time allocation. Some of his recent publications are "Time, efficiency and fitness in the Amazonian protein quest" and "Garden labor exchange among the Ye'kwana." He is Chair of the Department of Anthropology at the University of Nebraska-Lincoln.

Molly Bower Kux has served as an international environmental officer for the Department of State and the U.S. Agency for International Development since 1974. She has designed and managed technical assistance programs for developing countries in the areas of environmental planning and conservation of biological diversity. At present, she is Environmental Coordinator for the Bureau for Asia and Private Enterprise at the U.S. Agency for International Development.

Gary P. Nabhan is Assistant Director of the Desert Botanical Garden in Phoenix in charge of field conservation programs, and cofounder and board chairman of Native Seeds/SEARCH. **Donna House**, a member of the Navajo Nation, is a biologist working for The Nature Conservancy. **Humberto Suzan A.** and **Guadalupe Malda** are founding members of the Asociación Ecológica Tamaulipeca, researchers associated with the Rancho El Cielo Biosphere Reserve, and graduate students in botany at Arizona State University. **Wendy Hodgson** is a Research Botanist and Herbarium Curator at the Desert Botanical Garden. **Luis Hernandez S.** is former director of the Instituto de Ecología y Alimentos of the Universidad Autónoma de Tamaulipas, and a graduate student in botany at the University of Texas.

Margery L. Oldfield has conducted research on a variety of biodiversity conservation issues and has served as a biodiversity consultant for national and international organizations for more than 15 years. The author of *The Value of Conserving Genetic Resources*, she is currently an Assistant Professor in the Department of Biology at Trinity University in San Antonio, Texas. Recently, her research has focused on the role of animal behavior in extinction processes, and population dynamics of threatened mammal species.

Victor M. Toledo is a professor in the Centro de Ecología at the Universidad del Nacional Autónoma de Mexico in Mexico City. He has been actively involved in ecological research and rural development for two decades. His research has been focused on floristics, biogeography, biological conservation and ethnoecology of Mexico and Latin America. He has authored more than 70 publications, including four books. He is currently coordinating a research project on the economic potential of tropical rain forests, and is preparing a book on peasant appropriation of nature.

Garrison Wilkes has had a long standing interest in the evolution of cultivated plants under domestication, in particular, maize and its wild relatives. More than four decades of ethnobotanical studies and plant collecting in Meso-America and South Asia have enhanced his concern for plant genetic resources conservation by both *ex situ* management of gene banks and *in situ* conservation of landrace crops along with their surrounding native wild plants. Currently, he is a Professor in the Biology Department at the University of Massachusetts.

Foreword

E.O. Wilson

If all of humanity consisted of several million individuals dedicated to the maintenance of a stable population, we could afford to be what Raymond Dasmann calls "biosphere people." That is, humanity might draw down on the resources of the whole planet with a certain profligacy, confident that neither the climate nor the remainder of living organisms would suffer from our presence. The Earth, in other words, would be Eden, giving us life and absorbing our waste without complaint.

The truth is, however, that Earth's human population is in excess of five billion heading toward ten. The urban masses and their modern consumer culture are ecologically illiterate. Having supersaturated the land environment, we now coopt about 40% of the free energy captured from the sun by green plants. We are rapidly depleting the energy reserves in fossil fuels, changing the atmosphere into a new, hostile mix of gases, extinguishing a quarter of the other species on Earth, and setting the stage for our own decline and possible demise. In short, we are the first species in the history of life to go out of control on a global scale. Recognition of this bitter truth may turn the next hundred years into the "century of the environment."

We still have time to abate population growth and put our house in order. The multifarious essays in the present collection remind us that the adjustment will not be easy. It cannot come as the result of some technical fix such as a new energy source or a magically productive new crop. Rather, it will be achieved modestly as the summation of a large number of little steps, local in scale and tuned to biological exigencies and opportunity. It requires understanding that particular species and genetic strains of crops are optimally suited for particular places, provided they are raised in carefully constructed settings and by certain methods. Wild species are to be treasured as sources of genetic material evolved over thousands or millions of years in ways exactly tuned to individual habitats. Their Mendelian variety offers resiliency in the face of change, guarantees their growth and reproduction, and provides us with sustained production of goods and services.

The New Environmentalism of recent birth, adding the goals of economic progress to ecology, has led to a new emphasis on biodiversity. Ecodevelopment involves the promotion of methods to maintain and use biodiversity for the benefit of humanity. Western science will not provide all of the answers needed to achieve ecodevelopment. The knowledge of native peoples, evolved out of a kind of cultural Darwinism in which the bottom line was raw survival, is replete with practical leads on useful species and sustainable techniques of cultivation and ecosystem management.

The essays in this interdisciplinary book make the point that valuable, locale-specific knowledge is used by diverse, rural peoples who have relied for centuries upon the maintenance of biodiversity. On a global basis, human cultural diversity is associated with the remaining concentrations of biodiversity. Both cultural diversity and biological diversity are endangered. Modern cultures are undercutting traditional cultures, and modern knowledge is replacing traditional knowledge. Rural people are often classified as impoverished, ignorant masses, or treated as invisible. In the final analysis, however, they hold the key to successful biodiversity conservation. This book emphasizes the urgent need for detailed dialogue and collaboration between those of us who plan the future of biodiversity from afar and the culturally diverse rural people who know what it means to live with biodiversity for the long term.

The focus of the essays in *Biodiversity: Culture, Conservation, and Ecodevelopment* should be increasingly the focus of science and government policy, as well as the lodestar of the altruistic impulses of private businesses.

Acknowledgments

The editors are indebted to the Department of Wildlife and Fisheries Sciences of Texas A&M University for the financial and technical assistance that enabled preparation of the camera-ready copy of this book. We are especially grateful to Ms. Vivian Marie Gonzales, who helped Margery Oldfield to complete the book manuscript. M.L. Oldfield was supported by a NATO-NSF Postdoctoral Fellowship in Science. J.B. Alcorn received a travel grant from the Tinker Foundation through the Mesoamerican Ecology Institute at Tulane University; she also thanks her supervisors at the American Association for the Advancement of Science, U.S. Agency for International Development, and the Biodiversity Support Program at World Wildlife Fund for supporting her work on this book in 1988 and 1991. She also acknowledges the support from her parents, Bruce and Stella Bristol, who covered production costs for preparation of the figures.

This book began to take shape at the 64th Annual Southwestern and Rocky Mountain Division of the American Association for the Advancement of Science (AAAS-SWARM) meeting in Wichita, Kansas. AAAS-SWARM funded the symposium that brought most of the contributors together on March 30-31, 1988. The editors wish to acknowledge the many people who contributed to the conception, nurturing, and birth of this volume, especially: Marshall Alcorn, Michelle Balcomb, Sameer Bhatnager, Peter Brosius, John Browder, Jeff Campbell, Chris Franquemont, Cándido Hernández V., Molly Kux, Owen Lynch, Witoon Permpongcharaen, the late Deborah Rabinowitz, and Kirk Talbott. Janis Alcorn also wishes to acknowledge the people of Matala and Talmadgi, villages in Bidar District, Karnataka, India, who first taught her the meaning of biodiversity and its importance to rural economies during a prolonged drought and subsequent famine in 1970-72.

Chapter 3 was reprinted from *BioScience* (Vol. 37) with the permission of the American Institute of Biological Sciences. Other parts of the book are original, previously unpublished works. We thank the authors for their patience with our persistence, and for their willingness to brave the uncharted waters of interdisciplinary dialogue.

Biodiversity: Culture, Conservation, and Ecodevelopment

PART ONE

Biodiversity and Culture

Introduction

Margery L. Oldfield

> It is the story of all life that is holy and is good to tell, and of us two-leggeds sharing in it with the four-leggeds and the wings of the air and all green things; for these are children of one mother and their father is one Spirit.—*Black Elk*, as quoted by Neihardt (1961:1)

> (T)he cultural arrogance of the dominant society usually assumes that the subordinate society has nothing to offer, no contribution to make to it. The irony of this is that in this very cultural arrogance the dominant society sows the seeds of its own demise. (Appell 1975:1)

> It has almost been unanimously assumed . . . that the extinction of tribal life is natural, inevitable, and beneficial. Those who oppose these views are accused of romanticizing natives, and of attempting to deny them the benefits of civilization. (Bodley 1970:10)

Ever since the first homonids appeared on the face of the earth, our ancestors have existed as an integral part of nature. Tribal and aboriginal peoples have always been dependent on natural ecosystems and wild biota, and their oral histories indicate their cultural appreciation of the unity of all life. As *Standing Bear* related in an ancient prayer of the Lakota Indians: "All over the earth faces of all living things are alike. Mother earth has turned these faces out of the earth with tenderness" (DeMallie 1984:288). The same sentiments were recently echoed by the Kuna Indian delegate at the Fourth World Wilderness Congress, "For the Kuna culture, the land is our mother and all living things that live on her are brothers" (see Gregg, Ch. 13, this volume).

Since the Pleistocene, however, humans have constituted an increasingly powerful ecological component within natural ecosystems and have wielded more intense selection pressures on wild biota. When *Homo sapiens* began domesticating useful plants and animals, early agriculturalists increasingly transformed natural communities and ecosystems into man-modified landscapes. Even though our agricultural ancestors lived more apart from nature, they have continued to depend on biological resources obtained from wilderness areas and on the ecological and evolutionary services of nature to sustain their man-modified ecosystems and traditional life styles. In regions of traditional agriculture, indigenous cultures continue to influence the evolution of their domesticates as well as wild and weedy species (Harlan 1975). As a consequence, they have either purposefully or inadvertently conserved populations of both wild and domesticated species. For example, as Gary Nabhan (1987:31) has observed, "some cultural traditions have . . . been effective in conserving evolving gene pools *in situ* by acknowledging their own dependence on wild turkeys, wild corn and wild sunflowers."

Today, traditional cultures and biodiversity are both threatened by the continuing expansion of modern, industrialized societies. Modern cultures exist not as "a part of nature" but rather "apart from nature" due to their reliance on a rapidly expanding global economy (see Alcorn's Epilogue, this volume). As David Mayberry-Lewis (1977:56) has observed, the extinction of "small societies" is a "process that is as old as humankind itself" that has recently been hastened by the advent of Western technology: "The threat, either implicit or explicit, that they must die so that we may live is something we normally conceal from ourselves under comfortable phrases like 'the social costs of development,' or 'the price of progress.'" Moreover, the fate of vanishing human cultures is inextricably intertwined with the problem of vanishing wildlands and the accelerating loss of biodiversity–a situation that E.O. Wilson (1985) has termed the "biological diversity crisis." More than a decade ago, Altschul (1977:104) noted that "Many (plant) species are in danger of becoming extinct–as are the peoples and the folklore that harbor information about them–as advanced societies expand and disturb natural environments." Indeed, the "modern episode" of species extinctions "exceeds anything in the geological past" because current reductions of biodiversity are "destined to approach that of the great natural catastrophes at the end of the Paleozoic and Mesozoic eras–in other words, the most extreme in the past 65 million years" (Wilson 1988:11-12).

This dire global situation has led scientists from many different disciplines and perspectives to ask important questions about current and future efforts to conserve both biological and cultural diversity. How can humanity best conserve the earth's remaining wilderness areas, with their diverse ecosystems and communities of plants, animals, and microbes intact, while still meeting the needs of burgeoning human populations? To what extent is the fate of global biodiversity dependent upon the fate of vanishing aboriginal, tribal, or indigenous cultures? Should indigenous people be removed from newly established parks, nature reserves, and wilderness areas, even if they have resided in natural or man-modified ecosystems within the park or reserve boundaries for hundreds or thousands of years? Will parks and nature reserves, as envisioned by First World models, ever be successful in developing countries if indigenous people are denied access to the biodiversity protected within their national boundaries? Should particular indigenous cultures have access to the biological or other natural resources conserved within these parks or reserves, while most other people are denied access? Who should make these important decisions about conserving our global heritage of biological diversity? These are difficult questions that need to be addressed, and hopefully answered, within the next decade, particularly since "cultural diversity equates, at many important points, with ecological diversity . . . (and) is a basic and valuable attribute of the ecosystems of developing countries" (Bennett 1975:11).

It was this common purpose–to address the problems and prospects of integrating global conservation of biodiversity with efforts to ensure the survival and well-being of indigenous cultures–that initially brought together most of the contributors to this book in 1988 at the Symposium on "Traditional Cultures and Conservation of Biological Diversity" of the 64th Annual Meeting of the Southwestern and Rocky Mountain Division of the American Association for the Advancement of Science (AAAS-SWARM). This collection represents a multidisciplinary effort on the part of biologists, anthropologists and other scientists as well as resource management specialists and policy analysts, to provide a useful framework for evaluating efforts to identify and sustain viable traditional resource management strategies that will enable long term conservation of biodiversity.

Chapter 1 was originally presented by Raymond Dasmann as the Plenary Address for the AAAS-SWARM Meeting. Dasmann provides an overview of the importance of both cultural diversity and biological diversity, and makes a distinction between *ecosystem people*–"those who are dependent on and have learned to live in a sustainable manner within a natural ecosystem" and *biosphere people*–"those who . . . draw upon the resources of . . . the entire biosphere to maintain ways of life that are not necessarily sustainable and may be destructive to any one ecosystem." Finally, he provides a perspective on novel approaches that may allow the "simultaneous protection of . . . both biodiversity and ethnodiversity." Dasmann's contribution stands alone as a statement of the importance of biological and cultural diversity in relation to the aims and goals of this symposium volume.

It is imperative that those of us who exist within modern cultures and rely upon the global market economy reexamine the assumptions behind our collective thinking and renew our commitment to solving the dual problems of vanishing cultures and the "biological diversity crisis." The responsibilities of Neo-Europeans–Caucasians of European descent who today represent the cultural majority in lands far from Europe–now "require unprecedented ecological and diplomatic sophistocation: statesmanship in farm and embassy, plus greatness of spirit. One wonders if their comprehension of our world is equal to the challenge posed by the current state of our species and the biosphere" (Crosby 1986:307).

REFERENCES

Altschul, S.V.R. 1977. Exploring the herbarium. *Scientific American* 236(5):96-104.

Appell, G.N. 1975. Indigenous man: Chemotherapeutic explorations. *News from Survival International* 10:1-4.

Bennett, C.F. 1975. The advantages of cultural diversity. *Unasylva* 27:11-15.

Bodley, J.W. 1970. Tribal survival in the Amazon: The Campa case, AMAZIND/IWGIA Document No. 5. International Work Group for Indigenous Affairs, Copenhagen, Denmark.

Crosby, A. 1986. *Ecological Imperialism: The Biological Expansion of Europe, 900-1900.* Cambridge University Press, Cambridge, Massachusetts.

DeMallie, R.J., ed. 1984. *The Sixth Grandfather: Black Elk's Teachings Given to John G. Neihardt*. University of Nebraska Press, Lincoln.

Harlan, J.R. 1975. *Crops and Man*. American Society of Agronomy and the Crop Science Soceity of America, Madison, Wisconsin.

Mayberry-Lewis, D. 1977. Societies on the brink. *Harvard Magazine* (Jan.-Feb.):56-61.

Nabhan, G.P. 1987. Ethnobiology and conservation: Valuing diversity. *Environment Southwest* 519:28-31.

Neihardt, J.G. 1961. *Black Elk Speaks: Being the Life Story of a Holy Man of the Oglala Sioux*. University of Nebraska Press, Lincoln.

Wilson, E.O. 1985. The biological diversity crisis: A challenge to science. *Issues in Science and Technology* 2:20-29.

Wilson, E.O. 1988. The current state of biological diversity. Pages 3-18 *in* E.O. Wilson (ed.), *Biodiversity*. National Academy Press, Washington, D.C.

1

The Importance of Cultural and Biological Diversity

Raymond F. Dasmann

INTRODUCTION

The subject of this paper is one that could easily be addressed at a superficial level, but upon more detailed examination proves to be rather complex. The topic relates to a long-standing concern for nature conservation and to a more recent concern for the perpetuation of human ways of life that have been sustainable over past centuries. The two can go forward together and both be realized, although looking at today's world, this belief is obviously more an article of faith than a statement of any visible reality. This is not written from the perspective of an expert systematist on the biological side, nor that of a cultural anthropologist, but rather from that of a wildlife biologist whose occupation forces him to cross disciplinary boundaries into what often turns out to be thickets of controversy.

This paper will refer, in part, to tribal people. Even though it is not a popular term among some cultures, the word tribal is used herein following the lead of Robert Goodland (1982) in his World Bank publication, *Tribal Peoples and Economic Development.* The other terms–indigenous, aboriginal, autochthonous, native–are also found objectionable by some. In fact, every term used to describe original, traditional cultures has been tainted by its often contemptuous use by members of the now globally dominant cultures. The term *ecosystem people* will also be used to describe people who are dependent on and have learned to live in a sustainable manner within a natural ecosystem or group of closely related ecosystems. This form will be in contrast to that of *biosphere people*–those who, potentially at least, draw on the resources of the global economy or from the entire biosphere to maintain ways of life that are not necessarily sustainable and may be destructive to any one ecosystem (Dasmann 1974, 1984).

The term *biological diversity* has been worked over thoroughly in recent years, in particular by the many working groups and panels that produced the report of the U.S. Congress Office of Technological Assessment (OTA 1987), entitled *Technologies to Maintain Biological Diversity*. The term biodiversity refers to the totality of species, populations, communities and ecosystems, both wild and domesticated, that constitute the life of any one area or of the entire planet. It is a more impressive, all-encompassing term than "nature," but it conveys a similar meaning even though it specifically includes cultural modifications of the natural world.

MAINTAINING BIOLOGICAL DIVERSITY

The arguments for maintaining biological diversity are impressive and have been explored in such books as Margery Oldfield's (1984) *The Value of Conserving Genetic Resources* as well as in the OTA report on biodiversity. There is the obvious argument that we must allow the biosphere to function if humanity is to survive. For this we depend on the continuance of plant, animal, and microbial life. We are not in a position to say which species, varieties, numbers, or combinations thereof are essential. Yet we do know there is a risk involved in losing any species or community that is essential for ecosystem functioning. There is also a more economically based reason for maintaining biological diversity. The human population depends on the continued output of food, fiber, and other essentials derived from wild or domesticated life forms. The future of agriculture depends on our continued ability to develop new varieties of plants or animals resistant to diseases, pests, or climatic stress, and this in turn depends on having reservoirs of genetic materials to draw on. These, in turn, may include the wild relatives of existing domestic species, and the wild progenitors of future domestic species. The life-insurance value of biological diversity is potentially enormous, particularly as we go into a period where rapid climatic change or other severe planetary stresses are likely.

Apart from those arguments involving human survival are ethical and aesthetic reasons for conserving biological diversity. To many, these reasons are paramount. Some would argue that every species has an equal right to continued existence on planet earth, and that human pursuits that bring about the extinction of other species are the moral equivalent of genocide. Others would put it less dramatically, but still stress our ethical obligations to fellow passengers on planet earth. However, we must recognize that it is not within human power to preserve or maintain all of the biological diversity that exists or will exist. Extinction goes on in the absence of human agency. So does evolution. Our responsibility, if we will accept it, is to unnecessary increases in extinction rates or decreases in future rates of evolution. Furthermore, if we dig more deeply into the meaning of biological diversity, we run into problems. We do not know how many species of plants or animals there are on the planet. Are there three, five, ten, or fifty million? We do not know what past rates of extinction have been, nor what present rates are. Finally, we do not have and are not likely to have enough systematic biologists, ecologists, or ecosystem monitors to ever know the answers to these questions. How many systematic entomologists would it take, working for how many years, to describe, classify and catalogue 30 million insects? Who is going to monitor the rate of loss of genetic diversity among insects? How? Where? Obviously the best we can hope to do is to work with those forms that we can see and monitor, or those that appear to have key roles in ecosystems. We can then hope that if we take care of those, all of the others that we can't easily see or hope to monitor will also survive.

We must not get caught in the numbers game–guessing how many million species are likely to become extinct by the year 2020, or what the rate of loss per day, per hour, or per minute might be. Such guesses lack credibility. Let us concentrate on what we do know. Those facts are depressing enough. We do know that the number of many, if not most, of our larger vertebrates have been declining, and their ranges have contracted. Examples include the elephants, rhinos, lions, cheetah, whales, grizzly bears, wolves, quetzals, macaws, and condors. We know there is an overall decline in the abundance of wildlife globally. The great wildlife spectacles of the past have mostly gone, whether of bison and pronghorn on the American Great Plains, migratory springbucks in South Africa, passenger pigeons in Eastern America, or antelope in India. The actual loss of species is less impressive than the loss of faunal abundance. We also know that we are losing habitat at an appalling rate, with the tropical forests showing the most dismal record, but perhaps equally threatened are other dry land and wetland communities. With vegetation loss goes the animal life associated with rain forest, savanna, steppe or marsh. We know these things, and we try to do something about them. But either we are not doing enough, or we are going about it the wrong way, or both.

THREATS TO CULTURAL DIVERSITY

The threats to *cultural diversity* are certainly as depressing as those affecting biodiversity. All over the world, tribal people are being eliminated, or are having their cultures shattered by invaders of their territories. Some of these invaders are corporate or governmental representatives of the global economy. Others might be called refugees from the prevailing monoculture, people who could not fit in or find land to live on and are now seeking new land on the frontier. Some of these were once tribal people, who have lost their lands and have forgotten their old ways of living. This is an old story that has been told in many books, most recently in Alfred Crosby's (1986) *Ecological Imperialism*. Current struggles have also been described by Neitschman (1987) and in recent issues of *Cultural Survival Quarterly*. Apart from the sheer injustice and human suffering involved in the displacement or destruction (genocide) of tribal cultures, there is also the loss of indigenous knowledge of how to live sustainably on the land and the uses and values of wild species as well as the concomitant loss of the diversity of cultivated species and varieties that were once maintained in traditional systems. Regrettably, we are losing those who may be best equipped to aid conservationists and scientists in their efforts to maintain biodiversity.

A number of years ago, some American scientists present at an international meeting in Western Samoa pleaded that the first step toward establishing a system of protected areas in Western Samoa should be to bring in a team of (American) scientists to do a biological survey of the islands. It was argued that they would thus be able to tell the Samoans what species occurred where, and which areas were therefore most important for conservation. This was a particularly egregious case of scientific imperialism combined with cultural arrogance. The Samoans had lived on their islands for many centuries. Collectively they knew every nook and cranny of their lands. They not only knew what species grew where, but also the particular uses of those species. True, their species may not have been baptized with Linnean nomenclature, but their identities were known to Samoans. No outsider could learn as much as the local people knew without spending a very long time living and working with the people. However, we in the United States are so used to being strangers in our own land, where most of us are unable to identify most things that grow around us. We forget that elsewhere there are folks who are at home on their lands and know their biological neighbors very well.

We have learned to rely too much on the outside expert, or the trained scientist, to tell us what we should do. This has been reflected over the years in the ways we have used to provide for the conservation of nature. Most usually, national parks or nature reserves were set aside on the basis of advice from outsiders. Seldom were the tribal people who lived on the land consulted, even though they had cared for the wild plants and animals and had basically maintained the land in a wilderness condition deemed suitable for a nature reserve for many centuries. At best, they were told something about the government's plans; at worst they were simply moved out or told they could no longer use their land for their traditional lifestyles or purposes. Walter Lusigi (1978), a Kenyan biologist, has written at length about this process as it applies to Africa in his book, *Planning Human Activities in Protected Natural Ecosystems*.

Of course, there is a different way to protect biological diversity and nature. It involves learning the native language of the people who live in areas that require long-term protection. It involves spending a lot of time talking with them, telling them your concerns, goals, and objectives, and listening to theirs. It means seeking with them ways in which these goals can be jointly achieved. This is the approach Dan Janzen has utilized to attempt to establish the new Guanacaste National Park in Costa Rica (Allen 1988). But this approach takes time and patience. So, for the most part, we have gone ahead, producing new generations of poachers, surrounding nature reserves with alienated people and generations, instead of embedding them within such reserves and encouraging the development of a friendly human environment of stewards and protectors. But we are finally beginning to change our ways.

It is unfortunately true that most of the surviving tribal peoples occupy areas of the earth that have been the least accessible to the global cultures. So, they are found in the mountains and tropical forests, in the desert edges, boreal forests, and tundras. These are also the last remaining extensive wild areas that provide a home for the many wild species of fauna and flora that are adapted to such environments. They are areas of primary concern for the conservation of biotic diversity. Therein lies the dilemma. Should cultural diversity be sacrificed so that biological diversity may be preserved? Should the Amazonian Indians be chased out of areas designated as rain forest reserves? Should Eskimos be denied their traditional use of Arctic wilderness? The answer to these questions was "yes" when I first became involved with the national parks movement. Fortunately, we are changing our ways and recognizing that nature conservation and human use are not necessarily incompatible. But not everyone has been convinced.

RECONCILING CONSERVATION OF CULTURAL AND BIOLOGICAL DIVERSITY

One example of the way things are changing involves the Kuna Indians of Panama. Their homeland, Kuna Yala, includes the San Blas islands on the Caribbean coast and extends up to the top of the Cordillera, including a variety of rain forest habitats. This case study has been told often, but in brief it involves the establishment of a carefully protected reserve encompassing more that 60,000 ha of old-growth rain forest. To accomplish this, they have worked with many international organizations. Their fundamental goal is to protect their own territory against illegal settlement by outsiders. They have succeeded in incorporating ideas, concepts, and materials from the global economy into their own cultural context. The Kuna have maintained their cultural identity, yet are also protecting the biological diversity of their territory (Wright et al. 1988; see also Clay, Ch. 12 and Gregg, Ch. 13, this volume). In North America, the Indians and Eskimo people have been struggling for years to maintain lifestyles based on the use of wild animals. They have become extremely sophisticated in their dealings with the governments of Canada and the United States in the face of oil development, hydroelectric-power installations, and other forms of exploitation of their homelands. They are achieving the right to maintain their trapping, hunting, and fishing subsistence rights, and have accepted responsibility for conserving the species on which their continued livelihood depends. Now it seems that their greatest threat may come, not from exploiters, but from those who want to close down the fur business due to ethical or emotional concerns, and end all hunting and trapping (Anonymous 1986, Lack 1988).

In January of 1988, the World Wildlife Fund/U.S. approved funding for a project in Zambia that appears to be moving in a promising direction. Essentially, it involves turning the direct responsibility for management and conservation of wildlife back to the local people—villagers who still maintain most aspects of their traditional ways of life. The areas involved, called Game Management Areas, serve as buffer zones surrounding the national parks in which wildlife is legally protected. In Game Management Areas, game animals may be hunted under permit, whereas other land uses are not controlled. Hunting has primarily been done mostly by safaris, which originated from outside and paid high fees to the government. Local people had no right to hunt nor to acquire income derived from outside hunting. Naturally enough, they turned to poaching—either directly or in combination with outsiders who would share the meat. The government spent much money on anti-poaching patrols, but these were not really effective because during the wet season, they could not patrol remote areas. Under the new program, safari hunters must employ local people, and the income derived from safaris is shared directly with local communities. A share of income derived from the sale of wildlife-derived products is also directed to buffer zone villages. In addition, local people are now permitted to hunt within sustainable limits. Village scouts and rangers will be the protective force, and they have already demonstrated their ability to stop poaching by outsiders, wet season or dry, now that their incentive is the protection of their own wildlife and sources of income.

A similar program on an even larger scale is taking place in Zimbabwe, also supported by the World Wildlife Fund/U.S., among others. This derives in part from the game-ranching programs launched over 25 years ago (Dasmann 1964) that have proved to be successful on privately owned lands. The present scheme will be to extend the economic benefits of wildlife utilization to communally owned lands and the local people. In both Zambia and Zimbabwe, the government maintains ultimate responsibility and is still directly involved with management of the national parks which form the protective core of the system. However, benefits derived from uses of the parks will now flow outward to the surrounding communities. Local attitudes toward the parks are reported to be changing in a favorable direction (Child and Nduku 1986, Cummings 1987). Of course, the Zimbabwean, Zambian, and similar conservation programs in Africa depend on the ability of the local people to derive direct income benefits from the sale of wildlife products or hunting rights. These programs are threatened by the growing movement of opposition to the sale of wildlife meat, hides, furs, ivory, and other products, and to hunting in general.

The examples mentioned thus far, from Panama, North America, Zambia, and Zimbabwe represent a concept that combines strict protection of natural resources with consumptive human uses and sustainable development concepts. It is a step beyond the preservationist approach that featured only so-called nonconsumptive uses of national parks and strict nature reserves. This new approach allows for the simultaneous protection of nature and maintenance of culture–that is, of both biodiversity and ethnodiversity. Internationally, it fits in well with the concept of biosphere reserves, an ecosystem-oriented conservation technique developed by UNESCO Man and the Biosphere Program (MAB) in the early 1970s, (see Gregg, Ch. 13, this volume). Indeed, the Comarca de Kuna Yala reserve of the Kuna Indians has already been proposed as a new biosphere reserve. In Costa Rica, La Amistad National Park is part of a biosphere reserve designed to protect both the integrity of the local Indian culture and the unique flora and fauna of the Costa Rican Cordillera (Torres 1987). In México, the Sian Ka'an Biosphere Reserve is being developed as a means for protecting and assisting the local Mayan economy while also protecting nature (Tangley 1988a, 1988b). The biosphere reserve concept can be used in a pedestrian or imaginative manner. In the United States, it has been mostly an international overlay on an existing protected areas system, adding little except perhaps a greater research dimension. Fortunately, this is now changing. In México and some other countries, it has proved effective in enlisting popular support for an effort to protect both nature and sustainable lifestyles of traditional cultures (UNESCO 1986).

No doubt there are, or will be, other approaches better adapted to the tribal peoples or traditional cultures of other areas of the globe and the natural and man-modified ecosystems they maintain and depend upon for their continuing survival. In the future, these alternatives must be sought after and applied if we are to realize the importance of conserving both biological and cultural diversity.

REFERENCES

Allen, W.H. 1988. Biocultural restoration of a tropical forest. *BioScience* 38:156-161.

Anonymous. 1986. Alaska's frontier goes domestic. *Conservation Foundation Newsletter*, July-August. Washington, D.C.

Child, G. and W. Nduku. 1986. Wildlife and human welfare in Zimbabwe (Mimeo). African Forestry Commission, Working Party on Wildlife and National Parks. Food and Agricultural Organization of the United Nations, Rome.

Crosby, A. 1986. *Ecological Imperialism: The Biological Expansion of Europe, 900-1900*. Cambridge University Press, Cambridge, Massachusetts.

Cummings, D. 1987. A field study of multi-species indigenous wildlife utilization, Zimbabwe (Mimeo). World Wildlife Fund-U.S., Washington, D.C.

Dasmann, R.F. 1964. *African Game Ranching*. Permagon Press, Oxford, United Kingdom.

________. 1974. Difficult marginal environments and the traditional societies which exploit them. *Survival International News* 11:11-15.

________. 1984. *Environmental Conservation*, 5th Ed. John Wiley, New York.

Goodland, R. 1982. *Tribal Peoples and Economic Development*. World Bank, Washington, D.C.

Lack, L. 1988. All is not well in Cree country. *Whole Earth Review* 58:58-65.

Lusigi, W. 1978. *Planning Human Activities in Protected Natural Ecosystems*. Dissertations Botanicae 48, J. Cramer, Vaduz, Federal Republic of Germany.

Neitschman, B. 1987. The Third World war: Militarization and indigenous peoples. *Cultural Survival Quarterly* 11:1-16.

U.S. Congress, Office of Technology Assessment, 1987. *Technologies to Maintain Biological Diversity*. U.S. Government Printing Office,

Washington, D.C.
Oldfield, Margery L. 1984. *The Value of Conserving Genetic Resources.* U.S. Department of Interior, National Park Service, Washington, D.C.
Tangley, L. 1988a. Beyond national parks. *BioScience* 38:141-147.
________. 1988b. A new era for biosphere reserves. *BioScience* 38:148-155.
Torres, H. 1987. Conservation in the name of friendship: IUCN in Central America. IUCN Special Report Bulletin 18:6. IUCN, Gland, Switzerland.
UNESCO. 1986. *First International Congress on Biosphere Reserves,* UNESCO, Paris.
Wright, M., B. Houseal and C. de Leon. 1988. Kuna Yala: Indigenous biosphere reserve in the making. Pages 352-356 *in* J.H. Bodley (ed.), *Tribal Peoples and Development Issues*, Mayfield Publishing Co., Mountain View, California.

PART TWO

Genetic Resources and Culture

Introduction

Janis B. Alcorn and Margery L. Oldfield

We didn't fight two world wars to make the world safe for green vegetables. —Richard G. Darman, President's Budget Director, televized comment on lack of U.S. response to Global Climate Change (1990).

Public policy with respect to agriculture strongly reflects the attitudes of populations so removed from agriculture that they know little about how it operates. (Cox and Atkins 1979:3)

(T)o suggest seriously that a modest 10% of each country's national defense budget be reallocated to defending and protecting genetic and biological diversity gives one some idea of the magnitude of the conceptual, institutional, and political obstacles. (Dahlberg 1987:379)

During the past century, the shift from rural to urban-industrial attitudes and values in the world's most influential and affluent nations has been generally heralded as progress "toward human independence from nature"—even though the opposite is true (Cox and Atkins 1979:9). Plants and plant products still provide the primary foundation for all modern civilizations (Schery 1972). Therefore, agricultural genetic resources are critical to the well-being of people, the stability of governments, and ultimately to the modern way of life. Yet, as Mr. Darman's comment (above) illustrates, the role of plants in sustaining modern civilization and maintaining economic prosperity often goes unappreciated. Today, people in modern cultures have little interest in crop genetic resources. The average U.S. citizen cannot identify the plants or seeds of the major staple food crops, and has little understanding of crop genetics or the importance of conserving genetic resources for long term food security.

The narrow segment of modern society that is involved in crop breeding focuses on the creation of marketable seed stocks that will grow under controlled, uniform conditions over wide geographic areas (Simmonds 1991). Many scientists have acknowledged for decades that the prevailing crop production approach—monocultural agriculture—has led to the destruction of the very crop genetic resources that have enabled its continuing success. However, their voices have failed to stimulate significant investment in the long term security of these genetic materials through *in situ* reserves. As the relationship between traditional farming cultures and domesticated plants dissolves, scientists debate the appropriate solutions to secure the genetic resources on which modern agriculture is based (Anonymous 1990, Food and Agriculture Organization 1991). The contributed papers in Part Two provide insights as to how human attitudes about agricultural production processes affect the evolutionary relationship between agricultural genetic resources and humankind.

In Chapter 2, Dorothea Bedigian provides a case study of sesame–a representative crop. This case study illustrates the essential role of human cultural diversity in crop evolution, and the adverse impact of modernization on traditional farming practices that could otherwise maintain valuable crop genetic resources *in situ*. Over centuries, the Nuba of Ethiopia have created a tremendous diversity of native sesame crop varieties–also known as "landraces." But today, many Nuba farmers are no longer cultivating their genetically diverse landrace cultivars because urban merchants and buyers in the Sudan typically view sesame as a commodity and therefore do not value the genetic diversity contained within these traditional cultivars. Modern crop geneticists, who appreciate the value of sesame genetic diversity, have been urgently documenting and preserving samples of traditional sesame varieties in gene banks. However, *ex situ* storage of crop seeds in gene banks ecologically isolates crop genetic resources from their coevolving populations of wild pathogens and pests and the human cultures that once nurtured that diversity (see Oldfield and Alcorn, Ch. 3, and Browning, Ch. 4, this volume). Thus, *ex situ* conservation halts essential man-directed and natural evolutionary processes. Even though the sesame story has been told anecdotally for practically every major crop, there are very few published accounts of the transformation from traditional agriculture to modern agriculture within centers of crop origin and diversity (Plucknett et al. 1987). More case studies like Bedigian's are needed to characterize the level of crop genetic diversity still under farmers' management, the reasons behind the continued evolution and maintenance of this diversity, and the rates and specific causes of genetic erosion for particular crops in areas of traditional agriculture.

In Chapter 3, Margery Oldfield and Janis Alcorn provide a broad overview and summarize the core issues of *in situ* conservation of crop genetic diversity, some of which are addressed in detail by other contributors in Part Two. They introduce the ongoing debate over *in situ* (onsite) versus *ex situ* (offsite) conservation, discuss the major drawbacks of conserving seed stocks *ex situ* in gene banks, detail the importance of traditional agricultural practices in creating and maintaining crop genetic resources *in situ*, and discuss relevant rural development issues in relation to the *in situ* conservation strategy. Even though heightened awareness of and concern for endangered plant species have recently led some advocacy groups to take up the cause of *in situ* conservation of crop genetic diversity (Hoyt 1988), there is still little agreement on which crops should be the focus of such efforts. It is risky to make these important conservation decisions based on the detailed case studies that exist for a few crops. It is also risky to apply generalizations derived from studies of vegetatively propagated crops, such as potatoes (Brush 1986), to seed crops.

Plant pathologist and microbiologist J. Artie Browning presents a cogent argument in Chapter 4 for *in situ* conservation of coevolutionary processes facilitated by long term ecological associations between crop plants and their wild pathogens. Focusing on recent programs designed to conserve populations of wild crop progenitors and crop relatives within the indigenous natural ecosystems and agroecosystems of Israel, he explains why *in situ* conservation of crop germplasm resources is a necessary complement of conserving crop genetic stocks *ex situ* in gene banks. Crop plant-pathogen coevolution is impossible in modern agricultural systems, especially in Western-type farm fields from which crop seeds are not collected for subsequent replanting. However, if modern agriculture is to be sustained in the future, vital evolutionary processes must be sustained elsewhere in natural or man-modified agroecosystems. Browning describes major *in situ* conservation projects in Israel and the U.S.S.R.–modern societies that have given much higher priority to security of crop genetic resources via *in situ* conservation than the United States and most other modern, industrialized nations. These national conservation efforts will continue to sustain some of the safeguards that traditional agriculture once offered by default.

In Chapter 5, Garrison Wilkes, a proponent of *in situ* conservation since the early 1970s, focuses on the failure of the general public to appreciate the importance of plants to their quality of life. He argues that a public educated about "the primacy of plants" would be motivated to support global efforts to safeguard genetic resources. *In situ* conservation could be coupled with educational exhibits, such as "living historical farms," which would instill national pride in the plant domestication accomplishments of traditional farmers. At the same time, these displays would explain the evolutionary processes occuring in the fields as well as motivate the public to support cnsite conservation. Wilkes offers three other options for *in situ* conservation: a global network of *in situ* reserves of selected landraces, monitoring of existing populations of traditional cultivars and wild crop relatives, and improvement of landrace seed in ways that do not include complex plant breeding techniques (e.g., virus removal). He also reports on promising new *in situ* conservation projects in México and Guatemala—efforts which demonstrate that Third World culture-bearers recognize the importance of crop genetic diversity and are willing to assume world leadership in conservation of germplasm resources.

Despite the existence of techniques to conserve livestock genetic diversity, there is no cohesive or coordinated modern commitment to do so. Genetic resources of multipurpose livestock breeds are still poorly documented even though they are very important to and are primarily maintained and utilized by people in developing countries. In Chapter 6, Hank Fitzhugh and Ann Wilhelm compare traditional and modern livestock genetic improvement processes and explore the reasons for different livestock improvement goals in developed vs. developing nations, and discuss the erosion or loss of livestock genetic resources. They warn that increased productivity potential may not lead to improved human welfare over the long term if it is linked to loss of unique livestock breeds or genotypes adapted to particular diseases, extreme climates, and other environmental stresses common in developing countries.

The failure to appreciate the importance of crop and livestock genetic diversity for modern agriculture has resulted in insufficient investment in the maintenance of valuable genetic resources and evolutionary processes that were previously maintained by traditional cultures. The Congress of the United States recently initiated the Noah Project (Agency for International Development 1990), and thus, showed evidence of its recent concern that *ex situ* conservation facilities are grossly inadequate for the essential task of saving the wild species and crop genetic diversity now being lost.

In the past, genetic diversity for short term use in modern crop and livestock genetic improvement programs was freely harvested from traditional agroecosystems and natural ecosystems. Yet as we enter the 21st century, these natural genetic reservoirs are continuing to disappear, and they are, in many instances, no longer providing this service for free. Keystone Dialogue's Plant Genetic Resources Project has now brought this issue, and the attendant issue of intellectual property rights and farmers' rights, to national and international policy levels. Bilateral donors are beginning to support *in situ* conservation efforts (e.g., Cohen et al. 1991). National governments are investigating their options for *in situ* conservation of crop genetic resources (e.g., Lleras 1991). Traditional farmers are developing their own modern ways of maintaining their crop genetic resources (Nabhan 1989). Thus, today we see both signs of decay and signs of growth in our modern appreciation of the economic importance of crop genetic resources.

REFERENCES

Agency for International Development (A.I.D.). 1990. *Ex situ* conservation: Present status and future priorities. Report to Congress from the Agency for International Development. Bureau for Science and Technology, A.I.D., Washington D.C.

Anonymous. 1990. Germplasm experts study *in situ* conservation in Israel, visit Ammiad site. *Diversity* 6(2):13-14.

Brush, S.B. 1986. Genetic diversity and conservation in traditional farming systems. *Journal of Ethnobiology* 6(1):151-167.

Cohen, J., J.B. Alcorn and C. Potter. 1991 (in press). Utilization and conservation of genetic resources: Projects for sustainable agriculture. *Economic Botany.*

Cox, G.W. and M.D. Atkins. 1979. *Agricultural Ecology: An Analysis of World Food Production Systems.* W.H. Freeman, San Francisco, California.

Dahlberg, K.A. 1987. Redefining development priorities: Genetic diversity and agroecodevelopment. *Bulletin of Science, Technology, and Society* 7:367-382.

Food and Agriculture Organization of the United Nations (FAO). 1991. *International Code of Conduct for Plant Germplasm Collecting and Transfer.* FAO, Rome.

Hoyt, E. 1988. *Conserving the Wild Relatives of Crops.* IBPGR-IUCN-WWF, Rome.

Lleras, E. 1991. Conservation of genetic resources *in situ. Diversity* 7:72-74.

Nabhan, G.P. 1989. *Enduring Seeds.* North Point Press, San Francisco, California.

Plucknett, D.L., N.J.H. Smith, J.T. Williams and N.M. Anishetty. 1987. *Gene Banks and the World's Food.* Princeton University Press, Princeton, New Jersey.

Schery, R.W. 1972. *Plants for Man.* Prentice-Hall, Englewood Cliffs, New Jersey.

Simmonds, N. 1991 (in press). Selection for adaptation in plant breeding programs. *Theoretical and Applied Genetics.*

2

Genetic Diversity of Traditional Sesame Cultivars and Cultural Diversity in Sudan

Dorothea Bedigian

INTRODUCTION

This paper addresses the genetic diversity of traditional sesame cultivars in relation to Nuba cultural diversity in Sudan. Cultivars–or morphologically and genetically distinct varieties of crop plants–are the product of a complex, dynamic interaction between the selection pressures wrought by both human cultures and environmental factors. Sorghum and sesame are the major traditional crops grown for food in Sudan, and the local genetic variation of sesame and sorghum cultivars in Sudan is enormous. This is especially true in the Nuba Mountains, which are geographically and culturally isolated. There is also a significant level of geographical and cultural isolation among Nuba subgroups, and these factors that determine separation among the Nuba people have fostered both the creation and maintenance of traditional cultivar diversity. It is this interrelationship between human cultural diversity and the environmental diversity of particular geographical regions that accounts for the great diversity of sesame cultivars in Sudan.

Cultivars of sesame (*Sesamum indicum* L.) differ considerably from each other in their branching habit, flower color, size, shape and arrangement of capsules, and color and size of seeds (Bedigian et al. 1986). For example, there may be 1-3 capsules per axil, one or more axils per node, and the capsules may have 2-4 carpels per capsule. Seeds can range in color from white to black, with intermediate colors of beige, tan, red, brown, and gray.

There is a remarkably high level of morphological variation within cultivated sesame fields in the Nuba Mountains of Sudan (Yassin 1978, Bedigian 1988). Compared with the rest of Sudan where relatively few distinct sesame varieties occur, the Nuba occupied area harbors dozens of ancestral or traditional sesame varieties known as "landraces." The area may realistically be called a crop gene micro-center (Harlan 1951).

The Nuba people are culturally diverse in their religion, language, material inventory, and agricultural practices. For example, sixty-two distinct Nuba languages have been identified. Different Nuba peoples are defined by their distinct body ornamentation, speech, and the architecture of their houses and other structures. They are also distinguished by the distinct sesame and sorghum cultivars derived from their particular geographical area (Bedigian and Harlan 1983).

THE LINK BETWEEN NUBA CULTURAL DIVERSITY AND SESAME CULTIVAR DIVERSITY

The Nuba themselves identify specific cultivars or landraces of sesame with specific Nuba people and their particular inselberg hills (called *jebels* in Arabic). For example, they are aware of the particular oil content, seed color, and disease or insect-resistance qualities of various sesame cultivars of different Nuba subgroups. One sesame variety that has large gray seeds–'Morawi'–originated in the Moro Hills, and it still retains the name of its region of origin even though it is a successful cultivar grown today by Nuba outside this geographical area. One of the most popular landraces grown in Tukswana and Dubeibat, 35 km NE of Kadugli, is 'Abu Sindook', which means "father of containers"–a name suggestive of its shape, multi-carpellate form, and abundant seeds. Yet another landrace, 'Dinderawi', has gray seeds and capsules that are 50% longer than any other yet observed. The name of this long-capsuled cultivar is suggestive of the longneck or height of a giraffe; the nearby Dinder Game Reserve contains a large giraffe population.

Genetic diversity is maintained in traditional agroecosystems by cultural intervention as well as by natural selection. Observations of sesame cultivar diversity in the Nuba Mountains (Bedigian and Harlan 1983, Bedigian 1988) support the statement of Brush (1986) that crop genetic diversity is particularly marked in mountain areas. Mountainous terrains are characterized by well-marked environmental changes over small distances, marked attitudinal seasonality, and isolation between small crop production zones. Sesame cultivars selected for survival under arid conditions (*qoz*) exhibit morphological characteristics such as size reduction and early maturity. One example is 'Hirehir', which matures in just 70 days. Its short stature (90 cm), and small leaves with smooth leaf margins serve to reduce water loss. In contrast, on heavy clay soils with abundant water reserves, the late-maturing 'Tegil' "heavy" sesame cultivars require up to 180 days before harvest. These cultivars are taller (some exceed

1.5 m), more profusely branched, and have wide leaves with divided margins.

CULTURAL AND ECOLOGICAL FACTORS THAT AFFECT CULTIVAR DIVERSITY

The cultural and ecological diversity of traditional farming systems in ancient agricultural regions has shaped genetic variation within traditional crop varieties (Harlan 1975). Nuba farmers designate seed color and maturity as the principal characteristics for most sesame cultivars. For example, on clay soils, red-seeded varieties mature earlier than white-seeded cultivars while on sandy soils, white seeded cultivars mature earlier than red. Considering Altieri's (1987) suggestion that folk taxonomies should be more relevant in more marginal or "risky" environments, one could argue that the Nuba specify maturity of each sesame landrace to distinguish their crop genetic resources according to maximum benefit under environmental stresses. Particularly in the northern Nuba Mountains (north of Kadugli), farmers routinely plant, harvest, and store the next season's seed stock separately, based on maturity characteristics and seed color.

The red-seeded cultivars seem to be particularly well-adapted to the Nuba Mountains: they withstand high winds, heavy rains, and bacterial leaf spot or "blood" disease (*Xanthomonas sesame* Sabet & Dowson; *Pseudomonas sesame* Malkoff). At the sesame auction in the El Obeid main market depot, all red-seeded sesame varieties are called *jebeli*, which indicate they originated in the Nuba Mountains. The distinctive red-seeded sesame cultivar of Kordofan Province–'Kordofan Red'–is an unique form with high oil content grown primarily in the Nuba Mountains. The local name in the Nuba Mountains is *beledi*, which means native, local, regional. Kordofan Red is a 1.5-2 m tall, profusely branched, robust type with many large bicarpellate, coarse-looking capsules about 3-4 cm long; its seeds are brick-red, with reticulated, rugose seed coats that contain up to 60% oil. This cultivar type is grown specifically by farmers as an oil seed crop. A local name for this red-seeded variety is 'Denemet'–a term derived from dynamite, possibly due to the crackling sounds of its dehiscing capsules or its high oil content. Sesame capsules typically open with such explosiveness upon drying, that when standing in a sesame field at harvest time, one hears snapping, crackling, and popping sounds. This traditional cultivar is the most valuable cash crop in Umm Dorein in the southern Nuba Mountains region. Yet, it is not the only cultivar grown.

Nuba farmers who tend sesame fields of different sizes or who live in different geographical areas cultivate different types or combinations of sesame cultivars for a variety of reasons. For example, in Tebeldigo, farmers plant mixtures of cultivars if only a small plot of sesame is sown. On the other hand, if a larger field is sown, farmers prefer to keep different sesame varieties separate due to differences in maturity and specific uses. In the tiny village of Muheila south of Lagowa, white-seeded sesame varieties are considered to be superior for their oil content and flavor, but the oil of red seeded varieties is used specifically for stomach ailments by mashing cowpeas and red sesame seeds together as a porridge. On the *qoz* sands near Sunjukaya, three types of sesame are grown. The white-seeded 'Hirehir' is the earliest maturing variety. Another early type is a cultivar of short stature with tan-brown, reticulate seeds and small, thin-shelled capsules (1.5 cm) that shatter easily. Finally, a red-seeded type with coarse capsules called 'Jebrook' is the last to mature. In the eastern Nuba Mountains region, near the villages of Tukswana and Dubeibat, arabicized Nuba dwell along with a sedentary Arab population. Favored varieties include the early-maturing, high-yielding 'Abu Sindook' and 'Dinderawi' which, despite its lateness, is cultivated there because it has very long capsules that yield a large number of seeds. The sedentary, but formerly nomadic, Hallala Arabs in the region prefer 'Morawi'. The latter variety is attributed to the Moro Nuba; it has small, reddish-orange seed borne in short capsules. They also grow 'Denemet'. Each sesame variety is planted, harvested, and stored separately.

Such geographic and cultural isolation of farmers maintain sesame cultivar and genetic diversity. Nuba in the southern hills are more isolated from trade routes and markets, and maintain greater cultivar diversity within a single field. Consequently, fields of sesame in the southern *jebels* (south of Kadugli–in the region with less influence from the central government) often contain a great deal of genetic diversity. These growers maintain mixtures of genetic lines, all of which are reasonably well adapted to the region, but which differ from one another in date of maturity, seed color, and responses to prevailing diseases and insect pests. Southern Nuba farmers probably plant mixtures of seed from different sesame cultivars to "hedge their bets." This cultural method of crop insurance sustains sesame genetic diversity in this region. Some lines may be resistant or tolerant to certain races of disease pathogens, and some to others. The resulting genetic diversity confers at least partial resistance to crop specific diseases and thus allows farmers to exploit differences in microhabitats and to benefit from the available genetic variation of a given crop species (Harlan 1976a, see also Browning, Ch. 4, this volume).

Most sesame in the southern *jebels* is consumed locally. Therefore, southern Nuba people may have less incentive to obtain maximum yields for the export market. Rather, their primary goal is to obtain a reliable yield each season for local consumption. The southern Nuba store mixtures of sesame seeds and consume them as is, toast and pound the seeds to form a paste to which honey is added, or add the paste to a stew with cowpeas. Nuba closer to market centers in the North, and therefore to mechanized oil pressing facilities, often grow entire fields of high-oil content sesame that will yield them maximum profits.

Compared with the rest of the Sudan, the Nuba people grow many more varieties of sesame. In Gedaref in eastern Sudan, the acknowledged "capital of sesame production", one sees far fewer distinct cultivar types and all of these are high-oil or higher-yielding types recommended by the ARS (Agricultural Research Service). Luxurious growing conditions for sesame exist there on black, cracking clay soils, and sophisticated cultivar lines with longer capsules have recently developed here. Since farmers have easier access to the central market depot at Gedaref, this incentive may have encouraged them to experiment with a number of different sesame cultivars to grow as much as possible of the known, high-yielding varieties for export.

Many large agricultural tracts near Gedaref are leased by wealthy merchants who are absentee landlords. The varieties they favor are the 'Tajarup' or 'Ziraa' lines that have been improved and introduced by agricultural research stations. These are white-seeded varieties of intermediate maturity. Tajarup means "agricultural research station" in Arabic, whereas Ziraa means "agriculture." 'Ziraa-1' is another name for the Agricultural Research Service release 'Abu Naama A.1.10,' which was specifically named after the research station and its accession number. It was selected and released by Mahmoud A. Mahmoud at the Abu Naama research station nearly two decades ago. 'Ziraa-1' is a tall cultivar that matures fairly early (about 90 days). 'Ziraa-7' is also a favorite of the Semsem Agriculture Scheme. It is more profusely branched, and has a large number of capsules with a shorter distance (3-4 cm) between capsules that effectively increases oil-content yield. Both 'Ziraa-1' and 'Ziraa-7' have large white seeds with high oil content. Other favorites include 'Belia', a white seeded variety with a high oil content, 'Safra'–red-seeded, tall cultivar, 'Hirehir', the previously mentioned early maturing, white-seeded cultivar, and 'Aswad', a late maturing black-seeded cultivar that possesses the highest oil content. 'Galabat' is also favored for its very large white seeds in the Basunda area, but it is grown less frequently than other lines because it requires large amounts of water. Farmers around Gedaref said they have tried Abu Sindook, but have rejected it because it shatters too quickly once it ripens. In central Sudan, the most widely grown line is also Abu Naama A.1.10.

Owners of large tracts of land who are in the business of growing sesame to supply the oilseed trade or export market tend to seek and obtain advice primarily from the Agricultural Research Service. Conversely, sesame growers in very remote areas, who are isolated linguistically, culturally, geographically, and philosophically from the central government, or who are growing small sesame plots for household use, still retain their ancestral cultivar lines adapted to local environmental conditions, diseases, and pests.

THREATS TO TRADITIONAL NUBA AGRICULTURAL TECHNIQUES AND SESAME GENETIC RESOURCES

Many ecologically sound provide sustained sesame yields are evident in (Nuba farming practices) (Bedigian and Harlan 1983). These include carefully chosen agricultural methods that demonstrate adaptations to a variety of natural environmental conditions. Terraces control soil erosion, conserve water, and enable farmers to make maximum use of steep mountainous or hilly slopes for agriculture. The Nuba have developed genetically-diverse sesame cultivars appropriate for use in particular ecological microhabitats. Early maturing crops are chosen for terraced land, or for light, stabilized, sandy *qoz* soils where water is scarce. These crops include sesame, pearl millet (*Pennisetum americanum* (L.) Leeke), and roselle (*Hibiscus sabdariffa* L.) which provides *karkadeh*, the national drink of the Sudan. It is frequently intercropped with 'Hirehir', an early-maturing sesame cultivar that matures in 70 days. Selection of specifically-adapted cultivars for each environment or microhabitat (e.g., light, sandy soils vs. heavy clay, plains soils) facilitates more efficient use of water, sunlight, and soil. There is low risk of complete loss of crops in regions with unreliable rainfall due to the great diversity of genetic lines. The result is a high level of productive stability due to compensation by other varieties when one fails. The temperature and radiation extremes of the equatorial location, counteracted by increasing planting densities and mulching, are as seen in Muheila.

The bottom line for retention of sesame cultivar diversity in Sudan is continued availability. The inherent "conservation ethic" of farmers who live in geographical or cultival isolation is to continue the age-old practices of their forefathers because those agricultural methods "work." The introduction of Western ideas to African environments or cultures is not always appropriate. In fact, maintenance of successful sesame cultivars has probably occurred over the millennia as a consequence of experimentation by ancestral Nuba farmers who could not afford to fail. They adopted a "kit" of agricultural practices and cultivars that were successful in particular regions. Oldfield and Alcorn (1987) reviewed a number of cultivation practices that influence crop evolutionary processes, including terracing, fallowing, runoff channeling, soil mounding, mulching, use of fire, selective weeding, interplanting, trapping of particular crop predators, attention to individual plants, pollarding of selected wild trees in agricultural fields, and management of secondary successful vegetation. The Nuba have been observed to utilize all of these cultivation techniques (Bedigian and Harlan 1983). In this context, attempts to introduce Western ideas by Agricultural Research Service (ARS) "experts" has probably had a permanent impact on sesame cultivar diversity.

Isolation and poverty force the Nuba in remote hills to continue to use what they have. The remaining traditional sesame cultivars in specific regions are successful lines that are well adapted to the prevailing ecological conditions of each region. The Nuba still have little access to the "improved" varieties. Those who do have access are non-Nuba entrepreneurs who have leased land in the Nuba Mountains for a profit venture. These farmers usually seek assistance from and need the recommendations of the ARS, hence they grow seeds provided by the ARS. In the countryside close to El Obeid where non-Nuba have leased fields, Abu Naama A.1.10 is frequently grown. The goals of large-scale, non-Nuba growers, who plant ARS selections for urban markets, differ significantly from those of the Nuba farmers who primarily grow sesame for local consumption. Therefore, the adoption of recommended "improved" varieties is a response to market forces that is resulting in the long term loss of sesame cultivar diversity in Sudan. The Nuba continue to maintain their traditional sesame cultivars today, in part because of their isolation from market depots and the lack of adequate transportation systems (roads, vehicles) that prevent easy access to export markets.

There are no significant development projects in the region, even though plant breeders at the ARS station at Kudugli occasionally grow nurseries of traditional seed collections to select for desired characteristics such as higher yield, date of maturity, and disease resistance or drought tolerance, if applicable. The absence of significant development projects might actually serve to conserve the crop genetic diversity of Nuba groups. Often times, the university-educated agricultural "experts" have exhibited an arrogant contempt for the Nuba and a lack of appreciation of their demonstrated agricultural successes that have accrued from centuries of experimentation and trial-and-error testing. The "experts" have had no interest in learning about these ancient practices. Instead, they arrive on the scene with their own preconceived ideas, as representative of the dominant culture's unchallenged authority.

Nuba have observed that recommendations of the ARS staff or other outside experts are often incorrect. For example, Nuba farmers know that they must practice crop rotation, because if they do not, their fields will build up high densities of the parasitic pest, *Striga*, that lives on crop host species such as sorghum. Nuba cultivators rotate sorghum with sesame (Figure 2.1) to prevent the buildup of *Striga* parasite seeds (Bedigian and Harlan 1983). *Striga* is a root parasite that obtains its nutrients via the photosynthetic efforts of host plants. There are a number of different *Striga* species, and there is little host-specificity; over 100 different species of grasses serve as hosts for this parasitic plant. This is why *Striga* is such a terrible crop pest. It can easily spread from native savannah ecosystem grass hosts to gramineous crop hosts such as sorghum, and causes severe reductions in crop yields. Sesame and cotton are both examples of "trigger" crops that provide a suitable chemical environment for the germination of *Striga* seeds, even though they are nongramineous crop species that do not permit further growth or attachment of the parasite.

U.S.A.I.D. teams and university-trained ARS "experts" who are newcomers to Nuba lands prepared demonstration fields that consisted of sorghum monocultures. The mono-crops were grown on fields that were previously rotated, with a fallow period (see Figure 2.1) alternating with a three-crop cycle consisting of sorghum:cotton: sesame/groundnuts. The result of the sorghum monoculture during an eight-year period was a great increase in *Striga* seeds in the soil, which resulted in serious perennial infestations of the sorghum.

Control of both *Striga* on sorghum and the sesame pod borer, *Antigastra catalaunalis*, can also be accomplished by early planting of these crops. In this way, farmers can reduce the burden of labor and crop loss due to pests and parasites. Traditionally, pest populations have been minimized with overplanting, or by planting in times of low pest potential after the first rains (April-May). June is generally dry, and the rains usually begin in July. Hariq fires are often deliberately set after the first rains to remove germinated weed seeds, and then the desired crop seeds are sown. This greatly reduces the amount of weeding that needs to be done. However, useful weeds are permitted to remain in the fields. For example, wild okra is a prominent weed in sesame fields, and is tolerated and harvested by Nuba for food.

Cycles of crop rotation and fallow periods permit sustained, but relatively closed nutrient/pathogen cycles. Cowpeas are grown by Nuba farmers to restore soil fertility after more nutrient-demanding crops such as cotton and sorghum. A typical crop rotation cycle is cotton:sorghum:sesame:cowpeas followed by a fallow period of several years. Fallow periods generally become shorter as land pressure increases.

FIGURE 2.1 A sesame field at Kurgol, in the central Nuba Mountains, which is rotated for sesame and sorghum, but lies fallow for many years after five to eight years of continuous cropping. The trees, called *heglig* (*Balanites aegyptica*), are allowed to remain in the fields during farming cycles since they are used by many species of birds and produce a useful fruit that the Nuba consume as a laxative.

THE VALUE OF CONSERVING SESAME CULTIVAR DIVERSITY FOR CROP GENETIC IMPROVEMENT

Brown et al. (1989) and Oldfield (1989) have provided abundant reasons and numerous examples of the uses and values of conserving crop genetic resources. Oldfield documented specific crop genetic resources that have been used to improve particular crop species in various ways. Information on this subject, especially in relation to sesame, is sparse. For example, Harlan collected a broad sample of Turkish crop germplasm resources in 1948, as part of a U.S.D.A. expedition. He sent a total of 12,000 collections to the Plant Introduction Station in Beltsville, Maryland. However, these collections have not been well-maintained, and over the years much of the valuable ethnographic documentation has been lost. Today, much of what remains consists only of notations about the latitude and longitude of each collection. Nevertheless, the usefulness of these collections is undeniable. For example, one line of wheat collected by Dr. Harlan in a remote village (P.I. 178383) has been used extensively in breeding programs in the U.S. Pacific Northwest to introduce genetic resistance to five prevalent bacterial and fungal diseases. This single collection has contributed to such important releases as 'Moro,' 'Crest,' 'Elgin,' and 'Cardon' (Cunningham 1989). These discoveries were undertaken with much labor and expensive screening trials because wheat is a high cash value crop in world trade, and as such, research to improve this crop species is perceived as cost-effective.

It would be difficult to predict what precise genetic resources or "germplasm" might someday be derived from a particular Nuba cultivar to improve sesame varieties. At present, we have no research background from which to document this possibility, although this is a very typical dilemma in the case of minor crops. Oldfield and Alcorn (1987, see Ch. 3, this volume) observed that minor crops are generally most significant to cultures living at the subsistence level, yet, the extent to which minor crops are maintained off-site by *ex situ* technologies is minimal. Major industrialized countries have traditionally paid little attention to conservation and use of the genetic resources of minor crops used primarily in developing countries, even though preservation of such genetic resources may be of vital importance to local populations. In fact, many agricultural scientists in developing nations who have studied Western agricultural methods tend to reject the idea of utilizing the indigenous genetic resources of their own native crops.

In the case of sesame, support for this kind of research is nonexistent. No development programs, such as those sponsored by the U.S.A.I.D. or international agencies funded by the Rockefeller Foundation (such as CIMMYT and IRRI) are interested in sesame improvement. Sesame is a "poor man's crop." Therefore, mandates and assistance for sesame crop improvement research are nonexistent in agricultural assistance programs. Sorghum and cotton are the sole focus of attention in Sudan. Cotton, the most important cash crop that received attention during colonial rule, continues to retain its high priority status. Sesame has historically been regarded as of lesser significance than the dominant cereal crop in every region of the world. As a "poor man's crop," it requires much hand labor, and is not successfully harvested mechanically due to the tendency of its capsules to shatter (despite much effort by plant breeders to eliminate this undesirable trait). Even though sesame eludes the mandates of most crop genetic improvement programs, survival of rural people is a different concern than serving the world market. A different set of economic principles should operate when considering important subsistence crops like sesame. One reason this minor cereal crop has been successful in the hot, usually rainless conditions of Sudan is that its seeds are protected by the presence of antioxidants (sesamin and sesamolin); these chemical compounds allow sesame seeds to remain viable for many years in the absence of immediate planting.

Based on my studies of more than 350 sesame varieties from India in a nursery at the University of Illinois and on field observations and personal interviews conducted in India in 1979, a trend toward "modernization", is evident in India, and many small farmers are now cultivating improved sesame lines in the Indian center of sesame diversity. Among peoples in the developing world, the tendency is often to embrace the new. This will inevitably result in further losses of cultivar diversity, although growers do not appreciate what they are losing when they replace locally adapted landraces with "improved" cultivars. Similar trends may well influence the loss of sesame cultivar and genetic diversity in Sudan in the years to come.

Nevertheless, potentially useful genetic resources must be preserved for future generations. It is alarming that the world now relies on only a few major crop species (Harlan 1976b). This relatively recent phenomenon was not and is not characteristic of traditional subsistence agriculture. As this trend intensifies, human populations become more vulnerable to famine caused by natural disasters or pest outbreaks as their food supply increasingly depends on the genetic integrity and success of fewer crop species.

ACKNOWLEDGMENTS

Appreciation is extended to the Crop Evolution Laboratory, University of Illinois, officials of the Ministry of Interior of the Democratic Republic of the Sudan and at the Agricultural Research Service and Nuba Mountains Corporation in Kadugli. The author would also like to extend thanks to Roger Billington of Hunting Technical Services and his Sudanese counterpart Ahmed Khalid Showk, Mahmoud A. Mahmoud of the Agricultural Research Corporation, Wad Medani, and the people of the Nuba Mountains.

REFERENCES

Altieri, M.A. 1987. *Agroecology*. Westview Press, Boulder, Colorado.

Bedigian, D. 1988. *Sesamum indicum* L. (Pedaliaceae): Ethnobotany in Sudan, crop diversity, lignans, origin, and related taxa. Pages 315-321 *in* P. Goldblatt and P. P. Lowry (eds.), *Modern Systematic Studies in African Botany*, Vol. 25. AETFAT Monographs in Systematic Botany, Missouri Botanical Garden, St. Louis, Missouri.

Bedigian, D. and J.R. Harlan. 1983. Nuba agriculture and ethnobotany, with particular reference to sesame and sorghum. *Economic Botany* 37:384-395.

Bedigian, D., C.A. Smyth and J.R. Harlan. 1986. Patterns of morphological variation in *Sesamum indicum*. *Economic Botany* 40:353-365.

Brown, A.H.D., O.H. Frankel, D.R. Marshall and J.T. Williams, eds. 1989. *The Use of Plant Genetic Resources*. Cambridge University Press, Cambridge, United Kingdom.

Brush, S.B. 1986. Genetic diversity and conservation in traditional farming systems. *Journal of Ethnobiology* 6:151-167.

Cunningham, I.S. 1989. Jack R. Harlan: America's Vavilov. *Diversity* 5:34-35.

Harlan, J.R. 1951. Anatomy of gene centers. *American Naturalist* 85:97-103.

_______. 1975. Our vanishing genetic resources. *Science* 188:618-622.

_______. 1976a. Genetic resources in wild relatives of crops. *Crop Science* 16:329-333.

_______. 1976b. The plants and animals that nourish man. *Scientific American* 235:88-97.

Oldfield, M.L. 1989. *The Value of Conserving Genetic Resources*. Sinauer Associates. Sunderland, Massachusetts.

Oldfield, M.L. and J.B. Alcorn. 1987. Conservation of traditional agroecosystems. *BioScience* 37:199-208. (Ch. 3, this volume)

Yassin, T.E. 1978. Phenotypic variation in local sorghums in the Sudan Nuba Mountains. *Experimental Agriculture* 14:181-188.

3

Conservation of Traditional Agroecosystems*

Margery L. Oldfield and Janis B. Alcorn

INTRODUCTION

Much of the world's biological diversity is in the custody of farmers who follow age-old farming and land use practices. These ecologically complex agricultural systems associated with centers of crop genetic diversity include not only the traditional cultivars or "landraces" that constitute an essential part of our world crop genetic heritage, but also wild plant and animal species that serve humanity as biological resources.

To maintain the dynamic evolutionary processes involved in plant-human interactions, conservation of landrace varieties within their native agricultural landscapes, i.e., *in situ* conservation of traditional agroecosystems, has been proposed, but it is controversial. Crop genetic conservationists, who are keenly aware of the rapid and comparatively recent loss of traditional cultivars, usually consider such conservation programs to be impractical or impossible. At the same time, some wildlife conservationists consider traditional agriculture to be incompatible with *in situ* conservation of natural environments and wild taxa.

Both perspectives have influenced prevailing attitudes about conservation of biological diversity and traditional management of biological resources. Nevertheless, conservation efforts in the developing nations will ultimately fail if they are not integrated with the development needs of the people and prevailing land use practices.[1] We contend that both the human cultural resources, i.e., applied traditional knowledge, and the cultivated and wild biotic resources sustained within traditional agroecosystems are worthy of *in situ* conservation. In addition, the ecological-evolutionary role that traditional

*Reprinted from *BioScience* 37(3), with permission of the American Institute of Biological Sciences.

agriculturalists play in creating, maintaining, and enhancing crop genetic diversity should be acknowledged and incorporated within *in situ* conservation projects.

In situ conservation of ecosystems, with their indigenous species and genetic materials kept intact, has been termed the ideal (Frankel 1970) or preferred (IUCN-UNEP-WWF 1980) conservation strategy. Scientists and organizations have acknowledged the need for *in situ* conservation of crop genetic resources within traditional agroecosystems (Ugent 1970, Wilkes and Wilkes 1972, UNESCO-MAB 1974, 1984, Oldfield 1976, 1984, Brush 1977, Halffter 1985). The committee on Germplasm Resources of the U.S. National Academy of Sciences has recommended the United States adopt a policy to facilitate maintenance of traditional agroecosystems to conserve *in situ* the germplasm resources of major U.S. crops (NAS 1978).

In contrast, conservation of traditional agroecosystems has been considered impractical or unacceptable by some researchers (Frankel 1974, Prescott-Allen and Prescott-Allen 1982, Ingram and Williams 1984). The World Conservation Strategy (WCS), a proposal to link biological resource conservation with sustainable economic development, considers on site preservation impossible for domesticated species (IUCN-UNEP-WWF 1980, see WCS, sec. 6.5). Frankel[2] contends that preservation of traditional agricultural systems with their indigenous crop varieties intact is socially and economically impossible in the face of rapid technological advancements, especially within the developing nations. This view was echoed by the International Board for Plant Genetic Resources (IBPGR), which recommends *in situ* conservation primarily for wild crop relatives because "the interventions . . . required for the survival of landraces in agricultural landscapes would be too extensive to be considered" (IBPGR 1985).

The controversy over *in situ* conservation of traditional agroecosystems can be resolved only by examining carefully whether extant conservation policies and programs have achieved their goals; evaluating the oft-mentioned potential problems; and determining whether *in situ* conservation programs should be given the chance to demonstrate their feasibility and merits. In this paper we will address the following questions: What is the current conservation status of crop genetic resources? What can conservation of traditional agroecosystems achieve that *ex situ* methodologies cannot? What is the actual record of policy and program efforts in conservation of traditional agricultural systems? What considerations must guide the choice and maintenance of such *in situ* reserves, and what research efforts are needed for successful design and implemention of these conservation programs? Finally, what mechanisms are available for achieving traditional agroecosystems conservation?

STATUS OF CROP GENETIC RESOURCES

Genetic erosion–the extinction or depletion of traditional cultivars and distinct populations or species of wild crop relatives–has accelerated over the last half century. Despite decades of collecting activities (Hawkes 1983), much of the world's crop genetic heritage remains unsampled for *ex situ* conservation (Frankel and Soulé 1981, Lyman 1984). The remaining traditional agroecosystems serve, by default, as *in situ* genetic reservoirs for these unsampled crop genetic resources. The principal cause of genetic erosion has been the widespread adoption of modern crop cultivars within areas of ancient agriculture (Miller 1973, Frankel 1974, Harlan 1975, Oldfield 1984). But acculturation, replacement of mixed-crop subsistence farming by commercial agriculture, urbanization and industry, irrigation and drainage projects, tourism, and war have also contributed to genetic erosion (Wilkes and Wilkes 1972, Miller 1973, Brush 1977, Johannessen 1982, Oldfield 1984, Nabhan 1985).

The International Board for Plant Genetic Resources (IBPGR) was established in 1974 when the widespread loss of valuable crop genetic resources became so alarming that crop plant collectors, breeders, and conservationists recognized the need for an international network of *ex situ* crop germplasm repositories. The IBPGR organizes and funds cooperative collection expeditions with countries harboring major crop gene centers to facilitate *ex situ* conservation both within the country of origin and in specific facilities that form the IBPGR network of gene banks. This network provides plant breeders with the genetic resources necessary for the development of higher-yielding crop varieties that are more resistant to or tolerant of diseases, insects, herbicides, infertile and saline soils, drought, and temperature extremes (IBPGR 1975, Hawkes 1983, Plucknett et al. 1983). The IBPGR also designates priority crops and regions urgently in need of collection of crop germplasm resources (IBPGR 1976, 1981). However, present IBPGR policy dictates that future collecting expeditions will be limited to rescue operations urgently required for specific crop genetic resources (Holden 1984).

This restriction is unfortunate since so much of the world's crop genetic diversity is not well represented in *ex situ* reservoirs. The unsampled species include most of the endangered traditional cultivars and threatened wild relatives of wheat, sorghum, millet, potato, common bean, runner bean, cowpea, mungbean, Bambara ground nut, peanut, chickpea, and lupin (Lyman 1984). Half or more of the primitive cultivars of sweet potato, yams, finger and kodo millet, common bean, runner bean, Bambara ground nut, faba bean, and lupin have yet to be collected systematically. Moreover, significant proportions of the wild gene pool resources of many world crops, particularly wild rices, barley, millets, sorghum, sweet potato, cassava, yams, and most grain legumes, also remain uncollected; most of these wild species are important vegetational components associated with traditional agroecosystems.

ADVANTAGES OF AGROECOSYSTEM CONSERVATION

The primary advantage of *in situ* conservation of traditional agroecosystems is that crop populations continue to be influenced by the evolutionary processes that created and maintained the genetic resources now frozen in *ex situ* gene banks. Maintenance of crop plant-pest coevolutionary interactions, in particular, is a major objective, because traditional cultivars and their wild and weedy relatives continue to provide essential sources of genotypic resistance to disease pathogens, nematodes, insects, and other crop pests (Oldfield 1984). In many instances, genetic resistance has proved to be the only effective or applicable method of pest control (Leppik 1970). Moreover, studies of crop-pathogen interactions in traditional agroecosystems have facilitated our understanding of phytopathology, providing knowledge that has been beneficial to modern agriculture (Glass and Thurston 1978, Browning 1981).

Another advantage of *in situ* conservation of traditional agriculture is that it maintains the weeds and wild crop relatives associated with the cultivated species. These crop relatives have contributed to the rapid evolution of the cultivated species (Harlen 1969, de Wet and Harlan 1975). Weedy relatives also are one of the most significant reservoirs of germplasm for use both in maintaining primitive cultivars *in situ* and in improving modern cultivars (Harlan 1965, Ladizinsky 1985). In fact, traditional cultivators' potential for creating and maintaining genetic diversity within crop-weed complexes may be as significant for genetic resource conservation as their role in maintaining traditional cultivars.

The cultural and ecological diversity within traditional farming systems of the ancient agricultural regions has shaped the genetic variation within traditional crop varieties (Harlan 1975). Genetic diversity is maintained in traditional agroecosystems by cultural intervention as well as by natural selection. Cultural and socioeconomic factors that influence evolutionary processes include terracing, fallowing, runoff channeling, soil mounding, mulching, use of fire, selective weeding, interplanting, trapping of particular crop predators, attention to individual plants, pollarding of selected wild trees in agriculture fields, and management of secondary successional vegetation.

Unlike modern U.S. farmers who do not use the harvest from their fields for seed, traditional farmers are both seed stock producers and crop breeders. Seed selection and dispersal patterns, weeding regimes, harvest procedures, and storage practices create selection pressures. Because traditional farm management strategies often integrate livestock and wildlife production with crop cultivation, the farmers use complex production schedules. They also place the crops to create a patchy distribution of cultivated fields, different human-modified vegetation zones, and natural resource zones. Fallows, pastures, savannahs, riverbanks, forests, and rocky desert outcrops are included in traditional agroecosystems. This habitat diversity influences the organisms (e.g., insects and weeds) that enter the fields and interact with the crop plants.

The ecological complexity of traditional farms may seem surprising to those accustomed to modern temperate-zone farms. The complexity maintains high genetic variation due to temporal and spatial variation in selection pressures, intergenotypic and interspecific competition, and interactions with pests and pathogens.

Traditional farmers' fields are spatially heterogenous; each field contains multiple microhabitats created by the varying slopes and soils and different patches of intercropped cultivars and wild vegetation. There is also temporal variation, which may be cyclical, as when crops are regularly planted two or three times during the year and therefore are subjected to different climatic regimes. The temporal variation also may be unpredictable, as when individual farmers plant at different times each year (Alcorn 1984). Individual farmers are also likely to use slightly different farming methods or choice of seeds in one of several field types in any given year.

Some of the anthropogenic component of the spatial and temporal heterogeneity is deliberate; traditional farmers commonly use microclimate management as an agricultural tool. But some of the heterogeneity is simply a by-product of activities employed to another end. Different cultivars may be intentionally placed in specific niches within cultivated fields, or genotypes of one cultivar population may be scattered randomly throughout different niches so that individuals in such populations are subjected to entirely different selection pressures.

Crop plants are also involved in competition with individuals of the same species and in competition with neighboring weeds or with other crops cultivated in the same field. The effects of intergenotypic and interspecific competition vary according to microenvironmental factors and the densities of the interacting populations–factors that are influenced by site preparation, planting mechanics, and various other cultural practices.

A further source of heterogeneity is the individual variation and experimentation that are cultural aspects of traditional agriculture systems (Johnson 1972, Johannessen 1982, Alcorn 1984). Traditional farmers follow "scripts" of a general agriculture plan passed down from their ancestors, but these age-old patterns of land use change as each individual farmer chooses how to manage her or his land within the common cultural context.

Heterogeneous environments created within traditional agroecosystems not only provide a variety of selection pressures, but they also afford some crop plants protection from environmental stresses. Such protection is not provided by modern agroecosystems. For example, polyculture and other methods that use genetic diversity to create patchy environments (see Figure 3.1) affect crop plant-pest interactions (Janzen 1973, Glass and Thurston 1978, Browning 1981). Additionally, farmers may alter the community with which crop plants interact by protecting agronomically desirable, but weak, crop plants, cultivars, or useful wild vegetation. The protected elements differ from culture to culture, again creating different plant communities in different areas, thus enhancing overall crop genetic diversity.

Overlaid upon these effects is the diversity-enhancing influence of trade. Complex histories of trading networks among traditional farming cultures are believed to have played a major role in the crop evolution process (e.g., Harlan 1975, Brush et al. 1981, Johannessan 1982). While there is often strong selection for seed adapted to the conditions of a particular farmer's set of fields, opportunities also exist for introduction of lines from other local fields or more distant regions. The trading of seeds or other propagules permits gene exchange via hybridization among genetically distinct varieties, thus further enhancing the accumulated wealth of crop genetic resources (Harlan 1969, de Wet and Harlan 1975).

Another distinct advantage of agroecosystem conservation is that processes essential for the continuing evolution and maintenance of crop genetic resources can be conserved despite our present ignorance of these processes or their complex functions. It is currently difficult to make generalizations about how particular human activities or environmental conditions have influenced crop genetic diversity. The biology, breeding system, range of variability, genetics, and cytogenetics of a taxon influence the ways that cultural practices and ecological conditions affect evolutionary processes. In addition, there are tremendous differences among traditional agroecosystems around the world (Klee 1980); even within a nation there is often variation among agroecosystems supporting the same crop.

As traditional farmers abandon their landraces in favor of modern high-yielding varieties, they often discard or fail to maintain the seed stocks of their native cultivars. From a socioeconomic perspective, *in situ* conservation of traditional agroecosystems ensures that seeds of

FIGURE 3.1 Traditional farmers manage *in situ* reserves of genetic diversity within their polycultural fields, where crops continue to evolve in response to natural and artificial selection pressures. In this photograph, a Mexican Huastec Indian woman stands in a typical field, containing numerous species of wild and cultivated plants.

native crop varieties will be readily available in the event that subsistence farmers now growing modern cultivars need to rely upon their more traditional crops in the future.

Local varieties will yield a harvest dependably despite adverse agricultural or socioeconomic conditions. Because most peasant farmers live under unstable political and socioeconomic conditions, they are particularly vulnerable to disruption of the research, production, and trade networks that supply seeds of new improved cultivars (Plucknett and Smith 1986) and the mechanical and chemical inputs often necessary for these modern cultivars to produce an acceptable yield. If access to these networks is disrupted, if modern cash crops fail to produce dependable subsistence, or if use of genetically susceptible modern varieties results in widespread crop failures (Chang 1984), farmers may wish again to cultivate their traditional food cultivars. In such situations, seed stocks of traditional cultivars stored in distant institutions are not readily available to traditional or subsistence farmers.

Finally, maintenance of traditional knowledge of plant resources and farming practices within traditional agroecosystems ensures the conservation of a much greater variety of genetic resources than can possibly be conserved by conventional *ex situ* methods. Traditional farmers typically employ practices that enhance intraspecific, and in some cases interspecific, crop genetic diversity. Furthermore, they practice polyculture, cultivate minor crops, and maintain wild edible, medicinal, or industrially important plant species (Ugent 1970, Glass and Thurston 1978, Alcorn 1981, Oldfield 1984).

HISTORY OF CONSERVATION PROGRAMS

International programs for *in situ* conservation of traditional agroecosystems began more than a decade ago. In 1971, the UNESCO Man and the Biosphere (MAB) program established the biosphere reserve plan, under a project entitled "Conservation of natural areas and of the genetic materials they contain" (UNESCO-MAB 1974, Gregg and McGean 1985). The project's expert panel initially recommended that "samples of varied and harmonious landscapes resulting from long-established land use patterns should be maintained" in a manner that will conserve the ecological and cultural integrity and genetic diversity within these man-modified ecosystems.

The biosphere reserve management plan is unique among park plans in that it integrates wilderness conservation with rural development (von Droste zu Hulshoff and Gregg 1985). The biosphere reserves include resident humans as an integral component. Therefore the biosphere reserve program is, potentially, an important vehicle for international conservation of traditional agroecosystems and the adjacent natural environments with which they interact (Oldfield 1984, Halffter 1985, von Droste zu Hulshoff and Gregg 1985). Reserve areas are planned and administered by national agencies, which modify UNESCO-MAB guidelines to fit the social and ecological contexts of specific countries[3] (Cifuentes et. al. 1984, Halffter 1985).

The strategy of conserving traditional farming cultures and cultivars has been stressed only recently, although *in situ* conservation of wild crop relatives and a commitment to integrated rural development projects that use traditional knowledge and land use practices have long been objectives of this UNESCO-MAB project (UNESCO-MAB 1984, Halffter 1985). New biosphere reserves with potential for *in situ* conservation of traditional agroecosystems have been proposed or created in India (Government of India 1982), Costa Rica, and México.[4] To date, however, no model reserves in developing countries fully demonstrate the concept of a biosphere reserve (Gómez-Pompa 1984).

A major step in this direction is the current effort to incorporate the Kuna Yala Indigenous Reserve within the MAB biosphere reserve system (Houseal et al. 1985). The project provides a useful model for establishing *in situ* reserves containing indigenous agroecosystems. The Kuna Yala Indigenous Reserve, located on the northeastern coast of Panama, includes 60,000 hectares of tropical forests. The Kuna are applying ecologically sound resource management strategies and cultural development objectives consonant with those of the UNESCO-MAB biosphere reserve program (Houseal et al. 1985, Wright et al. 1985). The Centro Agronómico Tropical de Investigación y Enseñanza (CATIE) has advised the Kuna on economic development in the context of a park project. Funds for the project have been provided by a wide variety of contributors including the U.S. Agency for International Development (A.I.D.), the World Wildlife Fund, the Inter-American Foundation, the Smithsonian Tropical Research Institute, and the Kuna community. CATIE and the Smithsonian Institution also have provided a technical and scientific advisory group.

A 20-person Kuna staff is being trained in modern techniques to complement traditional methods for management and protection of Kuna natural resources. CATIE and Smithsonian scientists are in turn documenting Kuna ethnobotanical knowledge, agricultural techniques, and silvicultural management strategies to increase their understanding of the tropical forest ecosystem and its indigenous resources. The Kuna community is considering the option of biosphere reserve designation[5] and will make the final decision to accept or reject that option (Wright et al. 1985). If the Kuna accept the biosphere reserve designation, it means that they intend to follow the MAB guidelines and action plan–the "recipe"–for a biosphere reserve.[6] Acceptance does not guarantee the Kuna economic or technical aid for developing their reserve, but it will give them the status to apply for such aid (Gregg and McGean 1985).

The U.S. government's involvement with *in situ* conservation of crop genetic resources and traditional agroecosystems followed a report by the committee on Germplasm Resources of the National Academy of Sciences (NAS 1978). In November 1981, a three-day Strategy Conference on Biological Diversity was held at the State Department in Washington, D.C. This conference concluded that all countries should endeavor to maintain maximal genetic diversity of their indigenous biotic resources by both *in situ* and *ex situ* means and that national inventories of genetic resources should be conducted in areas of traditional cultivation as well as in protected natural areas and gene banks (U.S. Department of State 1982). In November 1983, the U.S. Congress amended the Foreign Assistance Act of 1961 to direct A.I.D. to collaborate with other national agencies to develop a national strategy for conserving biological diversity within developing countries. In compliance with this directive, A.I.D. organized the Interagency Task Force on Biological Diversity (ITFBD).

The task force's report to Congress (ITFBD 1985) noted a variety to U.S. government activities. The Departments of State, Interior, Agriculture, and Commerce, and the Environmental Protection Agency are participating in the UNESCO-MAB biosphere reserve program. A.I.D. provides substantial financial support to the IBPGR (IBPGR 1985). The Department of Agriculture has increased its interest in relatives of wild crops as sources of novel genes for breeding programs and is currently developing criteria for selection of natural areas for *in situ* conservation of crop species and their wild relatives. The Division of Biotic Systems and Resources of the National Science Foundation has designated characterization of the genetic and biological diversity of cultivated plants as one of its high priority areas for increased research support. And the Peace Corps has expanded its efforts in natural resources conservation and provided volunteer assistance for reserves and parks in developing nations.

The ITFBD also issued recommendations. When planning new agriculture projects, A.I.D. should encourage host governments (and provide the requisite funding) to evaluate the comparative potential of extant traditional agroecosystem production systems and more intensively managed, modern agroecosystem alternatives. The U.S. government, via A.I.D., should provide continuous support for ongoing training efforts in developing nations for *in situ* germplasm protection. The U.S. government, through A.I.D., various U.N.

programs, and multilateral donor banks, should provide additional long term support for the growth in the developing nations of research capabilities to create highly productive, sustainable agricultural systems that include localized subsistence economies and use of indigenous species. According to the recommendations, the U.S. government also should develop or strengthen the institutions concerned with both *in situ* and *ex situ* crop germplasm protection to assist the developing nations in preserving their indigenous plant species for possible agricultural uses (ITFBD 1985).

Government agencies of other nations have also made an effort to link rural development with traditional agriculture and *in situ* conservation projects. In México, for example, the National Research Institute for Biotic Resources (INIREB) has numerous programs aimed at integrating research,, biological reserves, and development of sustainable agroecosystems based on traditional management techniques (Gómez-Pompa and Giddings 1986). In Ecuador, the World Wildlife Fund is sponsoring a project that links conservation with development of traditional agriculture among the Siona Secoya people in the Ecuadorian national park Cuyabeno Reserve.[7] The framework of these projects contains elements necessary to support *in situ* conservation of crop genetic resources, although it is not an explicit goal.

Aside from reserves administered by national and international organizations, few other avenues have been proposed for conserving traditional agroecosystems. Some proponents of dynamic *ex situ* conservation recommend developing a system of village-level landrace custodians, e.g., to cultivate limited samples of endangered landraces native to a given region (Frankel and Soulé 1981). Iltis (1974) suggested "freezing the genetic landscape" by subsidizing farmers to continue traditional agriculture in selected zones from which agricultural and economic development would be deliberately excluded. Wilkes and Wilkes (1972) suggested setting aside 100 carefully chosen, sizeable (5 km x 20 km) strips of land under traditional agriculture around the world.

A few nonprofit citizen action groups have also attempted to promote *in situ* conservation of genetic resources. The U.S. organization Native Seeds/SEARCH based in Tucson, Arizona, for example, has complemented its *ex situ* conservation efforts by encouraging farmers to grow traditional crop cultivars. The organization provides the farmers with seed of varieties they have lost, and then purchases the farmers' excess production.[8] This organization has also promoted use of the tribal elders' knowledge of seed selection, seed storage, and traditional crop cultivation in local agricultural improvement programs (Nabhan 1985).

SITING AND MAINTAINING RESERVES

The process of establishing and managing *in situ* reserves is influenced by biological, logistic, socioeconomic, and political considerations. Policy makers need biological information to establish guidelines for requisite size, optimal boundary buffer zones, degree of isolation, proper genetic constitution of target crop plant populations, and maintenance procedures of *in situ* reserves. But biologists know little about the genetic makeup, demography, or population dynamics of local cultivar populations (Frankel and Soulé 1981).

This problem can be rigorously solved only by study of selected populations of the cultivars targeted for conservation; but most developing countries lack the strong scientific community necessary for undertaking such research projects (Gómez-Pompa 1984, Gómez-Pompa and Giddings 1986). However, with data on a particular crop and the traditional regimes under which it has been cultivated, informed choices can be made regarding which of the agroecosystem's principal elements should be retained within potential reserves.

One important biological consideration affecting siting and maintenance has been identified by Frankel: coadapted gene complexes must not be disrupted. Frankel (1970) has criticized proposals for scientifically controlled *in situ* reservoirs, i.e., maintenance of crop populations in experiment stations or substations located within the original environment. He contends that " . . . drastic changes in the environment and/or extensive introgression from alien cultivars may disrupt the genetic system [the cultivars] have acquired". Coadapted gene complexes within traditional cultivars could be broken apart by interbreeding with other varieties held in the same reserve, or with alien cultivars maintained in contiguous areas. Guidelines for selecting and maintaining *in situ* sites should include provisions for predicting the likelihood of radical changes in traditional farming practices within the potential site and for evaluating agricultural practices in bordering regions.

Achieving sufficient representation of diverse genetic resources is another consideration. There are biological and logistic problems in the process of sampling and maintenance of samples. Choices will have to be made on a country-by-country basis as to whether *in situ* conservation will focus on specific crops; on diversified traditional agroecosystems, and thereby on the array of crops found within them; or on both specific crops and a traditional agroecosystem. While it may be impossible to maintain all or even most crop genetic resources within an extensive network of small, isolated *in situ* reserves (Frankel 1970),[9] *ex situ* reserves are similarly plagued by problems.

The representation of crop genetic resources within *ex situ* reserves is limited by problems inherent in collection and conservation processes, rather than by site location problems. Among the greatest difficulties are maintenance of vegetatively propagated stocks and "recalcitrant" seeds that are not suited to conventional *ex situ* cold storage methods. Conservation of these resources typically requires living collections *ex situ*, which are very expensive in terms of labor and space, in part because they are often attacked by pathogens. Therefore, conserved stocks of recalcitrant seed, especially of many important tropical species (IBPGR 1985), and vegetatively propagated crops must be significantly smaller than those of other crops and are more endangered in storage (Frankel 1978, Lyman 1984). Yet more than one-fourth of the people living in the tropics rely on vegetatively propagated crops for their staple food (Coursey 1974). By focusing on recalcitrant seeds and vegetatively propagated crops, *in situ* conservation efforts could contribute more than *ex situ* reservoirs toward coverage of many tropical crop genetic resources.

Another problem with *ex situ* conservation is that the genetic integrity of all seed stocks deteriorates with time due to genetic drift and artificial selection pressures, such as those related to cold storage. Crop seeds are routinely withdrawn from cold storage for regeneration. During crop regeneration cycles, seed of particular crop cultivars is transported to a suitable agricultural environment, often in a tropical or subtropical climate, to be cultivated for purposes of stock maintenance. In these situations, unintentional outcrosses between distinct accessions can alter stock purity. Entire stocks have also been lost due to climatic conditions, natural hazards, and human error (Oldfield 1976, 1984, Roos 1984). Thus while in theory *ex situ* methodologies could achieve maximal coverage, in fact only limited samples of crop genetic diversity are being successfully maintained.

A difficult problem facing *in situ* conservation is the lack of national inventories of traditional cultivars and agriculture sites (U.S. Congress [OTA] 1986). Site selection requires analysis of data on sizes and distribution of cultivar populations; the number and proportion of indigenous cultivars, as opposed to introduced modern varieties, being cultivated in an individual farmer's fields in specific localities; how the cultivation of indigenous varieties is integrated with the cultivation of improved varieties; the specific kinds of traditional agriculture carried out in a given country; and how traditional cultivators are integrated into the national economy.

Many critics of *in situ* conservation projects emphasize socioeconomic problems. Frankel (1974) has maintained that *in situ* conservation of landraces is "an economic and social impossibility". How can policy makers encourage farmers to maintain viable populations of traditional cultivars or crop-weed complexes despite pressures to follow other agricultural paths? Conservation of traditional agroecosystems in the midst of agricultural modernization efforts in Third World countries is not only technically and scientifically complicated, but is also politically sensitive. Policy makers must establish that they are not attempting to freeze the peasantry in what may be perceived to be a backward way of life. Indigenous groups striving to maintain their cultural integrity are excellent candidates for inclusion in agroecosystem conservation programs.

A basic economic problem is that the production of staples does not usually offer great economic rewards; given the opportunity, farmers will cultivate cash crops when their subsistence needs have been met (e.g., Alvarez 1977, Barlett 1977, Brush et al. 1981). When any *in situ* conservation program limits or reduces perceived opportunities for cultivating cash crops, financial subsidies will be necessary. Incentives would have to be dependable and strong enough to ensure long term farmer participation in the conservation project.

However, introduction of subsidies may build unrealistic expectations and hasten socioeconomic change. The sudden infusion of money from outside sources unlinked to existing sociopolitical frameworks can have a dramatic effect on the social structure and farming decisions of relatively closed local economic systems. Can *in situ* conservation programs based upon community participation allow the continued social, cultural, and economic evolution of the community while ensuring conservation of traditional cultivars?

Another socioeconomic problem involves changes in land tenure. A program could lose its participants if land to be included in an *in situ* reserve shifts ownership from subsistence farmers to large-scale, commercial agriculturists. However, this problem is not likely to occur in marginal areas.

Political problems are interwoven with the socioeconomic problems. Site and maintenance decisions must reflect extraneous factors that affect successful implementation of any government-sponsored program (Grindle 1980). Social cohesion and a well-organized local political structure would be advantageous, perhaps necessary, for a farming community's participation in an agroecosystem conservation program. Because indigenous people can gain legal knowledge and political clout by cooperating with international conservation efforts (Brownrigg 1981), they may regard participation as politically advantageous. But national governments, may, for their own political and economic interests, oppose strengthened indigenous political clout, and may therefore resist the projects.

Until recently, conservation has been perceived as the domain of natural scientists. But the challenge of traditional agroecosystem conservation requires that policy makers include social scientists and area specialists among their advisors.

A PROPOSAL FOR FUTURE DIRECTIONS

What can be done to maintain optimally our genetic resources? We suggest a three-pronged approach consisting of *in situ* conservation linked with economic development efforts; *in situ* conservation and research centers; and *ex situ* reserves. We will always need *ex situ* genetic reservoirs as repositories for samples of the world's genetic resources, and research should continue on ways to upgrade *ex situ* methodologies (e.g., cryopreservation of meristem cultures). Nevertheless, *in situ* conservation programs could offset some of the major deficiencies of the *ex situ* approach (IBPGR 1985). These alternative *in situ* programs would seek to complement the *ex situ* reservoirs.

Modern crop breeders working in developed countries have shaped the prevailing view (Frankel and Soulé 1981) that traditional cultivars have two values for which they need to be conserved: as a bank of genes for insertion into modern cultivars and as a bank of gene complexes–adapted to specific environmental conditions–that could be improved by insertion of genes from other stocks. However, traditional cultivars have a third, more critical value for people in the developing nations. Locally adapted varieties of traditional crops provide dependable subsistence for many people in large areas where modern crop varieties attuned to local conditions are not available. Approaches linking *in situ* conservation with rural development efforts reflect a concern for all three of these values.

Efforts to integrate *in situ* conservation with rural development would probably be incapable of conserving landraces under traditional conditions. Development involves change. However, traditional farming has never been unchanging. The natural and cultural elements of the agroecosystem in which cultivars evolve have changed and will continue to change over time. But the new cultivar populations that develop would retain many of the adaptive complexes that are the mark of today's landraces. In addition, they would include many different genotypic combinations in dynamic interaction with evolving populations of pests and pathogens, unlike seed stocks conserved in *ex situ* reservoirs.

In situ reserves thus envisioned would contribute to Third World development in other ways. Modern cultivars created for use under ideal field conditions often have not met peasants' needs for predictable success when grown under the less than optimal conditions of peasants' fields. Furthermore, use of modern crop varieties that are not genetically resistant to or tolerant of specialized pathogens, parasites, or insects and other herbivores, has increased the vulnerability of agriculture production systems to major epidemics (NAS 1972, Wilkes and Wilkes 1972, Browning 1981, Brush et al. 1981, Johannessen 1982, Chang 1984). In contrast, the landraces propagated in conservation-development programs would be adapted to local conditions and therefore yield predictable subsistence.

Conservation-development programs could include research that may lead to scientifically based agricultural programs for the different natural and social conditions characteristic of tropical lands. However, *in situ* conservation sites should not be constructed as "cages," holding "primitive" farmers, crops, and agroecosystems for study by outside scientists. The participation of farmers educated in the goals and methods of the program would be necessary for its success (Gladwin and Murtaugh 1980, IPC 1984). The expertise of interdisciplinary research teams who have participated in small farmer development projects could also contribute to the success of such an approach.

The second approach would involve *in situ* conservation-research stations independent of private farmers. These research centers would both emulate the evolutionary processes occurring in farmers' fields and study evolution in traditional agroecosystems of the surrounding region. This approach has the advantage that it does not rely on freezing natural and social processes in traditional agricultural communities.

Any successful *in situ* research at the conservation stations will have to be based on careful studies of processes occurring in traditional agroecosystems. Such *in situ* research-conservation facilities will evolve as their operators learn their requisite roles and activities. From an emphasis on research about what needs to be done, there will be a transition period to the station's actual maintenance of genetic resources.

This approach also fulfills a conservation goal not being met by *ex situ* reservoirs. Intensive research on cultivar populations carried out at these stations could provide information on genetic differentiation and evolution of crop species (Frankel and Soulé 1981). Few *ex situ* collections can be used as sources of data to study evolutionary processes because some landraces are not represented or have been unevenly sampled; sampling procedures are not standardized and are often poorly documented; diversity in a given sample has been reduced by selecting a few individuals for preservation, while discarding others; the genetic integrity of the population's record has been altered by selection, hybridization, and genetic drift during reproduction for maintenance; information is lacking as to the collection's precise geographic origin; and the possible impact of modernization on the population at the time of collection was not recorded (Frankel and Soulé 1981).

Research-conservation stations ideally should be linked to a network of conservation-development program sites. Data gathered by one could be used by the other. In many ways, proposed MAB projects unite the two approaches. The program's initial indigenous culture biosphere reserves will reveal strengths and weaknesses of *in situ* conservation efforts. In conjunction with the United Nations Environment Programme (UNEP), UNESCO-MAB is now encouraging national governments to nominate specific sites as pilot rural development projects based upon indigenous knowledge. Emphasis is placed on integration of traditional and scientific knowledge of plant resources to further traditional uses.

With the involvement and encouragement of ethnobotanists, genetic conservationists, biologists, geographers, and anthropologists, *in situ* conservation of traditional agroecosystems in biosphere reserves may soon be realized. Researchers who have studied traditional cultivars, cultivators, and agroecosystems are urged to contact national MAB committees to provide information relevant to conservation of areas characterized by long-standing traditional land use and management practices. They should also nominate for biosphere reserve status promising areas for which sufficient information is available.

Any network of *in situ* conservation-development project sites and research-conservation stations will probably start out on so small a scale that some might say they would make an insignificant contribution toward solving the genetic resources conservation crisis. But during this stage, problems can be identified, and the programs can respond to these problems. Information from ongoing research will also contribute to the evolution of these programs. Today only speculations and opinions inform policy analysts and decision makers considering traditional agroecosystem conservation. Therefore, even data from early pilot programs will aid discussions about the feasibility of traditional agroecosystem conservation projects aimed at salvaging our crop genetic heritage.

ACKNOWLEDGMENTS

We thank Jim Barborak, J. Artie Browning, Bill Denevan, Dennis Glick, Arturo Gómez-Pompa, Bill Gregg, Gary Nabhan, Molly Kux, C.O. Qualset, Deborah Rabinowitz, and Hernán Torres for useful information and comments on earlier drafts of this article. Research for this paper was partially funded by U.S. Department of the Interior Contract No. CX0001-7-0017 to M.L. Oldfield and grants to J.B.

Alcorn from the Tinker Foundation awarded through the Mesoamerican Ecology Institute at Tulane University and from the American Philosophical Society. The authors, however, retain sole responsibility for the paper and any shortcomings therein.

NOTES

1. See Government of India, 1982, and also comments of Russell E. Train in *BioScience* 36:59 (1986).
2. O.H. Frankel, 1983, pers. comm. Division of Plant Industry, Commonwealth Scientific and Industrial Research Organization (CSIRO), Canberra City, Australia.
3. H. Torres, 1986, pers. comm. Asesor, Proyecto Parque International La Amistad, Centro Agronomico Tropical de Investigaciónes y Enseñanza (CATIE), Turrialba, Costa Rica.
4. J. Barborak, 1986, pers. comm. Wildlands Program, Centro Agronómico Tropical de Investigaciones y Enseñanza (CATIE), Turrialba, Costa Rica.
5. W.P. Gregg, 1985, pers. comm. U.S. UNESCO-MAB Secretariat, National Park Service, U.S. Department of the Interior, Washington, D.C.
6. J. Barborak, 1986, pers. comm. Wildlands Program, Centro Agronómico Tropical de Investigaciones y Enseñanza, (CATIE), Turrialba, Costa Rica.
7. D. Glick, 1986, pers. comm. Wildlands and Rural Development, World Wildlife Fund, Washington, D.C.
8. G. Nabhan, 1986, pers. comm. Native Seeds/SEARCH, Tucson, Arizona.
9. O.H. Frankel, 1983, pers. comm. Division of Plant Industry, Commonwealth Scientific and Industrial Research Organization (CSIRO), Canberra City, Australia.

REFERENCES

Alcorn, J.B. 1981. Huastec non-crop resource management: Implications for prehistoric rain forest management. *Human Ecology* 9:395-417.

__________. 1984. *Huastec Mayan Ethnobotany*. University of Texas Press, Austin.

Alvarez, J. 1977. *Traditional and commercial farm supply response in agricultural development: The case for basic grains in Guatemala.* Ph.D. dissertation, University of Florida, Gainesville.

Barlett, P.F. 1977. The structure of decision making in Paso. *American Ethnologist* 4:285-308.

Browning, J.A. 1981. The agro-ecosystem-natural ecosystem dichotomy and its impact on phytopathological concepts. Pages 159-172 *in* J.M. Thresh (ed.), *Pests, Pathogens and Vegetation*. Pitman Advanced Publications, Boston.

Brownrigg, L.A. 1981. Native cultures and protected areas: Management options. Pages 65-77 *in* International Union for the Conservation of Nature and Natural Resources (IUCN), *Conserving the Natural Heritage of Latin America and the Caribbean*. IUCN, Gland, Switzerland.

Brush, S.B. 1977. Farming the edge of the Andes. *Natural History* 86:32-41.

Brush, S.B., H.J. Carney and Z. Huaman. 1981. Dynamics of Andean potato agriculture. *Economic Botany* 35:70-88.

Chang, T.T. 1984. Conservation of rice genetic resources: Luxury or necessity? *Science* 224:251-256.

Cifuentes, M., C. MacFarland and R. Morales. 1984. Strategic planning of national or regional systems of biosphere reserves: A methodology and case study from Costa Rica. Pages 93-120 *in* UNESCO-UNEP, *Conservation, Science and Society*. UNESCO, Paris.

Coursey, D.G. 1974. Some culture-historical determinants of tropical agriculture research priorities. *Bulletin of the Tropical Production Institute* 1974:4-13.

de Wet, J.M.J. and J.R. Harlan. 1975. Weeds and domesticates: Evolution in the man-made habitat. *Economic Botany* 29:99-107.

Frankel, O.H. 1970. Genetic conservation in perspective. Pages 469-489 *in* O.H. Frankel and E. Bennett (eds.), *Genetic Resources in Plants*. Blackwell Scientific Publications, Oxford, United Kingdom.

________. 1974. Genetic conservation: Our evolutionary responsibility. *Genetics* 78:53-65.

________. 1978. Conservation of crop genetic resources and their wild relatives: An overview. Pages 123-149 *in* J.G. Hawkes (ed.), *Conservation and Agriculture*. Allanheld and Schram, Montclair, New Jersey.

Frankel, O.H. and M. E. Soulé. 1981. *Conservation and Evolution*. Cambridge University Press, New York.

Gladwin, H. and M. Murtaugh. 1980. The attentive-preattentive distinction in agricultural decision making. Pages 115-136 *in* P.F. Barlett (ed.), *Agricultural Decision Making*. Academic Press, New York.

Glass, E.H. and H.D. Thurston. 1978. Traditional and modern crop protection in perspective. *BioScience* 28:109-115.

Gómez-Pompa, A. 1984. The biological resources of biosphere reserves. Proceedings of the Conference on the Management of Biosphere Reserves, Gatlinburg, TN. National Park Service, U.S. Department of the Interior, Washington, D.C.

Gómez-Pompa, A., and L.E. Giddings. 1986. INIREB's new approach to applied research, development, and teaching. Pages 31-40 *in* *Journal* '86, The World Resources Institute, Washington, D.C.

Government of India. 1982. *Indo-U.S. binational workshop on conservation and management of biological diversity: Report and recommendations*. Department of Environment, Government of India, New Delhi.

Gregg, Jr., W.P. and B.A. McGean. 1985. Biosphere reserves: Their history and promise. *Orion Nature Quarterly* 4(3):41-51

Grindle, M.S., ed. 1980. *Politics and Policy Implementation in the Third World*. Princeton University Press, Princeton, New Jersey.

Halffter, G. 1985. Biosphere reserves: Conservation of nature for man. *Parks* 10(3):15-18.

Harlan, J.R. 1965. The possible role of weed races in the evolution of cultivated plants. *Euphytica* 14:173-176.

______. 1969. Evolutionary dynamics of plant domestication. *Japanese Journal of Genetics* 44(Suppl. 1):337-343.

______. 1975. Our vanishing genetic resources. *Science* 188:618-622.

Hawkes, J.G. 1983. *The Diversity of Crop Plants*. Harvard University Press, Cambridge, Massachusetts.

Holden, J.H.W. 1984. The second ten years. Pages 277-285 *in* J.H.W. Holden and J.T. Williams (eds.), *Crop Genetic Resources: Conservation and Evaluation*. G. Allen and Unwin, London.

Houseal, B., C. MacFarland, G. Archibold and A. Chiari. 1985. Indigenous cultures and protected areas in Central America. *Cultural Survival Quarterly* 9(1):10-20.

Iltis, H.H. 1974. Freezing the genetic landscape—the preservation of diversity in cultivated plants as an urgent social responsibility of the plant geneticist and plant taxonomist. *Maize Genetics Cooperation News Letter 48*:199-200.

Ingram, F.B. and J.T. Williams. 1984. *In situ* conservation of wild relatives of crops. Pages 163-178 *in* J.H.W. Holden and J.T. Williams (eds.), *Crop Genetic Resources: Conservation and Evaluation*. G. Allen and Unwin, London.

Interagency Task Force on Biological Diversity (ITFBD). 1985. U.S. strategy on the conservation of biological diversity: An interagency task force report to Congress. A.I.D., Washington, D.C.

International Board for Plant Genetic Resources (IBPGR). 1975. The conservation of crop genetic resources. IBPGR, Rome.

_________. 1976. Priorities among crops and regions. IBPGR, Rome.

_________. 1981. Revised priorities among crops and regions. IBPGR, Rome.

_________. 1985. Ecogeographical surveying and *in situ* conservation of crop relatives. IBPGR, Rome.

International Potato Center (IPC). 1984. *Breaking New Ground: Anthropology in Agricultural Research*. IPC, Lima, Peru.

International Union for the Conservation of Nature and Natural Resources, United Nations Environment Programme and World Wildlife Fund (IUCN-UNEP-WWF). 1980. *World Conservation Strategy: Living Resource Conservation for Sustainable Development*. IUCN, Gland, Switzerland.

Janzen, D.H. 1973. Tropical agroecosystems. *Science* 182:1212-1219.

Johannessen, C.L. 1982. Domestication process of maize continues in Guatemala. *Economic Botany* 36:84-99.

Johnson, A.W. 1972. Individuality and experimentation in traditional agriculture. *Human Ecology* 1:149-159.

Klee, G.A., ed. 1980. *World Systems of Traditional Resource Management*. Halstead Press, New York.

Ladizinsky, G. 1985. Founder effect in crop-plant evolution. *Economic Botany* 39:191-199.

Leppik, E.E. 1970. Gene centers as sources of disease resistance. *Annual Review of Phytopathology* 8:323-344.

Lyman, J.M. 1984. Progress and planning for germplasm conservation of major food crops. *Plant Genetic Resources Newsletter* 60:3-21.

Miller, J. 1973. Genetic erosion: Crop plants threatened by government neglect. *Science* 182:1231-1233.

Nabhan, G.P. 1985. Native crop diversity in aridoamerica: Conservation of regional gene pools. *Economic Botany* 39:387-399.

National Academy of Sciences (NAS). 1972. *Genetic Vulnerability of Major Crops*. NAS, Washington, D.C.

________. 1978. *Conservation of Crop Germplasm Resources*. NAS, Washington, D.C.

Oldfield, M.L. 1976. *The utilization and conservation of genetic resources: An economic analysis*. M.S. Thesis, Division of Man-Environment Relations. The Pennsylvania State University, University Park, Pennsylvania.

________. 1984. *The Value of Conserving Genetic Resources*. U.S. Department of the Interior, Washington, D.C.

Plucknett, D.L. and N.J.H. Smith. 1986. Sustaining agricultural yields. *BioScience* 36:40-45

Plucknett, D.L., N.J.H. Smith, J.T. Williams and N.M. Anishetty. 1983. Crop germplasm conservation and developing countries. *Science* 220:163-169.

Prescott-Allen, R. and C. Prescott-Allen. 1982. The case for *in situ* conservation of crop genetic resources. *Natural Resources* 23:15-20.

Roos, E.E. 1984. Report of the storage committee working group on "effects of storage on genetic integrity" 1980-1983. *Seed Science Technology* 12:255-266.

Ugent, D. 1970. The potato. *Science* 170:1161-1166.

United Nations Educational, Scientific and Cultural Organization, Man and the Biosphere Programme (UNESCO-MAB). 1974. *Final Report Prepared by the Task Force on: Criteria and Guidelines for the Choice and Establishment of Biosphere Reserves*. MAB Report Series No. 22. UNESCO, Paris.

________. 1984. Action plan for biosphere reserves. *Natural Resources* 20(4):1-12

U.S. Congress, Office of Technology Assessment (OTA). 1986. Staff paper on the role of U.S. development assistance in maintaining biological diversity in developing countries. OTA, Washington, D.C.

U.S. Department of State. 1982. *Proceedings of the U.S. Strategy Conference on Biological Diversity*. U.S. Government Printing Office, Washington, D.C.

von Droste zu Hulshoff, B. and W.P. Gregg, Jr. 1985. Biosphere reserves: Demonstrating the value of conservation in sustaining society. *Parks* 10:2-5.

Wilkes, H.G. and S. Wilkes. 1972. The green revolution. *Environment* 14(8):32-39.

Wright, R.M., B. Houseal and C. de Leon. 1985. Kuna Yala: Indigenous biosphere reserve in the making? *Parks 10(3):25-27.*

4

Conserving Crop Plant-Pathogen Coevolutionary Processes In Situ

J. Artie Browning

INTRODUCTION

Evolution of pathogen populations on homogeneous crop plant populations that are extensively grown and/or intensively managed has been called "man-guided evolution" (Johnson 1961). This reflects the extent to which pathogen evolution is under the indirect, but effective, control of crop improvement teams, farmers, and government policies that influence the genetic constitution and spatial distribution of crop production via breeding and husbandry of modern crop cultivars, factors that control the genetic constitution and spatial distribution of pathogen populations that mirror the crop populations. This unnatural situation, plus the time frame involved, precludes "conservation" of these evolutionary processes in modern agroecosystems. Therefore, the idea of "conserving crop plant-pathogen coevolutionary processes *in situ*" implies conservation in indigenous ecosystems and traditional agricultural ecosystems. This idea, while compelling, is one that must be accepted conceptually. Another important consideration is that our modern crop genetic improvement programs must still rely on *ex situ* reserves of seed stocks in order to obtain rapid access to new pathogen-resistant germplasm resources—resources that have not evolved in modern agroecosystems.

ACCEPTANCE OF THE CONCEPT BY NOBEL LAUREATE GEORGE BEADLE

In the early 1970s, few scientific or government leaders accepted the idea of *in situ*, or on-site, conservation of crop genetic resources. But some capable scientists did so when the idea was presented to them logically. For example, George Beadle, best known for winning the Nobel Prize for his outstanding work on *Neurospora* biochemical genetics, found the idea acceptable when I presented it to him in

1973. Upon retirement, Dr. Beadle returned to his first love, which was not genetics of *Neurospora*, but rather corn-teosinte cytogenetics and the evolutionary origin of maize (Beadle 1974). Dr. Beadle was invited to give a seminar on maize and teosinte at Iowa State University in 1973. During the discussion following his seminar, I queried him about means of preserving germplasm for future generations. Later, I wrote him a letter dated 12 April 1973, in which I stated the following:

> . . . I am the person who asked the question . . . on means of preserving germplasm for future generations. Your response that indigenous populations should be sampled systematically and preserved in germplasm banks like those at CIMMYT and Fort Collins certainly agrees with the opinions expressed by other good minds that have worked and thought deeply on this subject, like Norm Borlaug and Jack Harlan. I have little doubt but what this is the best overall approach, especially for land races of cultivated crops.
>
> For indigenous species that are progenitors of our cultivated species, however, I have real doubts. They frequently are less uniformly distributed, are more diverse in germplasm, and their seed dehisce severely. But worse, to me as a plant pathologist, when we store samples from these populations in germbanks, *we stop evolution*. You mentioned that teosinte contains factors for resistance to pathogens and insects that may be important for corn cultivars, and this probably is true for all crop progenitors and even land races. This resistance is a result of eons of coevolution of host and pathogen or insect. Yet when we store seed of the wild host and allow destruction of the wild host in the field, we stop evolution of our crop progenitors while allowing that of the pathogen or insect to continue on our commercial cultivars. This precludes further study of these interacting populations in a "natural" setting and of sampling them for germplasm to meet future needs. . . . We preserve large land areas for scenic and recreational value; why not also small "gene parks"? But first we need to know if the idea is sound. So few people agree, I'm beginning to think it isn't. If it is, then the political job of selling it can begin.
>
> I would welcome any further comments you would care to make on this subject, especially on the possible consequences of stopping the evolution in half of the host-parasite system, while allowing that in the other half to continue, and of how best to counter this if we might reasonably consider it not in the best interest of future generations.
>
> Sincerely yours,
> J. Artie Browning
> Professor of Plant Pathology

Dr. Beadle immediately accepted the idea and began to explore the possibility of setting up germplasm preserves where teosinte populations might interface with those of primitive corns in México. However, he was rebuffed by CIMMYT (The International Center for the Improvement of Maize and Wheat, in México). Therefore, he sought to avoid the México impasse, and tried to accomplish his goals by establishing teosinte preserves at the Pacific Tropical Botanical Garden in Hawaii. Figure 4.1 shows his letter of 26 July 1973.

NATIONAL *IN SITU* CONSERVATION EFFORTS

In 1984, Ingram and Williams emphasized that "*in situ* preservation . . . has become a stated international priority" (p. 163) and that "a global network of reserves . . . is an imperative" (p. 175). The International Board for Plant Genetic Resources (IBPGR) has also published guiding principles for its advisory work on *in situ* conservation (IBPGR 1985:72). Nevertheless, *in situ* reserves have not been adequately effected in international programs. Indeed, Ingram and Williams, who were IBPGR staff scientists, observed that "*in situ* conservation may be advancing more quickly at national levels, but no evidence that this is so is available to the authors" (p. 175). While IBPGR considers maintenance of wild species *in situ* preferable to *ex situ*, unfortunately *in situ* reserves are deemed beyond the scope of International Agricultural Research Centers (IBPGR 1990). Individual countries have acted, however. India established "the first gene sanctuary, for citrus" (U.S. Congress 1987:113). México has also established an *in situ* reserve for *Zea diploperennis*, the perennial maize relative, in the Sierra de Manantlan (Wilkes 1989). But the United States, which could reasonably be expected to take such initiative, has yet to include a single *in situ* crop genetic reserve in its otherwise comprehensive germplasm preservation system (U.S. Congress 1987, Shands et al. 1989). "The United States has not explored this mode either domestically . . . or internationally with such programs as those of A.I.D." (Wilkes 1989:29).

More recently, however, some U.S. legislators and other policymakers have begun to authorize actions to more directly address maintenance of biological diversity nationally. "With this new mandate, biologists, agricultural scientists, and conservation program managers *have started to develop* new ways to apply science to the problem of maintaining biological diversity on site" (my emphasis) (U.S. Congress 1987:102). This is good news, but I am unaware of any of the results portended by this announcement (see also Nabhan, Ch. 7, this volume). Minimally, I feel that the United States should establish *ecosystem reserves* that include *in situ* conservation of blueberry, cranberry, grape, pecan, and sunflower–all relatives of economically important crops that are indigenous to the contiguous 48 states. Additionally, national parks, wilderness areas, wildlife refuges, and national forests should be designated as *multiple use ecosystem reserves* also specified for conservation of indigenous crop *genetic* resources. These steps would have a positive influence on

GEORGE W. BEADLE
5533 DORCHESTER AVENUE
CHICAGO, ILLINOIS 60637

July 26, 1973

Dear Professor Browning:

Thanks to you we may have the first teosinte preserve under way — see enclosed copy of letter from Bill Stewart.

I tried to get CIMMYT to take some initiative with the response (from Ames grad): "Have US Gov propose to Mex gov." Muriel (Mrs B) says, "By then teosinte will be extinct."

That's what I like about Bill - no starting friction!

Regards
George

FIGURE 4.1 Letter dated July 26, 1973, from Nobel Laureate George W. Beadle to Dr. J. Artie Browning.

other nations that have been unable to accept the *in situ* idea conceptually, may question its feasibility, or may believe that the potential for *in situ* conservation of crop relatives is served adequately by the UNESCO-Man and the Biosphere (MAB) biosphere reserves project (see Gregg, Ch. 13, this volume). The USSR and Israel, however, have established them independently as part of an integrated national conservation strategy.

The effort of the USSR dwarfs all others. This country has established at least 127 *in situ* reserves for the protection of wild relatives of crops (Korovina 1980). For example, the Central Asiatic gene center east of the Caspian Sea has been divided into four provinces, 15 regions, and 57 ecogeographical areas. The USSR is also establishing reservations to conserve species valuable for breeding work in eight places where indigenous wild relatives of cultivated plants are most concentrated. So great is the Soviet effort that my Israeli colleague who was born in Russia, Professor Isaak Wahl, believes that one day the West will wake up to what the USSR has accomplished in conserving crop germplasm resources *in situ* and may consider their success as the equivalent of their having launched another sputnik—only this time, a biological sputnik.

The area of the Fertile Crescent that constitutes tiny Israel is home to wild progenitors of over 40 species of economically important crops (Zohary and Hopf 1988, D. Zohary, pers. comm.). These include the wild progenitors of wheat, oats, and barley and their foliar disease pathogens, which have been researched intensively. The germplasm resources of the wild grains have been collected and shared internationally. To facilitate this, Professor Wahl founded the Institute for Cereal Crops Improvement Through the Utilization of Germplasm of Their Wild Relatives at Tel-Aviv University (Anonymous 1982). This Institute includes the Lieberman Germplasm Bank, the world's first seed bank dedicated to the *ex situ* conservation of germplasm of wild grains, and the H.C. Murphy Laboratory for Rust Research. Additionally, the Institute is a major advocate of *in situ* ecosystem reserves of areas that include populations of wild grains. These natural ecosystems are their major laboratories, and conducting research on them is their major activity. The Israelis have, after a major educational campaign, acted to conserve all biotic components and their ecological interactions within these indigenous ecosystems. Their interconnected, member-funded Society for the Protection of Nature and state-funded Nature Reserves Authority and National Parks Authority oversee 49 national parks and 137 nature reserves. Protected areas set aside in Israel constitute 7% of the total country. Many of these are multiple use areas for conserving natural history and beauty, historic antiquities, and genetic resources in addition to their use for recreational purposes (Dinoor 1981, Paz 1981, Noy-Meir et al. 1988, Anonymous 1989, Shadur 1990, Anikster and Noy-Meir 1991, U.N. Safriel, pers. comm.). This is a conservation model which the Israelis are proposing for the entire Mediterranean Region (Ashri et al. 1989), and one that the United States should follow for its national parks and historic sites, national monuments, wilderness areas, and many state parks. The privately funded U.S. Nature Conservancy probably approximates Israel's Nature Reserves Authority in terms of philosophy. The California Gene Resource Conservation Program, that includes a goal of conserving natural populations, could also serve as a model for other states (Qualset and McGuire 1987).

THE CONCEPTUAL BASIS FOR *IN SITU* CONSERVATION

What is the conceptual basis for this idea that Dr. Beadle so readily accepted? The conceptual basis can be illustrated best by considering crop plants and their obligate parasites that have coevolved with them in natural ecosystems. Of these, the life cycle of the wheat stem rust fungus, *Puccinia graminis*, is most widely known. This obligately parasitic fungus has five spore stages on two hosts: the main or gramineous host–wheat or some other grass species–which harbors the dikaryotic stages of the fungus, and a dicotyledonous, woody alternate host–barberry–which harbors the monokaryotic stage. After a half century of studying the genetics of stem rust resistance in wheat, oats, and barley, much knowledge had been accumulated about the genetics of resistance to this obligate parasite on its gramineous hosts. However, little was known about the genetics of the alternate host, barberry. Consequently, very little had been learned aside from the immediate reactions of the economically important hosts to this potentially devastating fungus.

Then, H.H. Flor's research on flax and flax rust at the U.S. Department of Agriculture in Fargo, North Dakota, gradually impacted our science. Flax rust has many of the same characteristics as stem rust, except that it is monoecious, i.e., it has a single host–flax. Thus by understanding the genetics of only one host, he was eventually able to ferret out the genetics of all five spore stages of this obligate parasite-host interaction. From his research on flax and flax rust, Flor (1971) advanced his gene-for-gene hypothesis in 1942, which clarified the genetic relationship between avirulence/virulence in the pathogen and resistance/susceptibility in the host. Flor showed experimentally that each stage in the infection and pathogen development process is under full control of corresponding gene pairs in the host and pathogen–a marvel of coevolution. (For this, Flor should have won the Nobel Prize; however, it is not awarded in phytopathology.)

For the first time, Flor made it possible for coevolution occurring in indigenous ecosystems to be interpreted conceptually in terms of gene-for-gene interactions. The most relevant ecosystem to illustrate this phenomenon of host-parasite coevolution is that in Israel. Israel is a paradise for indigenous hosts and pathogens–not in the sense of an agroecosystem with recurring "boom and bust" cycles of cultivar resistance being overcome repeatedly by pathogen strains with increased virulence and/or aggressiveness (Harlan 1972)–but in the sense of the lamb dwelling with the wolf, the leopard with the kid, and the calf with the lion, with none predominating (Isaiah 11:6). Coevolved fitness traits require obligate parasites and their hosts to live together in an intimate ecological association in a state of evolutionary dynamic equilibrium.

The indigenous populations in Israel include the wild progenitors–*Avena sterilis*, *Hordeum spontaneum* and *Triticum dicoccoides* of cultivated oats (*A. sativa*), barley (*H. vulgare*), and wheat (*T. aestivum*), respectively. [Zohary and Hopf (1988) prefer *A. sativa* subsp. *sterilis*, *H. vulgare* subsp. *spontaneum*, and *T. turgidum* subsp. *dicoccoides*, respectively, as trinomials for these wild relatives. Other wild relatives of wheat that abound in Israel are *Aegilops* and *Agropyron* spp. (Manisterski et al. 1988).] Even though they differ taxonomically, the cultivated and wild forms are in the same respective gene pools. Furthermore, wild and cultivated barley and oats, respectively, which have the same number of chromosome sets and are therefore of the same ploidy level, intermate readily. The wild forms can be attacked by the same respective arrays of plant pathogens that have devastated farmer's cultivated crops recurringly for over a century and led to intensive investigations by plant scientists. Therefore, a century of accumulated mycology, plant pathology, genetics and breeding, and epidemiology literature can be brought to bear on the Israeli studies. This makes the Israeli results immediately meaningful and relevant to managing diseases in modern agroecosystems, and these indigenous ecosystems are therefore the most appropriate for study (see especially reviews by Browning 1974, 1980, 1988, Dinoor 1974, Anikster and Wahl 1979, Segal et al. 1980, Segal et al. 1982, Wahl and Anikster 1982, Anikster 1984, Dinoor and Eshed 1984, Wahl et al. 1984).

Interacting components of Israel's natural ecosystems are depicted in Figures 4.2-4.6. Figure 4.2 shows wild grains with the Sea of Galilee and the Golan Heights in the background. The indigenous natural ecosystem at Nahal Tut in Israel's Carmel Mountains is shown in 1969 in Figure 4.3. Historically, Nahal Tut is significant as the richest known source of specific genes for crown rust resistance, and such genes have been incorporated into virtually every contemporary oat cultivar grown worldwide. By 1978, however, Nahal Tut was wall to

FIGURE 4.2 An indigenous natural ecosystem (foreground) near the Sea of Galilee in Israel; this site is rich in both wild barley (*Hordeum spontaneum*) and wild oats (*Avena sterilis*).

FIGURE 4.3 Wild oats (*Avena sterilis*), the main host of oat crown rust (*Puccinia coronata*), is shown (center) next to Professor I. Wahl. At this site on the Plateau of Menashe, the alternate host, buckthorn (*Rhamnus palaestina*), is also present on the hill (background, right).

FIGURE 4.4 A dense cluster of wild barley (*Hordeum spontaneum*), the main host of the barley leaf rust fungus (*Puccinia hordei*), on the shore of the Mediterranean Sea, south of Caesarea, Israel. Wild grains such as these provided the basic food resources that enabled the rise of civilization in the Fertile Crescent.

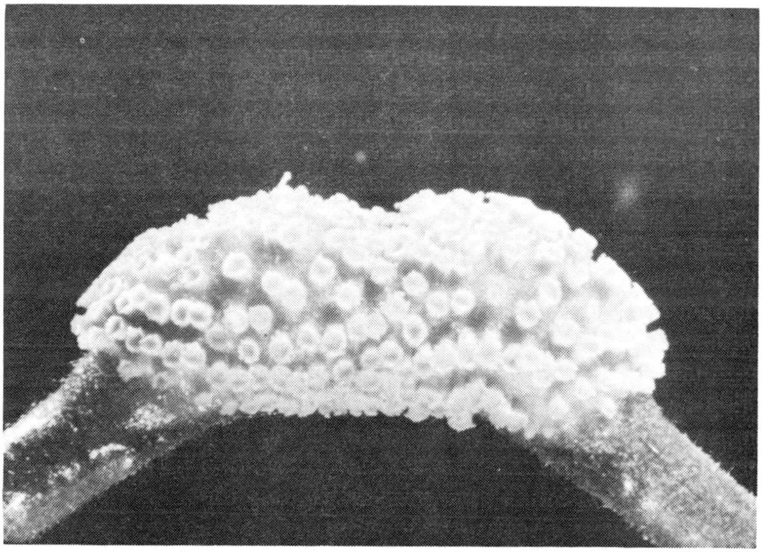

FIGURE 4.5 The aecial stage of the five-spore-stage oat crown rust fungus (*Puccinia coronata*) on buckthorn (*Rhamnus* sp.).

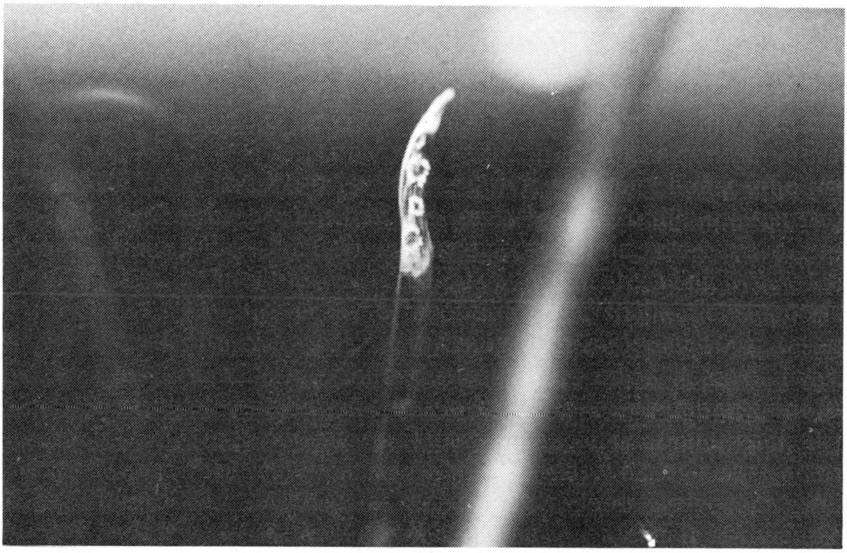

FIGURE 4.6 The aecial stage of the five-spore-stage barley leaf rust fungus (*Puccinia hordei*) on the Star of Bethlehem (*Ornithogalum* sp.), its liliaceous alternate host.

wall with beautiful orchards and vineyards, and this valuable crop germplasm was relegated to fence rows, the Lieberman Germplasm Bank, other germplasm banks, and breeding collections. The dense stand of wild barley shown in Figure 4.4 is on the shore of the Mediterranean Sea south of Caesarea, Israel. Such dense stands furnished grain for people and fodder for their flocks and herds of domesticated animals, and thus facilitated the rise of civilization in the Fertile Crescent. The flora depicted in Figures 4.2-4.4 is typical of Israel's oak park-forest ecosystem (Davis et al. 1969). The work most germane to this paper is that on coevolution and adaptation of rust species on their main and alternate hosts, including biogenic radiation to a wide array of plant hosts, coevolution of balanced polymorphisms, and genetic mechanisms of plant protection in natural populations in indigenous ecosystems. These topics are instructive for learning how to protect crops from pathogens in agroecosystems.

COEVOLUTION AND ADAPTATION OF RUST SPECIES ON MAIN AND ALTERNATE HOSTS

Coevolution of vascular plants and their obligate parasites occurs in areas where their centers of evolutionary origin and genetic diversity coincide. The evolution of the ability of rust fungi to parasitize new host species is spectacular, on the one hand, and to restrict themselves to only one host group, on the other—equally so! Dietel's Law states that "The transition to new hosts proceeds in the direction of the progressive evolution of the vascular plant, or to plants that originated at the same time as the source host, but not to plants that had evolved in earlier periods" (Anikster and Wahl 1979:369). Mycologists commonly accept that heteroecious rust fungi transferred from primary to secondary hosts by moving either the haploid or dikaryotic generation through an evolutionary process that Leppik (1967) has termed *biogenic radiation*. The *Avena-Puccinia coronata-Rhamnus* (Figures 4.2, 4.3, and 4.5) and the Gramineae-*Puccinia graminis-Berberis* host-parasite-host systems illustrate the shift of rust fungi to their main dikaryotic host from their alternate monokaryotic host, and the subsequent progressive coevolution of additional specialization on the gramineous dikaryotic host. In contrast, the barley-rust systems—*Hordeum-Puccinia hordei*-Liliaceae (Figures 4.2, 4.4, and 4.6) and *Hordeum-Uromyces*-Liliaceae—illustrate the opposite, i.e., a shift from the main dikaryotic host to the alternate monokaryotic host, with further specialization on the latter (Anikster 1989).

Survival over adverse periods in indigenous ecosystems is facilitated by resting teliospores (dormant, sexual spores in the rusts), woody or bulbous perennial alternate hosts, and by the strategy of evolving microcyclic correlated species with fewer spore stages. Through evolution, the rust fungus may become autoecious, homothallic, and lose two or more of the five spore stages, enabling the germplasm of the rust fungus to survive on a single perennial host as a *correlated species*. For example, *Puccinia mesnieriana* evolved as a short-cycled correlated species of *P. coronata* on *Rhamnus*. Its teliospores develop on *Rhamnus* spp. and are identical to those of *P. coronata* that develop on *A. sterilis*; when they germinate, they yield two binucleate, self-fertile basidiospores. *P. mesnieriana* lacks the pycnial, aecial, and uredial stages of *P. coronata*. Similarly, each macrocyclic *Uromyces* spp. on *Hordeum* is matched by a correlated microcyclic species on its respective, alternate liliaceous host. Microcyclic rusts have been bred in the laboratory from dioecious parents, tending to corroborate the evolutionary process of nature. The presence in Israel of the macrocyclic species and their respective correlated microcyclic species is considered by mycologists to be proof positive that Israel is in the center of origin and coevolution of these main and alternate hosts and their rust fungi.

The cereal rust fungi also have adapted the ability to cycle exclusively on the gramineous host, especially in agroecosystems. For example, the wheat stripe rust fungus (*P. striiformis*) has no known aecial host; the wheat leaf rust fungus (*P. recondita*) has *Thalictrum* for an alternate host, but it is non-functional over much of the wheat leaf rust range, and the stem rust (*P. graminis*) and crown rust (*P. coronata*) fungi have developed asexual populations (Roelfs 1985) that cycle over much of their crop's agricultural range because the alternate hosts have been eradicated or the crops are now grown where the climate or cropping season does not favor the sexual stage. The ability to parasitize two host species in turn, a single one for the microcyclic form, or a single one for the dikaryotic stage exclusively must be considered major, coevolved germplasm survival strategies for this group of opportunistic obligate parasites.

Great diversification has resulted at both the *formae speciales* level on host genera, and the pathogenic race level on host species and cultivars. *P. coronata* attacks several *Rhamnus* species (Figure 4.5) in Israel, and *P. hordei* is specialized on seven different *Ornithogalum* species (Figure 4.6). Isolates from all *formae speciales* of *P. hordei* are also virulent on two wild species, *Dipcadi erythraeum* and *Leopoldia eburnea* (Anikster 1989). Significantly, the last two hosts grow in desert regions, whereas *P. hordei* does not; this indicates that presence of the pathogen in a favorable environment fosters coevolution of avirulence/virulence in the pathogen and resistance/susceptibility in its liliaceous host. Manisterski et al. (1986) found this true also for *P. hordei* on *H. spontaneum* accessions. Four heteroecious *Uromyces* spp. develop on wild barleys, and have aecia on the liliaceous genera *Bellevalia*, *Muscari*, and *Scilla*, yet each fungus has its respective microcyclic species. All *formae speciales* of *P. hordei* and both macrocyclic and microcyclic forms of *Uromyces spp.* are compatible with *Leopoldia eburnea*, making this a universal host that has a unique role in the evolution of these species and *formae speciales* of rust fungi (Anikster 1989). This universal host is central to further study of this coevolved complex.

Possibly, the tremendous genetic diversity and complexity that can result from continuing host plant-pathogen coevolution is displayed best on gramineous hosts. For example, 75 species from 36 genera of grasses were attacked by *P. graminis avenae*, including cultivated oats and four species of wild oats (Gerechter-Amitai 1973). Nine species of *Bromus*, *Hordeum*, and *Lolium* were susceptible to both *P. graminis tritici* and *P. graminis avenae*, facilitating crossing among these *formae speciales in nature*. The alternate host most important to crown rust, *Rhamnus palaestina*, enables virulent and prevalent pathotypes of *P. coronata* to survive unfavorable weather conditions and a diversity of pathotypes with new virulence combinations to arise. As a result, two to three times as many pathotypes commonly occur in aecial isolations from *Rhamnus* or uredial isolations from oats near *Rhamnus (Figure 4.3)*, as compared to uredial isolations distant from *Rhamnus*.

PROTECTION OF PLANT POPULATIONS IN INDIGENOUS ECOSYSTEMS

Truly the area that includes Israel favors the evolutionary development of both hosts and their pathogens, yet the natural ecosystem remains healthy (Figures 4.2-4.4). From our modern agroecosystem perspective, this may seem inconsistent. Israel, with its diversity of crops, weeds, wild species, and favorable climate is a paradise for pathogen evolution, development, and survival, because there is greater opportunity for the evolution and maintenance of more different pathotypes on genetically diverse wild hosts than would occur on genetically uniform cultivars. Vavilov (1957:212 [201]) stressed that in the Mediterranean region, varieties of wheat, barley, legumes, and flax "are distinguished by their large seeds, large ears, great height, excellent grain quality, breadth of leaves and bushiness." He stated that this is a result of both natural and artificial selection since the beginning of ancient agriculture in this area, and that both species and crop varieties are characterized by extensive resistance to the most widespread fungal diseases, fruit-flies, and Hessian-flies, and by an overwhelming concentration of multiple (resistance). Vavilov

(1957:212 [201]) concluded that the "ancient local types of the Mediterranean area stand out globally (for) their extremely valuable properties" and that the "Mediterranean agroecological region ranks first in the world for high quality grain and valuable biological and physiological properties." This seeming discrepancy between the diversity of pathogenic strains on infected hosts, and the evolution of robust, vigorous, and productive wild progenitor species and indigenous crop cultivars, is the crux of the issue.

A plant scientist familiar with a more linear relationship between disease epidemics and crop damage in a genetically uniform agroecosystem would be amazed and probably overwhelmed with the dynamic stability one can observe in an indigenous ecosystem where such plant-pathogen populations still interact ecologically. For instance, a site at Nahal Tut (Figure 4.3) on Israel's Plateau of Menashe was especially rich in crown rust resistance genes. The very diverse flora of Nahal Tut included *A. sterilis* as a conspicuous component. The alternate host, *R. palaestina*, abounded on the surrounding hills (Figure 4.3), and the climate favored the development of crown rust (Figure 4.5) and other foliar diseases. In this small but diverse area, *A. sterilis* plants were located that thrived both with and without crown rust. Even though some plants were infected with the powdery mildew fungus (*Erysiphe graminis*), they remained green and productive. A few oat plants, but not many, were systemically infected with the destructive oat loose smut fungus (*Ustilago avenae*). Some oat plants, but again not many, showed symptoms of systemic infection by the barley yellow dwarf virus. And most plants had some stem rust, but they too seemed green and productive. Indeed, the ecosystem and its plant populations were judged healthy. What is the explanation of the evolution of excellent yield, high protein percentage and other desirable agronomic traits, and the extensive host range despite the host specificity of these obligate parasites and evidence of severe disease development?

Obligate parasite-host interactions that involve several cereal rust fungi and powdery mildews are characterized by broad host ranges coupled with great host specificity and gene-for-gene control of the interactions. One aspect of the apparent paradox can be explained by examining the evolutionary history of these obligate parasites. Many host tribes, genera, and species have evolved from a common ancestral type, yet each fungus has evolved specific adaptations that facilitate specialization for a different host or set of alternate hosts. This adaptive radiation, which is a characteristic of many plant pathogens, accounts for the seemingly broad range of adaptation to various plant hosts within a genus or tribe. Thus, many host species contain, in varying proportions, gene combinations obtained from several seemingly divergent ancestors that actually trace their ancestry back to a common progenitor. As a result, individual cultures of some pathogens are capable of simultaneously expressing their recognized host- or cultivar-specificity while still attacking different plant species or even plants in remote genera or tribes.

The other aspect of the apparent paradox can be understood from examination of extensive pathogen-host studies conducted by Israeli scientists over more than three decades. One important conclusion from these studies is that populations of Israel's wild small grain progenitors possess identifiable protective mechanisms. These are primarily the pathological-ecological phenotype related to the host population's resistance or tolerance genotype, and alternatively, the plant pathogen's pathogenicity or aggressiveness. These generalized features are frequently polygenically inherited. Polygenic host resistance/susceptiblity as well as pathogen avirulence/virulence probably became superimposed during evolutionary time over the basic polygenic systems that were primarily responsible for host-pathogen homeostasis in indigenous populations. Oligogenic systems can and do contribute in a major way to maintaining homeostasis where both polygenic and oligogenic resistance occur and where protective population structures coevolved. Nature uses one resultant type of *epidemiologic* resistance–dilatory or rate-reducing resistance–to protect the population; but it uses many types of *genetic* resistance and population structures to achieve it. This has been called the "*protection of indigenousness*" (Browning 1974) and it can be emulated in agroecosystems. The challenge that faces agricultural science is to utilize specific and general genetic resistance to create crop cultivars or spaciogenically stable crop populations with dilatory epidemiologic resistance or tolerance, and to incorporate these crops within managed agroecosystems to emulate the protection of indigenousness.

"Sustainable development" may defy accurate definition and may not even be attainable, but the development of sustained yield agroecosystems (SYAEs) is possible to achieve (Greenland 1975)! Our opportunity and obligation as concerned agricultural scientists is to develop them! People demand not just food, but food from stable, dependable systems–SYAEs that could be developed, in part, via application of knowledge of how wild crop progenitor populations protect themselves. To accomplish this, "libraries" of genetic information will have to be employed that were only recently "learned" from studying indigenous natural ecosystems in areas of traditional agriculture or crop progenitor genetic diversity. A major constraint to achieving SYAEs in the past has been excessive reliance on genetic uniformity. Genetically uniform crop populations in modern agroecosystems are vulnerable to the more variable pathogen populations that possess great epidemic potential (Harlan 1972). "If uniformity be the crux of genetic vulnerability, then diversity is the best insurance against it" (NAS 1972:295). A suitable compromise between legitimate economic and agronomic needs for genetic uniformity and alternative benefits of genetic diversity to counter biotic and abiotic sources of stresses is exemplified by the three-cultivar mixtures developed and studied in England (Wolfe 1985, Browning 1988), which minimally, but adequately, mimic plant population structures in natural ecosystems, and thus can achieve substantial protection of indigenousness. This alternative offers an excellent compromise and represents a suitable extrapolation from lessons learned from indigenous *in situ* populations of wild crop genetic resources.

OTHER ADVANTAGES OF *IN SITU* CONSERVATION OF INDIGENOUS ECOSYSTEMS

This discussion has concentrated on cereal rust fungi and their main and alternate hosts in an area where all are indigenous species. This alone constitutes a compelling conceptual basis for *in situ* conservation of coevolutionary processes. The real *pièce de résistance,* however, is consideration of the other components and other associations of the indigenous ecosystem. It is relevant to ask whether Flor's gene-for-gene hypothesis is one of the most significant biological hypotheses, and if so, what other associations within these ecosystems may be under gene-for-gene control as well? Ellingboe (1979) has stated that more than 90% of the genetic variability reported in the literature is consistent with the gene-for-gene hypothesis. The same pattern appears to hold for interactions between plants and bacteria, fungi, insects, nematodes, and possibly viruses. It may also hold for interactions between animals and non-antigenic disease-causing entities. Thus, Ellingboe (1979:15) concluded that the gene-for-gene hypothesis "seems to be quite universal in interorganismic genetics."

The many trophic levels of ecological organization that invoke interactions among organisms in natural ecosystems are described succinctly by an anonymous ditty that states:

> Dogs have fleas
> And fleas have fleas
> On their backs to bite 'em.
> And little fleas have lesser fleas,
> And so *ad infinitum*.

The same ecological relationships hold for plants and animals and their pathogens or pests—whether fungal, bacterial, viral, or arthropod.

Conceptually, the gene-for-gene hypothesis may also apply to these relationships as well as those between a pathogen and its pathogens or hyperparasites, the hyperparasites' hyperparasites, and so, *ad infinitum*. Plants did not evolve in a germ-free environment, but also along with beneficial microbes. Furthermore, the relationship between plants and beneficial microorganisms is probably gene-for-gene. This would include nodulating bacteria, root-extending mycorrhizal fungi, biocontrol agents, and other beneficial microbes.

These often ignored microorganisms and their interactions among themselves and with plants may be integral parts of the supporting infrastructure that sustains agriculture and makes it and, therefore, life as we know it, possible. Conservation of supporting germplasm of beneficial microbes in the ecosystem, therefore, is imperative. Indeed, as Jong (1989:242) has pointed out, their role "is fundamental to the proper functioning of ecological systems and to the very existence of plant and animal life. The interdependency of this relationship is such that a loss of microbial species would result in a loss of animal and plant variety."

CONSERVATION OF GERMPLASM *IN SITU* IN NATURAL ECOSYSTEMS

Who is prepared to say that some day we may not need germplasm of any one of these—or other yet unknown—taxonomic entities? Certainly, to conserve germplasm of any one of them, germplasm of all of them must be conserved, because the ecological links among the several trophic levels must be maintained. Many organisms—arthropods, nematodes, and microorganisms—that may prove to be ecologically important are yet taxonomically or otherwise unknown. Thus, we do not presently know how to collect and preserve them *ex situ*, particularly in the case of many multicellular organisms. It is better to conserve them *in situ* in ecosystem reserves under conditions of benign neglect than to sample and preserve only a small fraction of the ecosystem *ex situ*—namely samples of germplasm resources of selected wild crop progenitor or crop cultivar hosts in seed banks.

Other vitally important ecological associations (Hunter and Aarssen 1988) are also lost when germplasm resources are stored away in gene banks. For example, plant-competitor, plant-herbivore and predator-prey associations that are not under gene-for-gene control would be precluded, even though these are no less vital to the ecosystem and the populations involved in these coevolutionary associations. Some plants associate with particular plants but repel other plants through allelopathy; these associations would be lost. Early farmers developed and perpetuated specific crop associations, such as legumes, maize, and squash in Mesoamerica. These may be intercropped in traditional agroecosystems, but may not be preserved *ex situ* storage in seed banks. Another association vital to traditional agriculture is that between people and their plants (see also Oldfield and Alcorn, Ch. 3, Wilkes, Ch. 5, and Gliessman, Ch. 10, this volume). Many food, feed, fiber, and medicinal plants have coevolved with humanity over the centuries. Plants and mankind are mutually interdependent, and both the plant resources and the ancient knowledge of their husbandry merit conservation. Mankind also merits survival, yet to the extent that we have endangered and even extinguished the valuable plant resources with which we are interdependent, we also endanger our own existence (Browning 1972).

The philosopher of plant pathology and father of modern plant disease epidemiology, J.E. Vanderplank (1968:140) once stated that: "The conservation of nature's vast inheritance of genes is probably this shortsighted world's most urgent single genetic problem." Faced with mankind's perceived imperative to alter natural habitats to meet short-term needs for food, feed, fiber, and firewood for a rapidly expanding human population, this genetic problem has become and is still becoming ever more urgent. A major response to the recently accelerating loss of natural habitats is the drive to document biological diversity. For example, Miller et al. (1989:3) have proposed that "the first order of business in biological diversity research must be to make a complete inventory of the world's biota. This task is essential to nearly all conservation and germplasm plans for the future." Biological diversity should be documented. Yet strangely, many scientists dedicated to this effort offer no support for *in situ* conservation. Their focus is more that of a "Manhattan-" or "human-genome-sequencing" -size project that will require massive funding and political commitment to the education and training of armies of taxonomists and systematists, ecologists, and other naturalists. This is an admirable endeavor, yet time simply precludes postponing the establishment of *in situ* reserves until all of the rapidly disappearing components have been identified and counted and definitive data exist on where to "site" them. This is akin to arguing against securing the door to a bank vault until all the money has been sorted and counted. It is certainly wiser to secure the door first and then sort and count the money at one's leisure. In a similar vein, some representative ecosystems should be secured as *in situ* ecosystem reserves; then at a later time, their flora, fauna, and microorganisms should be identified and studied at our relative leisure. Anyway, while "continued improvements in ecosystem classification and assessment would facilitate better decisions on where protection is most needed, . . . descriptions of threatened ecosystems are now adequate in every country to undertake effective programs for conserving biological diversity" (U.S. Congress 1987:102).

Some argue that *ex situ* methods, e.g., seed banks, offer the only feasible means to preserve crop germplasm resources, and that they are adequate in their own right. Yet, *ex situ* resources are to a vast ecosystem what zoos are to animal conservation in the wild: small nonrandom samples of a few spectacular species that have been singled out for preservation. Preserving them as working collections is highly justified, but judging them adequate as the sole means of long term conservation is not. *Ex situ* gene banks freeze evolution of that germplasm while simultaneously ignoring vital associated germplasm—possibly relegating it to eventual extinction. Even germplasm resources stored in seed banks are not secure. For example, much of the USDA World Collection of Small Grains was lost during the Great Depression of the 1930s when budget constraints precluded timely seed increases (H.C. Murphy, pers. comm.). Much of the collection of the Vavilov Institute for Plant Industry was lost during the Nazi's siege of Leningrad during World War II. And Nicaragua's *ex situ* maize collection was lost twice—once due to an earthquake and, later due to an electrical failure (Plucknett el al. 1987). These disasters could strike again and could affect any governmental or other organizational unit at some future time. Even now, some germplasm in the National Seed Storage Laboratory (NSSL) is in jeopardy because budgetary constraints may force a partial repeat of the 1930s experience (U.S. Congress 1987, Wilkes 1989). Also, according to a 1981 Government Accounting Office study, at least one-fourth of the crop germplasm resources held by the National Plant Germplasm System may be lost due to storage in inadequate containers, undesirable climates, or because they were not sent to the NSSL (U.S. Congress 1987:235). I personally witnessed the destruction of a sizeable percentage of the world collection of potatoes of the Rockefeller Foundation at Bogota, Colombia due to systemic infection with viruses. Many germplasm collections, especially cultures of microbes, have the life expectancy of their founder and curator, and are terminated upon his/her retirement. These examples illustrate that germplasm resources are more significant than and should survive the budgetary whims of individuals or nations. They represent a continuous thread of life derived from the past that has continued until the present and should project on into the future. In contrast, budgets, individuals, and entire nations are relatively ephemeral.

In all these cases, I would feel more secure knowing that the germplasm resources of small grains progenitors were adequately preserved also in the *wadis* of the Fertile Crescent, or those of potatoes in the isolated valleys of Peru, where they are subjected to benign neglect as the main conservation philosophy. Each ecosystem reserve would constitute a biological insurance policy for germplasm resources within that natural ecosystem. In such *in situ* reserves, all ecological components and evolutionary processes would be allowed to continue for future coevolution, study (including investigations with the marvelous modern tools of biotechnology), and collection to meet the needs of future generations. It seems to me that the acme of scientific arrogance is to say to any young person that "Neither you, your children, nor your grandchildren will ever need this germplasm or its indigenous ecosystem, either for spare genetic parts or for future study, therefore we don't need to save it." We must be willing to save some natural ecosystem reserves with their wild crop progenitors intact for future generations!

The questions are: "How and what size?" Discussion of nonscientific problems like political, social, and financial considerations is not the purview of this paper. The Israeli experience was previously described as one the United States should emulate. Merely designating our national parks as *in situ* genetic reservoirs for all ecosystem components in addition to recognizing their scenic and recreational significance would cost nothing, yet would make a tremendous philosophical statement of an enlightened national commitment. The National Park Service is already deeply involved with the UNESCO- MAB Programme (see Gregg, Ch. 13, this volume). One cannot help but be thrilled at that program and its objectives, one of which is Project No. 8 on "Conservation of Natural Areas and the Genetic Material They Contain" (U.S. Congress 1987: 264). It may be the best approach to conserving traditional agroecosystems and their germplasm (see Oldfield and Alcorn, Ch. 3, this volume). In many cases, man-crop-pathogen-hyperparasite interactions have continued in traditional agricultural fields for 300-500 years or more, and may have resulted in the evolution of balanced polymorphisms–not unlike those found in pathogen-host systems in indigenous natural ecosystems. Coevolution is dead in modern agroecosystems–but it continues in crop fields of traditional agriculturalists and in indigenous ecosystems. Therefore, representative traditional agroecosystems should be conserved *in situ*. The major difference between indigenous natural ecosystems and traditional agroecosystems is the presence of coevolved human-plant associations in the latter. MAB biosphere reserve projects may eventually prove to be the best approach for conserving traditional agroecosystems, including their necessary human and rural development elements.

However, MAB biosphere reserves are too few, sometimes too large, and too structured to adequately serve as indigenous ecosystem reserves for conservation of crop genetic resources *in situ*. If migratory animals are not involved, *in situ* ecosystem reserves need not be large to preserve plants and their coevolved ecosystem components. Wilkes (1971) considered that small *in situ* reserves could suffice for conservation of teosinte in México. A three-acre virgin tall grass prairie ecosystem is thriving on the grounds of the Ames High School in Iowa. Many nature reserves in Israel are small. Some protection and/or management (e.g., grazing or controlled burning) will probably be required, depending on the predominant flora and the neighbors. But in many cases, the areas need not be large. Israeli scientists have a relevant, on-going multidisciplinary project on *Triticum dicoccoides*, in part to answer the questions such as: What size area is required to conserve genetic diversity of this autogamous species *in situ* where it is indigenous, and how should these areas be managed? (Noy-Meir et al. 1988, Anikster and Noy-Meir 1991). Additionally, areas that have never been plowed, or have never had agricultural fertilizers and/or pesticides applied have become invaluable as ecosystems for obtaining baseline ecological data for many scientific purposes.

CONCLUDING REMARKS

Nature and mankind accumulate and carry over three kinds of information: 1) geologic/biologic (e.g., tree rings; radioactive carbon, etc.), 2) genetic, and 3) "language" or recorded information (e.g., music, computer, and spoken languages). The last two, genetic information and language, both of which involve "libraries," are relevant to *in situ* conservation. Clearly the Israeli ecosystem includes an extensive genetic library of knowledge on how to stabilize host-pathogen interactions.

Why conserve germplasm *in situ* in ecosystem reserves? For the germplasm per se and for information and acquisition of knowledge about that germplasm and its ecosystem. Germplasm per se should be preserved for both present and <u>future</u> collection of "spectacular species" of animals and plants, and of both taxonomically known and unknown, potentially valuable species we have yet to conserve or to learn how to conserve properly. It also should be preserved to facilitate the <u>continuing</u> coevolution of both known and presently unknown species. The acquisition of knowledge is only slightly less important than germplasm conservation per se. It includes information from <u>future</u> studies, e.g., as inventories are made of higher organisms and many beneficial as well as deleterious microorganisms and arthropods. Information from <u>future</u> studies of ecological interactions and genetic interrelations among interacting populations of higher and lower organisms and knowledge of how plants in indigenous ecosystems buffer or protect themselves from sources of stresses such as diseases, arthropods, fire, and drought, should lead to superior management practices for all agroecosystems.

Recently, a summary of results and implications for plant breeding and germplasm conservation was published from long term studies of host-pathogen coevolution between two genetically variable barley populations and the barley scald pathogen (*Rhynchosporium secalis*). The conclusions were consistent with those from long term Israeli studies, emphasized the benefits of *in situ* over *ex situ* conservation of coevolving genetic resources, and they illustrated the flow of valuable germplasm resources and knowledge that can be obtained from such coevolving populations (Allard 1990).

Thus, as George Beadle had recognized, the logic behind *in situ* conservation of germplasm in ecosystem reserves is compelling (see Figure 4.1). Why then have so few reserves been established? I feel that it is because most scientists and other leaders still do not appreciate the significance of continuing coevolution. Several authors recognize evolution of the target crop alone. Wilkes (1989:29), however, clearly recognized coevolution; he advocated "evolutionary gardens where crops and their wild relatives hybridized and the resulting variation is in dynamic evolution with evolving pest and pathogens." Unexpectedly, sociologists Busch et al. (1989:52) also recognized coevolution: "Nature preserves permit evolution to go unhindered . . . and . . . the continuation of the complex interactions that occur both within and between species. Perhaps most importantly, they permit the preservation of the enormous, perhaps infinite, diversity that is found only in nature." If, as Ingram and Williams (1984:169) have claimed, "preservation without prospects of utilization is a nonstarter" and, as Wilkes (1989:28) has remarked, "genes in (an *ex situ*) bank are like candles under a basket" unless they are used, germplasm resources of all biota conserved *in situ* in an ecosystem reserve are akin to assets in a bank that, because of continuing coevolution, continue to appreciate at a compound interest rate. We must conserve these resources to make them available for future study, collecting needs, and use because "nature preservation will in the end be a protection of ourselves and our future" (Paz 1981:30).

ACKNOWLEDGMENTS

I acknowledge with gratitude: Neil Jensen and George Kent who, while I was in graduate school, challenged me to consider intravarietal diversity; Isaak Wahl and associates (especially Yehoshua Anikster, Nissan Binyamini, Uzi Brodny, Amos Dinoor, Zahir Eyal,

Zeev Gerechter-Amitai, Jacob Manisterski, the late Adrian Segal, and Daniel Zohary) who instructed me and whose counsel and concern for indigenous natural ecosystems forced me to look at both natural ecosystems and agroecosystems differently; K.J. Frey, M.D. Simons, H. Tachibana, and R.A. Frederiksen for intellectual and collegial support; I. Wahl for Russian translations and, with Yehoshua Anikster and Jacob Manisterski, for reviewing and making valuable suggestions for completion of this manuscript; E.M. Rosenthal and Isadore Roosth for their help with the Texas-Israel Scientist Exchange Program; and the Council for International Exchange of Scholars, Tel Aviv University, and Texas A&M University for a Fulbright Fellowship which enabled me to complete this contribution in Israel.

REFERENCES

Allard, R.W. 1990. The genetics of host-pathogen coevolution: Implications for genetic resource conservation. *Journal of Heredity* 81:1-6.

Anikster, Y. 1984. The *formae speciales*. Pages 115-130 *in* W.R. Bushnell and A.P. Roelfs (eds.), *The Cereal Rusts. I. Origins, Specificity, Structure, and Physiology*. Academic Press, Orlando, Florida.

Anikster, Y. 1989. Host specificity versus plurivority in barley leaf rusts and their microcyclic relatives. *Mycological Research* 93:175-181.

Anikster, Y. and I. Noy-Mier. 1991. The wild wheat field laboratory at Ammiad, Israel. *Israel Journal of Science* 41:(in press).

Anikster, Y. and I. Wahl. 1979. Coevolution of the rust fungi on Gramineae and Liliaceae and their hosts. *Annual Review of Phytopathology* 17:367-403.

Anonymous. 1982. BioBriefs. *BioScience* 32:763.

Anonymous. 1989. A time to build, and a time to preserve. *Eretz Magazine* 4(2):36-47.

Ashri, A., I. Noy-Meir, D. Zohary and U.N. Safriel. 1989. Dynamic conservation of wild progenitors of cultivated plants: Israel as a model for the Mediterranean Region. Research proposal submitted by The Hebrew University of Jerusalem and The Nature Reserves Authority to the Council of Europe. Strasbourg, France.

Beadle, G.W. 1974. Recollections. *Annual Review of Biochemistry* 43:1-13.

Browning, J.A. 1972. Corn, wheat, rice, man: Endangered species. *Journal of Environmental Quality* 1:209-211.

__________. 1974. Relevance of knowledge about natural ecosystems to development of pest management programs for agroecosystems. *Proceedings of the American Phytopathological Society* 1:191-199.

_________. 1980. Genetic protective mechanisms of plant-pathogen populations: Their coevolution and use in breeding for resistance. Pages 52-75 *in* M.K. Harris (ed.), *Biology and Breeding for Resistance to Arthropods and Pathogens of Agricultural Plants*. Texas Agricultural Experiment Station Miscellaneous Publication No. 1451. College Station, Texas.

_________. 1988. Current thinking on the use of diversity to buffer small grains against highly epidemic and variable foliar pathogens: Problems and future prospects. Pages 76-90 *in* N. W. Simmonds and S. Rajaram (eds.), *Breeding Strategies for Resistance to the Rusts of Wheat*. CIMMYT, México, D.F.

Busch, L., W.B. Lacy and J. Burkhardt. 1989. Culture and care: Ethical and policy dimensions of germplasm conservation. Pages 43-62 *in* L. Knutson and A.K. Stoner (eds.), *Biotic Diversity and Germplasm Preservation, Global Imperatives*. Kluwer Academic Publishers, Dordrecht, The Netherlands.

Davis, P.H., P.C. Harper and I.C. Hedge (Eds.). 1969. *Plant Life of Southwest Asia*. Botanical Society of Edinburgh, Scotland.

Dinoor, A. 1974. Role of wild and cultivated plants in the epidemiology of plant diseases in Israel. *Annual Review of Phytopathology* 12:413-436.

_________. 1981. Dynamic preservation of germ-plasm in nature reserves. *KIDMA Israel Journal of Development* 6(4):22-25.

Dinoor, A. and N. Eshed. 1984. The role and importance of pathogens in natural plant communities. *Annual Review of Phytopathology* 22:443-466.

Ellingboe, A.H. 1979. Inheritance of specificity: The gene-for-gene hypothesis. Pages 3-17 *in* J.M. Daly and I. Uritani (eds.), *Recognition and Specificity in Plant Host-Parasite Interactions*. University Park Press, Baltimore, Maryland.

Flor, H.H. 1971. Current status of the gene-for-gene concept. *Annual Review of Phytopathology* 9:275-296.

Gerechter-Amitai, Z.K. 1973. Stem rust, *Puccinia graminis* Pers. on cultivated and wild grasses in Israel. Ph.D. Dissertation, Hebrew University, Jerusalem. (Hebrew with English summary)

Greenland, D.J. 1975. Bringing the green revolution to the shifting cultivator. *Science* 190:841-844.

Harlan, J.R. 1972. Genetics of disaster. *Journal of Environmental Quality* 1:212-215.

Hunter, A.F. and L.W. Aarssen. 1988. Plants helping plants. *BioScience* 38:34-40.

International Board for Plant Genetic Resources (IBPGR). 1985. *Annual Report 1984*. IBPGR, Rome.

_________. 1990. *Annual Report 1989*. IBPGR, Rome.

Ingram, G.B. and J.T. Williams. 1984. *In situ* conservation of wild relatives of crops. Pages 163-179 *in* J.H.W. Holden and J.T. Williams (eds.), *Crop Genetic Resources: Conservation and Evaluation*. George Allen and Unwin, London.

Johnson, T. 1961. Man-guided evolution in plant rusts. *Science* 133:357-362.

Jong, Shung-Chang. 1989. Microbial germplasm. Pages 241-273 *in* L. Knutson and A.K. Stoner (eds.), *Biotic Diversity and Germplasm Preservation, Global Imperatives*. Kluwer Academic Publishers, Dordrecht, The Netherlands.

Korovina, O.N. 1980. Wild relatives of cultivated plants and regional division of the Central Asiatic Gene Center. *Bulletin of Applied Botany, Genetics, and Plant Breeding* 68:3-14. (Russian with English summary)

Leppik, E.E. 1967. Some viewpoints on the phylogeny of rust fungi. VI. Biogenic radiation. *Mycologia* 59:568-579.

Manisterski, J., A. Segal, A.A. Levy and M. Feldman. 1988. Evaluation of Israeli *Aegilops* and *Agropyron* species for resistance to wheat leaf rust. *Plant Disease* 72:941-944.

Manisterski, J., L. Treeful, J.R. Tomerlin, Y. Anikster, J.G. Moseman, I. Wahl and R.D. Wilcoxson. 1986. Resistance of wild barley accessions from Israel to leaf rust collected in the U.S.A. and Israel. *Crop Science* 26:727-730.

Miller, D.A, A.Y. Rossman and J. Kirkbride. 1989. Systematics, diversity, and germplasm. Pages 3-11 *in* L. Knutson and A.K. Stoner (eds.), *Biotic Diversity and Germplasm Preservation, Global Imperatives*. Kluwer Academic Publishers, Dordrecht, Netherlands.

National Academy of Sciences (NAS). 1972. *Genetic Vulnerability of Major Crops*. NAS, Washington, D.C.

Noy-Meir, I., Y. Anikster, M. Waldman and A. Ashri. 1988. Population dynamics research for *in situ* conservation: Wild wheat in Israel. FAO/IBPGR Plant Genetic Resources Newsletter 75/76:9-11.

Paz, U. 1981. *Nature Reserves in Israel*. Massada, Ltd. Givataim, Israel, Volume 1. (Hebrew)

Plucknett, D.L., N.J.H. Smith, J.T. Williams and N.M. Anishetty. 1987. *Gene Banks and the World's Food*. Princeton University Press, Princeton.

Qualset, C.O. and P.E. McGuire. 1987. California moves to save imperiled collections. *Diversity* 12:9-10.

Roelfs, A.P. 1985. Wheat and rye stem rust. Pages 3-37 *in* A.P. Roelfs and W.R. Bushnell (eds.), *The Cereal Rust. II. Diseases, Distribution, Epidemiology, and Control*. Academic Press, Orlando, FL.

Segal, A., J. Manisterski, J.A. Browning, G. Fischbeck and I. Wahl.

1982. Balance in indigenous plant populations. Pages 361-370 *in* H.M. Heybroek, B.R. Stephan and K. von Weissenberg (eds.), *Resistance to Diseases and Pests in Forest Trees*. Pudoc, Wageningen.

Segal, A., J. Manisterski, G. Fischbeck and I. Wahl. 1980. How plant populations defend themselves in natural ecosystems. Pages 75-102 *in* J. G. Horsfall and E.B. Cowling (eds.), *Plant Disease, an Advanced Treatise. V. How Plants Defend Themselves*. Academic Press, New York.

Shadur, Joseph. 1990. *The Society for the Protection of Nature in Israel, SPNI*. Society for Protection of Nature in Israel, Tel Aviv.

Shands, H.L., P.J. Fitzgerald and S.A. Eberhart. 1989. Program for plant germplasm preservation in the United States: The U.S. National Plant Germplasm System. Pages 97-115 *in* L. Knutson and A.K. Stoner (eds.), *Biotic Diversity and Germplasm Preservation, Global Imperatives*. Kluwer Academic Publishers, Dordrecht, The Netherlands.

U.S. Congress, Office of Technology Assessment. 1987. *Technologies to Maintain Biological Diversity*. OTA-F-330. U.S. Government Printing Office, Washington, D.C.

Vanderplank, J.E. 1968. *Disease Resistance in Plants*. Academic Press, New York.

Vavilov, N.I. 1957. *World Resources of Cereals, Leguminous Seed Crops and Flax, and Their Utilization in Plant Breeding*. Academy of Sciences, USSR. Moscow, USSR. (Translated from Russian by M. Paenson and Z.S. Cole, Israel Program for Scientific Translations).

Wahl, I. and Y. Anikster. 1982. Role of the alternate host in evolution of some cereal rusts. *Garcia de Orta* 9:21-27. (Serie de Estudos Agronomicos, Revista da Junta de Investigacoes Cientificas do Ultramar, Lisboa, Portugal)

Wahl, I., Y. Anikster, J. Manisterski and A. Segal. 1984. Evolution at the center of origin. Pages 39-77 *in* W.R. Bushnell and A.P. Roelfs (eds.), *The Cereal Rusts. I. Origins, Specificity, Structure, and Physiology*. Academic Press, Orlando, Florida.

Wilkes, G. 1989. Germplasm preservation: Objectives and needs. Pages 13-41 *in* L. Knutson and A.K. Stoner (eds.), *Biotic Diversity and Germplasm Preservation, Global Imperatives*. Kluwer Academic Publishers, Dordrecht, The Netherlands.

Wilkes, H.G. 1971. Too little gene exchange. *Science* 171:955.

Wolfe, M.S. 1985. The current status and prospects of multiline cultivars and variety mixtures for disease resistance. *Annual Review of Phytopathology* 23:251-273.

Zohary, D. and M. Hopf. 1988. *Domestication of Plants in the Old World*. Oxford University Press, New York.

5

<u>In Situ</u> Conservation of Agricultural Systems

Garrison Wilkes

INTRODUCTION

In situ conservation of economically useful plants and animals is an "idea whose time has come." Interestingly, there is growing public awareness of *in situ* conservation–not from the perspective that it is more advantageous than *ex situ* conservation, but rather from an increasing realization of the public service functions served by natural environments. *In situ* conservation is seen as an alternative to development fostered by unlimited growth. Unlimited growth patterns have caused environmental impoverishment and instability and have reduced the human carrying capacity of the global environment. Initiation and maintenance of *in situ* conservation will serve to enhance our environmental awareness and sensitivity to the remaining natural systems around us.

Only a century ago, human populations were less densely packed, our demands on natural environments were less harsh, and communication was more localized. We have only recently developed a concern for genetic conservation in perpetuity and begun to talk in terms of global carrying capacity (Altieri and Merrick 1987, Altieri et al. 1987, Hoyt 1988, see also Oldfield and Alcorn, Ch. 3, this volume).

OPINIONS FOR BIODIVERSITY CONSERVATION IN PERPETUITY

Biological conservation in perpetuity can take three somewhat distinct but overlapping levels or pathways. First, entire biomes or major types of natural ecosystems can be conserved. This entails preservation of vast tracts of land or water areas with *in situ* conservation of animal and plant populations or species within them–a continuum from wilderness areas that are almost totally devoid of humans, to unique natural habitats that require considerable human intervention, management, and protection. Preservation at the biome level has typically entailed the removal of humans from the natural ecosystem picture; therefore, it has not, and does not, promote humans as the pivotal point of natural systems. This type of conservation will have little impact on regeneration of perturbed habitats exploited as a resource base.

The second level or pathway is that of onsite, or *in situ*, conservation of useful plants and animals in evolutionarily dynamic interactions–an option that does not necessarily lead to decreased carrying capacity of the environment for human populations. This pathway also involves a habitat continuum of idealized habitat types from near–wilderness tracts necessary to sustain hunter-gatherer societies, to the cultivated fields of shifting agriculturalists, to more intensively cultivated fields where local landraces or native crop cultivars may still hybridize with their weedy or wild relatives along field margins. The essential elements of this option are that the plants and animals are useful to humans, and human population density has not increased to the point where degradation of the natural resource base has occurred. In truth, these idealized habitat-types have not been, and will not be, easy to achieve. But such *in situ* conservation efforts will give us more time to understand and appreciate how human-modified agroecosystems evolved. Considerable potential exists for creating institutional arrangements to facilitate such *in situ* conservation options, especially for cultural and historical sites in developing countries. The widespread cultivation of "heritage varieties" by members of the Seed Savers Exchange, a grassroots, national heritage *in situ* seed conservation effort that is widely dispersed across the entire country, is another unique form of *in situ* conservation. This concept, seldom realized by the agricultural research establishment, has enhanced the value of the "heritage varieties" collection (Wilkes 1983).

Finally, the third pathway is *ex situ* (offsite) preservation of plant seeds, animals, or in vitro cell lines that are isolated from the natural or human-modified environments in which they evolved. These primary modes for preserving genetic resources of direct economic importance are evolutionarily static systems that freeze the genetic landscape, e.g., by storing seed in genebanks, clones in clonal gardens, and animal semen/ova in liquid nitrogen. Such systems pull genes out of circulation and therefore, to be useful, require documentation and evaluation so that their potential value will be known to animal and plant breeders. Information management is therefore as important as the physical arrangement of storage space for seed, gametes, etc. Genebanks slow down crop and domestic animal evolution and necessarily require intervention of plant or animal breeders to make the *ex situ* preservation option useful. *Ex situ* preservation has three aspects: 1) exploration, collection, and banking of genetic resources; 2) evaluation and documentation; and 3) breeding for genetic enhancement via hybridization and subsequent selection.

Twenty years ago, at the Third Technical Conference on Crop Genetic Resources in Rome, Erna Bennett invited me to give a presentation on *in situ* conservation along agricultural zones where maize and its wild relative–teosinte–hybridize (Wilkes 1971). The subsequent discussion was short and to the point–"good idea, but it won't work". At the time, no one, including myself, asked "Why?" The system had been working on the biological side for millennia, so it must have been on the economic side that it "would not work." Since priorities often establish funding availability, it must be acknowledged that if one's priorities are for *ex situ* genebanks, then *in situ* conservation will be out of the question–i.e., "it won't work". Since the early 1970s, crop genetic conservation priorities have changed. In 1981, at the Fourth International Technical Conference in Rome, I chaired a panel on *in situ* preservation; yet this "idea whose time has come" still has yet to be realized (Jackson 1984). Because of the large number of accessions (two million) in major *ex situ* crop collections, there is no way that *in situ* conservation could be employed for most landraces or traditional crop cultivars. In essence, such a mandate would force landraces to persist in the genetic landscape of peasant farming, and would place high social costs on peasant farmers.

A limited number of selected landraces in key sites would be feasible, practical, and appropriate. These key sites would be especially useful if located in "hot spots" where the diseases, pests, and pathogens of the crop were also evolving along with it under the pressures of both natural and artificial selection (see also Browning, Ch. 4, this volume). These selected *in situ* stations would highlight the issue of conserving crop genetic resources because of their location and educational function. To be effective, such *in situ* conservation of traditional crop cultivars or landraces needs to be coupled with both tourism and research. A visit to the Maya Ruins at Chichen Itza along with a nearby reconstruction of a *milpa* corn-field (carefully researched by the Mexican Escuela Nacional de Agricultura) would impress well-heeled tourists about the primacy of crop plants. The same could be done for the potato and other tubers in the Peruvian highland agricultural region at Inca Ayacucho. In the Old World, an Indian village at the fertility temple sites around Khajuraho would be seen by hundreds of thousands of tourists annually. In the United States, the apple orchard and fields at Sturbridge Village, Massachusetts (circa 1837), or gardens at Williamsburg, Virginia, provide suitable examples of living historical farms that combine tourism and research values (see also Nabhan, Ch. 7, this volume).

These "living farms" and their heritage germplasm of landraces conserved *in situ* would not preserve the genetic landscape as much as they would instill pride in the plant domestication accomplishments of indigenous peoples in the past. Establishment of such farms would encourage people in other regions to do likewise, perhaps at local agricultural colleges. For example, at the Punjab Agricultural University in Ludihana, India, there is a reconstruction of a small landowner's compound–the brain child of a former Punjab Agriculture University chancellor inspired by a visit to Williamsburg, Virginia. The frequently visited buildings are the pride of the campus. To broaden the educational impact of such farms, timeline 1870 genotypes or domesticated animal breeds and plant cultivars should be included in the farmyard and fields.

These living farms would not focus on just one crop, but rather on a mixture of different crops; they would also demonstrate typical cropping patterns. This is what *in situ* conservation projects can do best–whether it be a demonstration of how the crop hybridizes with its wild relatives, how two different genotypes of the same crop interact, or how two distinct crops may interact in the same field. Until a few years ago, it was possible to find two distinct barley varieties–one early-maturing and the other late-maturing–that had identical stature and were often planted together in the same fields in Nordic countries. Presumably, seed stocks of these are now stored in the Nordic Genebank, but the knowledge of which two accessions go together has been lost! There is a pea and barley combination that has also been lost. In México, the corn/bean/squash coadapted *milpa* crop systems are in place, but they will probably decline in area given current economic priorities.

There is much optimism about México progressing in *in situ* conservation because it is a very "plant-sensitive" culture (as witnessed by the expression "my mother maize"). México already has a national *in situ* reserve–the first of it's kind–for the only perennial wild relative of maize, *Zea diploperennis* in the Sierra de Manantlan! At this site, a sizeable acreage protects this recently discovered, rare endemic plant species along with its associated vegetation. The reserve's research laboratory is also studying local agricultural diversity and other endemics–both plant and animal– found on this mountain. Overall, they have combined conservation with research, but they still need to add education and tourism to futher justify the value of conservation by communicating the primacy of plants both as a source of food and as a part of natural systems.

Pioneering efforts in *in situ* conservation under local control are being initiated in Guatemala. During a 1991 expedition to revisit sites where teosinte populations existed 30 years ago, I found that populations had been reduced to 25% of their earlier distribution.

The situation was discussed with local farmers and ranchers who agreed that teosinte was disappearing and recognized that it was important to maintain teosinte wherever it was still found. Because of their concern that teosinte be kept in their ecosystems, they volunteered to cooperate with the Guatemalan Institute for Science and Technology in Agriculture (ICTA) to establish and protect *in situ* reserves of teosinte. ICTA will support these local initiatives with workshops and formal recognition, but responsibility for the management of the reserves remains in the hands of the landowners (a rancher and two communities). Thirty years ago Sir Otto Frankel maintained that *in situ* conservation was impossible. Guatemalans and Mexicans are showing us that it is possible!

IN SITU CONSERVATION: AN IDEA-GATHERING MODEL

To be blunt, if something is not used, it is not worth conserving. This perspective may run against the grain of many, but it should not. The real value of conservation is that others, notably future generations, can have comparable opportunities and resources to those that we enjoy now. Numerous crop seed samples are now being stored in genebanks in the hopes that some might be useful in the future; but in truth, most of the seed now stored by such *ex situ* methods will not prove to be useful. The reason these seed stocks are stored as such now is that in the absence of better genetic evaluation than currently practiced, we don't know which seed is "most" useful and which is "least" useful. Indeed, this is a major problem with *ex situ* genebanks—they are useless until their contents have been evaluated.

In contrast, *in situ* conservation generates its own evaluation with every change of the seasons. Seed in genebanks have to be planted in the ground for evaluation, whereas seed stored *in situ* is already connected with or rooted to the soil. *In situ* conserved populations are dynamic, because they continue to evolve in response to pest and pathogen populations, whereas gene banks freeze the seed population sample in its present generation (see Browning, Ch. 4 this volume). Great care is necessary to prevent the loss of rare alleles when regenerating seed stored in *ex situ* banks.

Human population growth cannot be stopped. Nor can we halt the simplification of ecosystems by cultivation of crop monocultures. But these conditions do not have to prevail everywhere. *In situ* reserves are thresholds to what the world once was, and hopefully they can be models to show us how conservation might be achieved in the future. This does not mean that buffalo (bison) hunting will be reinstituted, nor that the Eastern half of the United States will again become four-fifths reforested. But it does mean that by preserving some of the remaining, human-sustaining biological systems *in situ*, there will still be models to guide us. The development and success of dwarf stature wheat led to the idea of developing dwarf stature rice. Likewise, the traditional "gardens of chaos" begot modern relay planting of crops now successfully deployed in Asia. The multiline crop cultivars consisting of individual varieties of the same stature, but with different dates of maturity, pest resistance genes, or stress-tolerance genes, are all modeled in gardens of chaos as though the garden had no creator! Coadapted complexes comprised of different taxa are models reminiscent of what biotechnologists talk about when they suggest cell fusion across taxa lines; yet, such systems already exist in many village kitchen gardens and dooryard agroecosystems. These *in situ* "extended kitchens" possess considerable information of agricultural and cultural significance, and their preservation is best justified in that they should be well studied.

LINKS BETWEEN RURAL PEOPLE AND LEVELS OF *IN SITU* MANAGEMENT

Different levels of *in situ* management are appropriate for different classes of genetic resources (see Pluckett et al. 1987). Four major groupings, in terms of decreasing intensity of management, by people, include:

- Cultivated plants and domesticated animals, conserved on living historic farms and national heritage sites rich in historical or cultural significance.
- Wild relatives of domesticated and animals, found in evolutionary gardens and "chaotic" gardens in centers of crop diversity and crop origin; the latter often show peasant farming at its best, allow *in situ* monitoring of wild populations, and might also include "hot spots" for coevolution of selected crop plants with their pest and pathogens.
- Wild species used directly by people, found in "kitchen gardens" of peasant farmers that may extend into nearby forest; even though the species may be wild, their survival is actively managed by people.
- Wild species of no current use to people, which may be useful someday even though they are still not currently recognized.

The strongest case for *in situ* conservation of useful plants in a self-sufficient peasant society is its value as a teaching tool to reestablish the primacy of plants to people from industrial societies who are urban dwellers entrained on the supermarket. These individual decision makers have forgotten that civilization depends on an assured food supply or have never really known that crop yields and harvests are affected by changes in weather, pests, and pathogens. Thus, they do not understand that genetic variability is essential for maintaining a steady supply of food on market shelves. *In situ* conservation of self-sustaining agroecosystems is one of the best ways to demonstrate that environmental impoverishment affects the urban dweller.

IN SITU CONSERVATION IN A NATIONAL CONTEXT

The indigenous *milpa* agriculture of México was based on hoe and dibble-stick cultivation, and is ideally suited for *in situ* interpretative farming. There were no large domesticated mammals to affect or dominate the agricultural landscape. Seed sowing was not done in plowed fields by broadcasting seeds, but rather by hilling up small mounds or making several holes with a pointed stick in undisturbed soil after slashing and burning a forest. Into each hole, or mound, seeds of maize, beans, and squash were placed, and the soil smoothed over. The maize grew tall and upright, the beans twined around the maize stalk, and the squash formed a ground cover with its large leaves, thus decreasing competition from weeds. What is seen here is not just one of the fifty indigenous landraces of maize in México, but a coadapted maize/bean/squash crop complex with interacting genetic systems. In modern terminology, the crop complex was "locked in" on evolution under domestication through artificial selection by local cultivators.

This cultivator-to-plant–"I-thou" relationship of indigenous people and their crops is in part responsible for the tremendous genetic diversity of native crops found in México. Subsequent hybridization between genetically distinct, often geographically isolated taxa found in domesticated mountainous terrain has actually increased the rate of evolution of these crops. Because crop plants are harvested one plant at a time and the cultivator has eye contact with individual plants, artificial selection by humans has resulted in the accumulation of a tremendous diversity of size, shape, intense color, and taste variants. In this way, *in situ* conservation can also serve as an "artificial selection window" to past evolutionary events.

Beyond our dependence on plants for their high caloric yields to meet the needs of expanding human populations, there is a second aspect of the human-crop plant symbolic relationship that is well illustrated by the classic Mexican diet. In any subsistance agricultural system where cultivators must depend on what they grow, there is natural selection for balanced nutritional intake. Over the millenia, if human nutrition was well-balanced, the cultivators had sufficient energy to tend their plants to increase crop yields. This promoted cultivation of crops that possess complementary plant proteins. A New World example is that of maize (corn) which is deficient in the amino acids lysine and methionine and is typically consumed with beans which are deficient in cysteine. An Old World example is that of wheat—deficient in lysine and isoleucine and also usually consumed with cysteine-deficient foods like curded milk or lentils. In addition, cooking styles that alter amino acid ratios, increase the amount of digestible protein, or add essential minerals to the everyday diet were adopted as well (Katz et al. 1984). A good example of the latter is the soaking of maize kernels in lime before processing them into tortillas. The Mexican people did not coevolve with milk-producing animals. Consequently, their diet tended to be more calcium-deficient. By soaking maize in lime, calcium was added to the diet; this practice basically allows an adequate nutritional balance similar to that found in human diets elsewhere in the world that include milk, curds, or cheese (Wilkes 1977a).

Although maize is the most important food grain and crop plant in México, it does not provide a complete nutritional package. Even with the addition of beans and squash, only the vitamins A and B, and part of the protein and caloric needs of the population are met. Yet, by the simple addition of tomatoes and chili peppers that provide vitamin C and fatty acids, and avocados that provide fatty acids and protein, indigenous agriculturalists of México were able to sustain one of the healthiest, most densely settled populations in the world at the time of the Spanish Conquest in 1520. By use of *in situ* reserves or farms such as a circa 1520 *milpa* at the National Museo de Anthropologia in Central México City, all of this useful information could be communicated. Such a "Self-Sufficiency Garden" would provide a living history of what the Museo Nacional de Cultures Populares has so beautifully described in the 1982 book *El Maiz*. Such an achievement could lead to the establishment of other *in situ* living farms or gardens in other Latin American museums.

IN SITU CONSERVATION ON THE FARM

The Hungarians also have living farms, staffed by retired public school teachers from natural sciences, for demonstrating the value of their national heritage of crop germplasm resources. These farms, often located on national forest lands, are currently maintaining many traditional vegetable varieties by use of a carefully managed system of geographic isolation that prevents outcrossing to modern commercial cultivar lines. Closer to home is Kent Whealy's Seed Saver Exchange in the United States.

The elements these *in situ* programs have in common is a present-day living connection to the past. Live crop plants on display represent a shared heritage from the past for all people living today. These are the crops that have and still do feed us, support our cities, and make it possible for humans to live in high density populations. In making a connection between the past and present, we thus retain our identity with the past. México can be proud of maize, its contribution to the world community; India–for its contribution of sugarcane and rice; and China–for soybeans, rice, and tea. The list goes on for about 500 significant food or cash crop species that now support a human population of more than 5 billion.

The most significant service of *in situ* landrace preservation on heritage farms is to demonstrate the primacy of plants. This facilitates work to establish the efficacy of our germplasm collections through increased evaluation and breeding for plant improvement, while still holding us to the principle of free and open gene exchange. If people could value plant genetic resources through the *in situ* interpretative demonstration mode, this would be a fantastic accomplishment.

EVOLUTIONARY GARDENS

Another opportunity for *in situ* conservation is evolutionary gardens–the great natural resource base from which our useful crops have developed. These are the agricultural areas where crops and their weedy or wild relatives on the margin of fields continue to hybridize and where dynamic evolution occurs through introgressive hybridization (Wilkes 1977b) or via changes in ploidy levels (numbers of chromosome sets) of crop species. The teosinte-maize system in México is a classic example. These systems are analogous to a still where the boiling of the mixture is the process of hybridization and the distillation is achieved by both artificial selection (on crop species) and natural selection (on both the crop and its wild or weedy relative). This results in the dynamic generation of genetic refinements in crop plant populations. In current practice, seed of landrace crop cultivars and their wild or weedy relatives derived from these locations is usually not cross-referenced when preserved in *ex situ* genebanks. One of the few crop plant collectors who has been very sensitive to this problem is Angel Kato-Y. in México. Nevertheless, *in situ* conservation of these sites would allow the continuation of future evolution of the crop genetic system under the selection pressures of both pests and pathogens (Wilkes 1971).

We know little about domestication process for most major crops. Therefore, another important function of *in situ* zones of crop origin and evolution is that they serve as windows to the past that are now being closed by changes in land use. For example, Mexican maize fields harbor teosinte mimics of maize that are not weeded out from the crop population. Maize fields which I previously used in 1962-64, 1971, and 1978 to obtain baseline information for studying maize evolution (e.g., see Wilkes 1977b, 1987) are now part of a Volkswagon automobile dealership on the outskirts of Chalco. Future generations might thank us immensely for the worldwide preservation of only 500 of these remaining agricultural systems for future study.

Now we come to the other side of the evolutionary gardens, where the focus is not on single crop species, but rather on all plant species used by indigenous inhabitants in the landscape. This also represents an evolutionary sequence, because some of the plant species are full domesticates, some are incipient domesticates, and others are gathered regularly or only occasionally. The ethnobotanist's dream is to find a stable society that utilizes vast knowledge of its natural surroundings. This lifestyle exists in non-cash societies–most of which have already disappeared or are now doomed.

The information potential in these "gardens of chaos" named by Edgar Anderson (1952) is tremendous. Agricultural extension officers consider it their responsibility to set gardens in straight rows and generally "clean-up the act." But Anderson correctly saw everything in chaotic gardens as being in their place, balanced together like a miniature natural ecosystem where ecological relationships among species are more often defined by webs rather than chains or straight lines. Often these peasant or kitchen gardens have no clear lines of demarcation where native vegetation begins. Indeed, the wild plants in these gardens are usually managed for useful products. Academic books can list the useful species found there, but they rarely connect all the lines that describe the intertwined web of life between these "self-sustaining" peoples and their domesticated plants and animal cohabitants of the "hearth". Here, evolution occurs as a consequence of the ecology of the entire habitat, and only *in situ* conservation can capture this for purposes of long term genetic conservation.

IN SITU MONITORING: THE EXAMPLE OF TEOSINTE IN MEXICO

Over the past 20 years the maize program of CIMMYT (The International Center for the Improvement of Maize and Wheat in México) has maintained the world's largest collection of maize landraces along with germplasm resources of the wild relatives of this crop in *ex situ* genebank cold storage. Recently, it has taken on the self-imposed task of monitoring wild populations of annual teosinte in México and Guatemala in cooperation with the respective national program staff of each country. A highly variable wild plant, teosinte is represented by both annual and perennial species. The seed-producing annuals are all diploid (2N=20), and are found in eight geographically isolated population clusters–six in México and two in Guatemala. Some teosinte populations are small. The smallest occupies less than 1 km², whereas the largest covers hundreds of square kilometers around the Rio Balsas, primarily in the state of Guerrero. Yet at many sites on México's central plateau, teosinte has died out in recent times due to intensive land use. By best estimates, the current distribution of teosinte is about half of what it was in 1900. The disappearance of remaining teosinte populations will accelerate as more roads are built and land use continues to intensify. Just 20 years ago, a third of the teosinte populations was far from any road, and another third were accessible only by dirt roads that could be travelled with difficulty. Now all these populations are accessible by roads that can be easily negotiated by automobile.

Due to threats to extant annual teosinte populations, the CIMMYT Maize Germplasm Bank has cooperated with national program staff to initiate an annual *in situ* monitoring effort designed to watch for potential status changes of wild teosinte in natural habitats. This work should not to be confused with *in situ* conservation, which is a much more time-consuming and complicated endeavor. Yet, maize appears to be the first major crop plant for which any *in situ* monitoring of the status of its naturally occurring wild relative has been undertaken. For CIMMYT staff, most teosinte monitoring sites are less than one day's drive from the Central headquarters. Monitoring by periodic visits is the most flexible and efficient approach that provides a sufficient guarantee that the resource material will be closely watched. If the status of a recognized population appears to be changing or has been placed in immediate danger of extinction, there will be time for program staff to respond in cooperation with appropriate national organizations. In the past, teosinte populations went extinct with no realization by either national program staff or international organizations. Such extinctions will probably no longer occur with annual teosinte–the closest relative of maize. *In situ* monitoring is an activity that should be extended to the wild relatives of all major crops as well as to locally important minor crops (Wilkes 1987).

RECENT DEVELOPMENTS: FARMER'S CHOICE

As suppliers of agricultural seed become fewer, there is a generalized fear that commercial seed will become even more genetically uniform, and farmers will be in even greater risk.[1] Many developing nations have not allowed multinational seed companies to conduct business or have limited their operations within their borders. With prevailing patent laws and the utility patent, there are real questions about ownership of the seed. Another question is whether farmers will be able to retain their freedom to plant the seed and crops of their choice (Fowler and Mooney 1990). Industrial farmers find these trends less frightening than peasant farmers. In some cases, considerable mistrust has developed–witness the FAO undertaking of 1983 and the subsequent lack of agreement on how to proceed with use and conservation of crop plant genetic resources on a worldwide level. One breakthrough in this cycle of mistrust came with the 1988 Keystone Dialogue concept of "farmer's choice" which is directly related to *in situ* conservation of landrace cultivars.

Farmer's choice should not be confused with farmers' rights. They are separate and distinct concepts. The idea of farmers' rights is a vague notion that farmers in centers of origin (i.e., where a crop or notable landrace variety has evolved) should be compensated somehow for the plant breeding efforts of their ancestors. This concept of farmers' rights is an antidote or countermeasure to more recent recognition of European plant breeders' rights and U.S. patent protection laws. Farmers' choice, on the other hand, is a pledge to maintain distinct seed stock options for farmers in the marketplace. This commitment applies to commercial seed suppliers in open market economies as well as to those in the public sector in centrally planned economies. Simply stated, seed suppliers must provide improved seed uat a reasonable price for both high input and low input agriculture.

Low input seed, unlike high input seed, must be of a kind which farmers can regenerate to maintain their own seed supply.[2] High input seed is typically envisioned as hybrid seed stocks that cannot be regrown by farmers. Supplying high input hybrid seed is a familiar role of commercial seed companies; but the idea of supplying improved, low input seed is a relatively unfamiliar role. In essence, this was what seed companies did in the last century. In the 1860s, the American Shaker communities had an eager market for their garden vegetable seed which was well adapted, free of off-type seed, and had desirable flavor, storage, and other kitchen qualities. The Shakers supplied good commercial seed so that farmers could regenerate themselves, i.e., it wasn't hybrid seed and was not covered by variety protection legislation or other forms of "ownership".

The concept of crop genetic improvement is central if farmer's choice is going to work. Obviously, high input hybrid seed is genetically improved, or farmers would not be willing to pay more for it in comparison to cheaper, open-pollinated seed stock. The rule of thumb has been that farmers will pay up to one-third the value of increased crop yields for hybrid seed. Therefore, the meaning of "improved seed" needs some clarification at the low input end. Any additions of quality or quantity to the yield component would constitute improvement that may or may not involve selective breeding. Plant breeding with repeated backcrossing and yield trials would not pay for itself with low input seed. This is basically the reason seed companies have not served this market. But there are many improvements that can be made that would not involve extensive plant breeding programs. Recently, Guatemalan landrace beans were improved in Europe by removing viruses from seed stocks; the virus-free seed was returned to the New World where it has demonstrated a yield performance improvement of 20%. The local environmental adaptation, flavor, and cooking qualities remain the same as before because there was no plant breeding and no artificial selection. Another suitable example is the creation of a "synthetic" from the extremes of a landrace cultivar. Such synthetics often exhibit increased yields in farmer-produced seed for a period of three to five years. This has been done in México with the maize landrace Tuxpeño.

Cost-effectiveness of crop breeding for adaptation to marginal habitats and farmer generated-seed will only exist for specialty crops. There are many other ways to improve seed stocks that have been explored and will be encouraged by the acceptance of farmer's choice. Large multinational firms with good hybrid seed stock to sell might contract with a small company to improve an "orphan crop" that may be nationally important even though it may never return a profit for the seed supplier. Under this scheme, landowners of agricultural land would have access to improved seed stocks, but local farmers of marginal agricultural land would have access to improved seed as well.

It is a romantic concept to believe that landraces of crops never change, but some advocates of *in situ* conservation have held this view (Fowler and Mooney 1990). The landraces planted by peasant farmers are participants in a dynamic evolutionary process that is best characterized by continuing changes in gene frequencies. The central threat to most landraces grown by peasant farmers is that they will be abandoned for other higher-yielding seed or for other crops that have a higher market value. The concept of farmer's choice would, by supplying incentives for improvement of landrace cultivars, actually slow down the process of landrace abandonment. This would therefore reduce the rate of genetic erosion, because improved landrace seed would give higher yields despite low input. The best source of germplasm resources for most marginal habitats and many minor, but regionally important, crops is found in landrace or local varieties. Much of this material could be genetically improved by current technologies that would be cost-effective for seed companies if hinged to the sale of more profitable, high input (often hybrid) seed. Since peasant farmers will produce their own landrace cultivar seed in subsequent years, the contentious problem of seed ownership is a moot argument. This formal recognition of the value of seed for low input agriculture has the potential to conserve more *in situ* landrace germplasm resources than all of the other modes of *in situ* conservation. Therefore, the concept of farmer's choice is worthy of encouragement and support by the crop conservation community.

CONCLUSIONS

Our need for biological resources has not been diminished by industrialization—only our options. It takes active support of the idea of *in situ* conservation in both developed and developing nations if we are to capture the value and importance of these indigenous agroecosystems before they are lost forever. Extinction of an entire species, a particular genetic line, or a species/genetic complex in an agroecosystem represents an irreversible loss of a unique biological resource. Increasingly, we consume food, use medicines, and employ industrial materials that owe their production to genetic resources derived from biological entities. Given the uncertainty of future generations, genetic resources should be acknowledged as one of society's most valuable raw materials. Any reduction in such resource diversity narrows society's scope to respond to new problems and opportunities.

To the extent that we cannot be certain what needs may arise in the future, it makes sense to keep our options open. This conservation rationale applies to the earth's endowment of useful plants more than almost any other category of natural resources. It is difficult to visualize a challenge more profound in its implications, yet less appreciated by the general public than conservation of plant genetic resources. By maintaining representative samples of some living agroecosystems *in situ*, the best teaching tool for demonstrating the primacy of plants and plant genetic resources would be maintained for both present and future generations.

NOTES

1. Advocates of *in situ* conservation often cite the heterozygosity of landrace cultivars as a positive attribute of peasant farming systems. The reasoning is that homozygous or genetically uniform plantings of modern cultivars promote genetic vulnerability to disease pathogens or pests. (Even though modern hybrid cultivars are, by definition, heterozygous, they are genetically identical within a given cultivar or variety.) Therefore, there is presumably greater safety in crop genetic diversity for traditional farmers who maintain landrace cultivars. There are elements of truth in

both assertions, but neither is always true. First, landrace cultivars are not necessarily more heterozygous than modern cultivars. Second, genetic homozygosity does not necessarily initiate genetic vulnerability to pathogens or pests; instead, genetic uniformity leads to genetic vulnerability if susceptablity to a pest or pathogen is present.

2. This concept should include consideration of seed stocks that have to be repurchased every few years, e.g., when "synthetics" lose their yield components or once "virus-free" stocks accumulate a virus load over a number of years in farmers' fields.

DEDICATION

I would like to dedicate this paper in memory of the late Professor Efraím Hernández Xolocotzi, Escuela Nacional de Agricultura and Centro de Botanica Colegio de Postgraduados, Chapingo México, who opened my eyes to the vastness of variation found in farmers' fields and the power of folk-selection in cultivated crops.

REFERENCES

Altieri, M.A. and L.C. Merrick. 1987. *In situ* conservation of crop genetic resources through maintenance of traditional farming systems. *Economic Botany* 41:86-96.

Altieri, M.A., M.K. Anderson and L.C. Merrick. 1987. Peasant agriculture and the conservation of crops and wild plant resources. *Conservation Biology* 1:49-58.

Anderson, Edgar. 1952. *Plants, Man, and Life*. University of California Press, Berkeley.

Fowler, C. and P. Mooney. 1990. *Shattering Food Politics and the Loss of Genetic Diveristy*. University of Arizona Press, Tuscon.

Hoyt, E. 1988. *Conserving the Wild Relatives of Crops*. IBPGR/IUCN /WWF. 45pp.

Jackson, W. 1984. A new threat to agriculture and a window of opportunity. *The Ecologist* 14:119-124.

Katz, S.H., M.L. Hediger and L.A. Valleroy. 1984. Traditional maize processing techniques in the New World. *Science* 184:765-773.

Museo Nacional de Cultures Populares. 1984. *El Maiz*. 2nd Edition. García Valades Editores S.A., México, D.F.

Oldfield M.L. and J.B. Alcorn. 1987. Conservation of traditional agroecosystems. *BioScience* 37:199-208 (see also Ch. 3, this volume).

Plucknett, D.L., N.J.H. Smith, J.T. Williams and N.M. Anishetty. 1987. *Gene Banks and the World's Food*. Princeton University Press, Princeton, New Jersey.

Wilkes, G. 1971. Too little gene exchange. *Science* 171:955.

__________. 1977a. Native crops and wild food plants. *The Ecologist* 7:312-317.

__________. 1977b. Hybridization of maize and teosinte, in Mexico and Guatemala and the improvement of maize. *Economic Botony* 31(3):254-293.

__________. 1983. Current status of crop plant germplasm. *CRC Critical Reviews in Plant Science* 1:133-181.

__________. 1987. Conservation of the wild relatives of maize. *CIMMYT Research Highlights* (1986):33-39.

6

Value and Uses of Indigenous Livestock Breeds in Developing Nations

Hank Fitzhugh and Ann E. Wilhelm

INTRODUCTION

Today, most of the world's people and most of the world's livestock are located in developing countries. This correlation is not due to chance. For the majority of the rural poor in developing countries, livestock improve their quality of life, offer a means to escape poverty, and in some instances, provide the means to survival. Animals are an important component of many traditional farming systems in developing countries. Such complex interdependencies have led to a longstanding relationship between livestock and humans.

Although most important livestock species were first domesticated in the Middle East and Asia, relatively few world "breeds" are truly indigenous. Livestock have moved (indeed often provided the transport) as humans migrated under the influence of agroecological changes and sociopolitical pressures. For example, most of the important New World livestock breeds are relatively recent imports. Only the capybara (*Hydrochoerus capybara*), guinea pig (*Cavia porcellus*), turkey (*Meleagris gallopavo*), llama (*Lama glama*) and alpaca (*Lama pacos*) appear to have originated in the Americas. On occasion, introductions of livestock have stimulated major changes in traditional cultures. The impact of horses and wool sheep on the North American Indian culture are good examples. Without their introductions to the American southwest by Spanish conquistadors, exploits of Navaho weavers and Apache warriors might never have become part of American history.

Domestication of livestock was a major step in the evolution of primitive humans. As humans alter the environment, develop new technologies and change their needs and desires, the nature of their interdependence with animals also changes. Thus, the values and uses of livestock continue to evolve.

DOMESTICATED ANIMALS

Natural selection is the principal directing force behind genetic change in wild animal species. Interactions between an animal's genotype and its environment determine its phenotype–which influences its overall fitness or lifetime reproductive success. The most "fit" individuals are those that survive until sexual maturity and leave more offspring to populate the next generation than most other individuals in the population–often because they are best adapted to their environment. Over time, superior genotypes in given environmental conditions will predominate. Natural selection processes shaped the form and function of animal populations until humans intervened by imposing their own preferences through artificial selection pressures.

Today, practically all domesticated animal species and incipient domesticates or "candidates for domestication" are vertebrates from the phylum Chordata, as shown in Figure 6.1. Mammals (Figure 6.2) –and to a lesser extent, birds (Figure 6.1)–predominate. The first wild animals domesticated by humans were probably the dog and pig –both of which date back about 12,000 years. Whether the original intent of domestication was to provide humans with food or to fulfill sacred functions is still debatable (Peel and Tribe 1983). However, early domesticates became an integral part of primitive human culture–providing meat during periods of scarcity and serving sacrificial or ceremonial roles. Both wild dogs and wild pigs were already adapted for scavenging the campsite wastes of hunter-gatherer tribes. Although their natural ability to survive on human food aided in their domestication, it also placed them in direct competition with humans. Subsequently, herbivores (ruminants, equines and cameloids) were favored for domestication because they survive on low quality forage, and thus, convert the grasses and browse that humans do not eat into meat. Humans that had ventured into more inhospitable habitats, such as deserts, high mountains, and frozen tundra, eventually domesticated wild species that were already well adapted to these harsh environments, such as the camel (*Camelus* spp.), yak (*Bos grunniens*), llama, and reindeer (*Rangifer* spp.).

Additional products and purposes for early domesticates soon became apparent. Female animals could be milked and adult animals could be bled to provide daily nutrient supplements to human diet. Dogs protected the campsite, provided companionship, and joined in the hunt, helping to control the flight of game (not unlike their eventual role in herding livestock). Loads were lifted from human shoulders to be carried on the backs of animals or to be pulled on sledges behind them. Eventually, animals were carrying people as well as their packs and pulling plows as well as sledges.

Along with the domestication of animals came the ability to control their reproductive process and determine which individuals

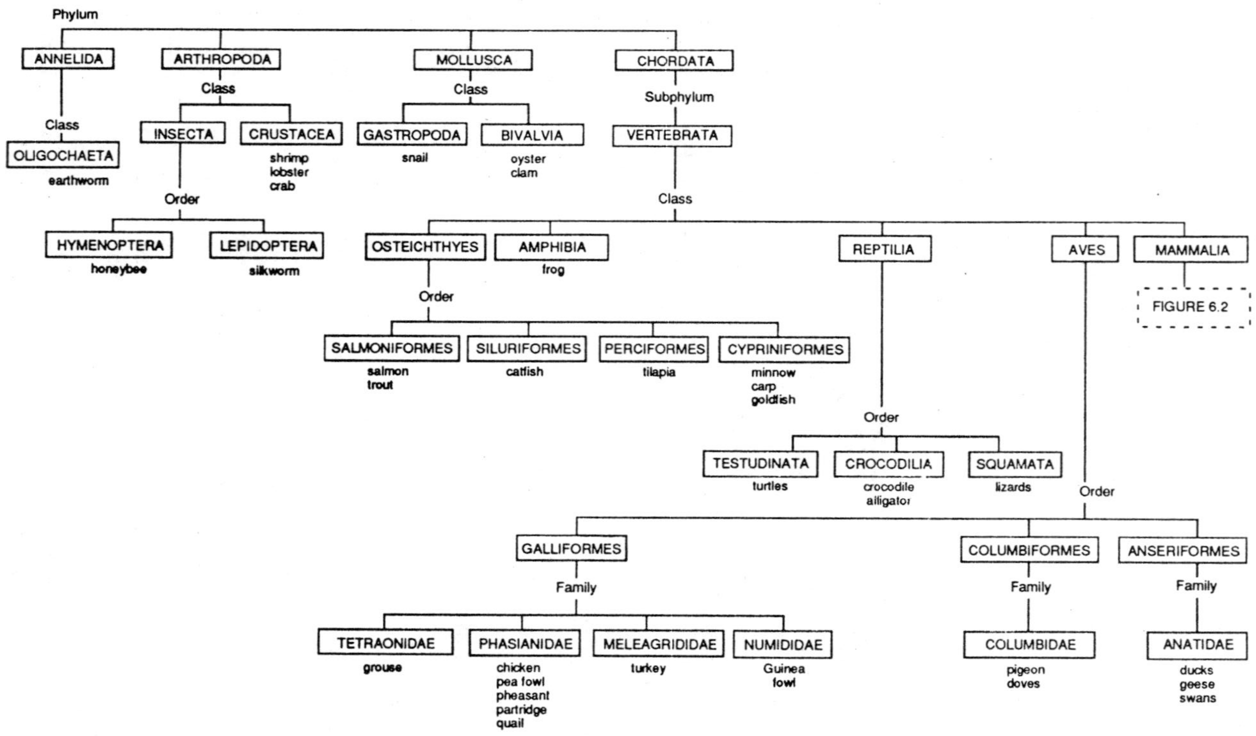

FIGURE 6.1 Domesticated animals in the animal kingdom, including candidates for domestication.

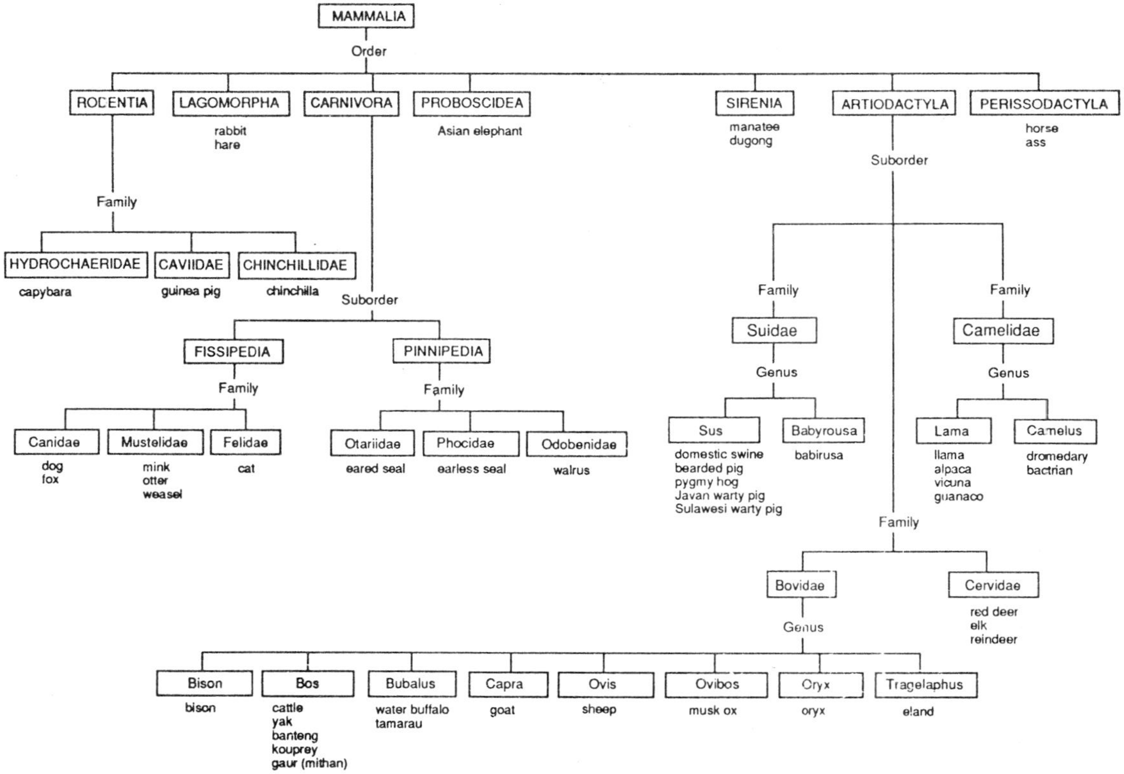

FIGURE 6.2 Domesticated animals in the Class Mammalia, including candidates for domestication.

produced the next generation. Through artificial selection, humans began to modify livestock species–molding them to meet their needs. When human selection preferences differed from those favored by nature, humans eventually modified the production environment to allow genotypes less fit for the natural environment to be able to survive. These modifications included the harvest and storage of feed to provide winter requirements and construction of housing to provide protection from the elements. Today, humans continue to modify the environment of livestock by providing vaccinations against diseases, and even caesarean sections to enable the birth of oversized progeny. Through environmental alterations and intensive selection, humans have developed highly specialized livestock breeds, such as the Holstein dairy cattle capable of 10-month milk yields of more than 15,000 liters.

Even though most domesticates and candidates for domestication are mammals and birds (Figures 6.1 and 6.2), invertebrates provide many important exceptions–especially insects such as the honeybee (*Apis mellifera*) and silkworm (*Bombyx mori*), crustaceans such as shrimp (*Crangon* spp.), and mollusks such as snails (*Triodopsis* spp.). Today, most principal candidates for domestication are generally similar to already domesticated types, i.e., uses or markets for their products or purposes are already well established. Little effort will be expended to domesticate new breeds as long as there are domesticates which already fit the available product/market niche–even if the candidate potentially fits better than the domesticate. However, sometimes these candidates are well adapted to an environmental niche not filled by an important domesticate.

Some candidates for domestication include the capybara from the wet lowlands of South America; the manatee (*Trichechus* spp.) and dugong (*Dugong dugon*) from subtropical waterways; the seal (*Callorhinus* spp.) and walrus (*Odobenus* spp.) from cold temperate waterways; several types of pigs, cattle, and buffalo indigenous to tropical Asia (National Academy of Sciences 1983); and game species, such as the eland (*Taurotragus* spp.) and oryx (*Oryx* spp.) grazing semiarid African rangelands. Particular attention has been paid to the economic potential of several species in the family Cervidae. Farming of red deer (*Cervus elaphus*) in New Zealand has become a major commercial industry annually exporting millions of dollars of venison, hides, antler velvet, and slaughter by-products (Drew and McDonald 1976).

Other candidates for domestication are animals which are hunted and are now in danger of extinction, either from overhunting, destruction of their environment, or both. If humans continue to desire products from these animals, production systems must be devised in which endangered species become managed resources. The iguana (*Iguana iguana*) is an example. In areas of Central and South America, iguana meat has traditionally been eaten and iguana eggs are considered a delicacy. However, as pressures from hunters and egg collectors increase and as tropical forests are cleared for more intensive agriculture, iguanas are threatened with extinction in some areas. Under managed production systems, iguanas are initially reared in small enclosures, and later tattooed and released into small patches of forest on the farmer's land. In addition to improving the lives of rural people by providing them an additional source of food and income, this new type of farming promotes the reforestation of fragile lands and the conservation of an endangered species (Stoney 1987).

VALUES AND USES OF LIVESTOCK

In developed regions, livestock and livestock products are sometimes criticized as symbols of conspicuous, even health threatening, overconsumption. Livestock may compete with humans for resources, and the highly specialized production systems can be wasteful. Beef production systems may support a cow for an entire year for the production of a single calf, while bull calves born on dairy farms have little value. In contrast, the livestock in developing countries are usually multipurpose (Table 6.1). For example, nomadic pastoralists depend on their herds for milk, which provides a substantial portion of their daily nutrients, and for transport of family and household goods as they move from place to place. The animals in turn depend on their herders to guide them to pasture and water.

Often unrecognized is that livestock have traditionally facilitated sustainable systems of agriculture. Draft animals–both bovine and equine–frequently provide the power needed to plow fields, thresh crops, run irrigation pumps, and carry products to market. This power is provided at a relatively low cost when compared to fossil fuel driven machinery. While fossil fuels are expensive and can impose a large drain on national economies, draft animals often live on crop residues or roadside fallow, materials which might otherwise have little use. It's been estimated that tractors become uneconomical on farms less than four hectares in size. With over 100 million farms in developing countries under four hectors in size, the demand for animal power will not disappear (Vietmeyer 1982). However, the importance of animal power in traditional farming systems has been neglected. The relatively low productivity of African farmers can be attributed in part to dependence on human labor for field cultivation. The addition of draft power can double, even triple, crop productivity per unit of land cultivated (Falvey 1982). Very little is understood

TABLE 6.1
Domesticated Agricultural Mammals and Birds: Products and Uses

Domesticated Animals	Food			Fiber	Skins, Hides, Fur & Leather	Transport & Traction	Other Products or Uses
	Meat	Milk	Eggs				
RUMINANTS:							
Cattle	X	X			X	X	Manure (fuel, fertilizer), pharmaceuticals, horn
Sheep	X	X		X	X		Manure (fuel, fertilizer), pharmaceuticals, horn
Goats	X	X		X	X		Manure (fuel, fertilizer), pharmaceuticals, horn
Buffalo	X	X			X	X	Manure (fuel, fertilizer), pharmaceuticals, horn
Yak	X	X		X	X	X	Manure (fuel, fertilizer), pharmaceuticals, horn
Reindeer	X	X		X	X	X	Manure (fuel, fertilizer), pharmaceuticals, horn
Musk Ox	X	X		X	X		
CAMELOIDS:							
Camels	X	X		X	X	X	Manure (fuel, fertilizer), pharmaceuticals
Llama	X			X	X	X	Manure (fuel, fertilizer), pharmaceuticals
Alpaca	X	X		X	X	X	Manure (fuel, fertilizer), pharmaceuticals
Vicuna				X	X		
SWINE:	X				X		Manure (fuel, fertilizer), pharmaceuticals, bristles
EQUINE:							
Horse	X	X			X	X	Manure (fuel, fertilizer)
Ass	X	X			X	X	Manure (fuel, fertilizer)
Mule	X				X	X	Manure (fuel, fertilizer)
ELEPHANT:					X	X	Ivory

TABLE 6.1 (Cont.)

Domesticated Animals	Food			Fiber	Skins, Hides, Fur & Leather	Transport & Traction	Other Products or Uses
	Meat	Milk	Eggs				
CARNIVORES:							
Dog	X				X	X	Herding, hunting, companionship
Chinchilla					X		
Mink					X		
RODENTS:							
Rabbits	X			X	X		
Guinea pig	X				X		
POULTRY:							
Chicken	X		X				Feathers, manure (fertilizer)
Turkey	X		X				Feathers, manure (fertilizer)
Ducks	X		X				Feathers, manure (fertilizer), weed control
Geese	X		X				Feathers, manure (fertilizer), weed control
Quail	X		X				Feathers, manure (fertilizer)
Guinea fowl	X		X				
FISH:	X						Fertilizer, pharmaceuticals

Sources: McDowell (1977), Fitzhugh et al. (1978).

about the physical requirements of draft animals, and harness designs often remain unimproved. Because of the vital role that the draft animals play in most developing countries, there is a tremendous need for research on the physiology, nutrient requirements, genetic improvement, and management of draft animals.

Animals provide numerous other products and services to humans. Their manure is an important means of recycling essential plant nutrients. Approximately 40% of the world's farmers use animal manure to fertilize their soil. Additionally, manure is depended on as a source of fuel in many developing countries, with more than 200 million metric tons burned annually (McDowell 1979). Grazing ruminants allow economic returns from pasturelands too steep, too dry, too high, or too cold for cultivation. The altiplano, which extends from northern Ecuador to the borders of Peru and Bolivia, ranges in altitude from 3,000 to 5,000 meters. Temperatures may drop to 15 degrees below zero centigrade and strong winds and storms occur frequently. Cameloids, such as the alpaca, help human survival in the harsh, inhospitable environment of the Andean altiplano by providing meat, milk, leather, and wool (Cimó 1988). Unlike sheep and goats that have hard hooves that damage fragile soil, the hooves of cameloids are padded so they would do less damage than the goats and sheep.

Perhaps, the most important values of livestock stem from the strong market is the demand for animal products. Increasingly, limited resource families in the developing countries struggle to cope as the economy shifts from meeting basic needs directly to dependent on cash to purchase goods for basic needs and more. Livestock are often their only means to convert available but low opportunity cost resources (feed, labor) to high value market goods (DeBoer and Fitzhugh 1983). Animal food products of meat, milk, eggs, and cheese frequently are an important source of income. Animal fiber, skins, horns, hooves, and bones are important in the production of clothing, shoes, blankets, carpets, and other handicrafts which support cottage industries and provide income to many rural families.

The full values realized from the domesticated livestock are frequently unrecognized. Moreover, substantial potential exists to increase these contributions. Relatively little research has been devoted to improving the productivity of the indigenous animals and the traditional farming systems in developing countries. Potentials of the camel whose physiological adaptations allow it to survive in the arid regions prone to drought and heat, and the yak, an animal uniquely suited to the high mountain climate of the Himalayas, remain largely untapped. The importance of animals such as these and their integral role in traditional agricultural systems deserve greater recognition.

LIVESTOCK PRODUCTIVITY

Although the vast majority of the world's livestock are found in developing regions, their productivity is generally low. For example, livestock in developed nations produced almost three-fourths of the 46 million metric tons of meat and more than four-fifths of the 458 million metric tons of milk produced globally in 1985, even though they accounted for approximately one-third of the world's 1.3 billion cattle (Fitzhugh 1988). Productivity of livestock in developing regions is comparatively low for two main reasons. First, emphasis is placed on multiple products and uses; thus productivity of any single product tends to be lower. Obviously, a cow capable of pulling a plow is unlikely to produce as much milk as one accustomed to standing all day at the feed trough. With development has come more emphasis on specialization. However, it has only been the last few decades that livestock in today's developed regions were specialized. At the turn of the century, most cattle breeds were triple purpose: milk, draft, and meat. Second, investments in improving the production environment are often not cost effective in many developing countries. The North American (or Australian) rancher has a ready and generally profitable market for cattle, plus ready access to transport by road or rail. Profits provide incentive to improve the production environment. The North African pastoralist, however, has substantially less incentive to invest in an improved environment for his cattle herd. Transport to market is not available, so cattle must be trekked long distances. Thus, the abilities of cattle to survive droughts and disease have more value to African pastoralists than does growth rate or meat quality.

The low productivity of livestock in developing countries tends to reflect the low value of animal products relative to costs of inputs. The principle involved is illustrated by the relationship of outputs to feed requirements:

$$\text{Feed Efficiency} = \text{Outputs/Inputs}$$

$$= \text{Outputs}/(I_M + I_R), \text{ where}$$

I_M = feed resources required for maintenance, and
I_R = remaining feed resources available for reproduction or to support production of meat, milk, draft power.

When I_R equals zero, the animal can only maintain itself; outputs (other than manure) are negligible, and feed efficiency is essentially zero. When the value of livestock products increases substantially above the costs of inputs, producers usually move quickly to improve the quantity and quality of inputs provided their livestock.

The global spread of intensive, confinement broiler operations illustrates this responsiveness of producers to economic incentives. In many developing regions, chicken is the most expensive meat and is generally saved for special occasions (not unlike the chicken for Sunday dinner tradition in the United States prior to the 1950s). Offtake from household flocks is low; perhaps, only four or five chicks survive per hen per year. However, the level of inputs required to maintain household flocks is also low. Predation, disease, and the limited nutrition available from household wastes, seeds, and insects combine to lower productivity. In contrast, intensively managed, grain-fed, confinement flocks exceed 200 chicks marketed per hen per year. Efficiency of production is high because only a small proportion of inputs is required for flock maintenance. The fast growing, modern broiler strains approach biological efficiency ratios of a pound of gain per pound of grain fed. Thus, using technology transferred from Europe and North America, chicken is now the least expensive meat in some developing countries.

This transfer of efficient "poultry package" has lowered the price of poultry meat paid by urban consumers, but has had little impact on the income of small-scale farmers. These farmers lack the resources to invest in the levels of inputs required to make production efficient. They have neither the capital or access to credit needed to build housing, provide feed and medicine, and purchase improved genetic strains. Thus, commercial poultry production tends to be in the hands of a relatively few wealthy investors. Similarly, efficient, profitable confinement swine and dairy cattle technology packages are increasingly dominating commercial traditional pork and milk production in developing countries. This example illustrates that efficiencies are attainable through high levels of inputs, often using "improved" breeds introduced from developed regions. However, increased productivity is not always associated with improved human welfare over the long term, especially when increased productivity leads to loss of genotypes adapted to disease, climate, and other environmental stresses prevalent in developing regions.

PRODUCTIVITY VERSUS ADAPTABILITY

Modern meat and milk production "packages" tend to involve livestock breeds selected in developed, temperate regions with emphasis on genetic capacity for high production volume. Almost by definition these "improved" breeds are poorly adapted to the low inputs and poor production conditions of the tropics. Thus imported exotics are not very productive under developing country environments and may not even survive unless they are protected from the stresses of disease, climate, and poor nutrition. Government subsidies and short-term profitability may pay costs of this protection for a time. Problems arise if profits decline and subsidies cease. Resulting financial losses to investors are not as great a concern as the loss of genetic diversity when exotics have replaced traditional adapted breeds. The extent and value of this lost genetic diversity are largely unknown because little effort has been made to characterize traditional breeds in developing countries. Fortunately, a few dramatic examples have illustrated the potential values resident in these populations. These examples include trypanotolerant cattle in western Africa (ILCA 1979) and prolific hair sheep capable of raising large litters of lambs within a tropical environment (Fitzhugh and Bradford 1983).

CONSERVATION OF GENETIC DIVERSITY

Concern for the erosion of genetic diversity has stimulated several international conservation studies. These include the FAO Standing Panel for Conservation and Management of Animal Genetic Resources (Hodges 1984) and the U.S. National Research Council Committee on Global Genetic Resources, which addresses both plants and animals (Pino 1986). These and other studies have given highest priority to the characterization and conservation of genetic resources in developing regions.

Criteria for assigning conservation priorities (Simon 1984) include:

- The population constitutes an original, independent, closed breeding group not closely related to other nonendangered or already conserved groups.
- The population carries a specific genetic value (e.g., superiority for a production trait, a major gene, a character of potential importance, or expectation of substantial heterosis in crosses with other populations).

Unfortunately, use of these criteria presume that populations at risk of losing genetic diversity are reasonably well characterized. This is not often the case for livestock populations in developing regions. For example, the babirusa (*Babyrousa babyrussa*) is a pig-like animal whose stomach more closely resembles that of a ruminant rather than the monogastric stomach of pigs (Davis 1942). Development of the babirusa could prove useful since its unique anatomy allows it to live on fibrous materials that pigs cannot normally digest. However, the babirusa remains little known and its numbers are dwindling.

For species which are not well characterized, the best remaining option is to conserve a representative random sample of livestock populations. For some species, though, it may even be too late. The kouprey (*Bos sauveli*), an animal native to Kampuchea and Laos, has several features which could prove valuable to tropical domestic cattle. It has a large, pendulous dewlap, giving it greater evaporative surface for body cooling, and it is believed to have resistance to rinderpest. However the population is known to be dangerously close to extinction, and the last sighting was of five animals in 1982 (National Academy of Sciences 1983).

Efforts to conserve genetic variation generally involve one of two approaches: 1) living banks that are important for maintenance of small samples of endangered breeds or types in zoos, parks, or private herds, or for establishment of randomly intermating gene pools of several breeds or lines; and 2) storage of gametes and/or embryos from endangered breeds or types. Results from genetic and economic analyses (Smith 1984a, 1984b) favor frozen storage of gametes and embryos as compared to maintenance of living banks. However, storage is effective only for those species for which cryopreservation techniques for sperm (cattle, sheep, goats, pigs, rabbits, horses) or embryos (cattle, sheep, goats, rabbits, horses) are well established (Renard 1984). Among major domestic species, poultry are a notable example of species for which practical techniques for frozen storage of eggs remain to be developed. Nevertheless, techniques for freezing gametes or embryos for all domestic species can probably be developed if sufficient resources are devoted to the task. Obtaining and conserving representative samples of uncharacterized populations native to developing regions should be technically feasible, even if practically difficult. At issue, however, is who will pay the costs. Few developing countries have the resources or inclination to make conservation a priority.

CONCLUSIONS

Livestock are likely to become increasingly important to the welfare of people in developing regions. Regardless of their socioeconomic background, most people aspire to a better standard of living for themselves and their children. Measures of quality of life include opportunities for education and health care and freedom from the drudgery of hard labor. Livestock can provide the means for meeting these aspirations. Problems arise, however, as growing populations of humans and their livestock compete for scarce resources. There is need to increase the supply of animal products demanded by the growing urban populations. However, this increase should be achieved by expanding production efficiency, so that relatively fewer inputs are used for maintenance than for production, not by increasing livestock numbers.

Populations of livestock in developing countries have evolved under conditions of environmental stress. On the presumption that human populations in developing regions must aim for food self-sufficiency, livestock adapted to tropical climates and low value feedstuffs will be essential. Furthermore, adapted breeds which have evolved under stressful conditions may serve as pools of genetic diversity to be tapped through genetic engineering. Thus, the value and uses of indigenous livestock breeds in developing nations are most likely to be even greater in the future. The challenge is to enact practical conservation programs so that this potentially useful genetic diversity will be available when needed.

REFERENCES

Cimó, P. 1988. Alpaca, a symbol and source of hope in the Andes. *Ceres* 21(1):32-37.

Davis, D.D. 1942. Notes on the anatomy of the babirusa. *Field Museum of Natural History* (Zoological Series) 22:363-411.

De Boer, A.J. and H.A. Fitzhugh 1983. *Sheep and Goats in Developing Countries*. World Bank Technical Paper. The World Bank, Washington, D.C.

Drew, K.R. and M.F. McDonald, eds. 1976. *Deer farming in New Zealand: Progress and Prospects*. Editorial Services Limited, Wellington.

Falvey, J.L. 1982. *An Introduction to Working Animals*. Australian Print Group, Melbourne, Australia.

Fitzhugh, H.A. 1988. Relevance of animal science research: International perspectives. *Journal of Animal Science* 66(8):2110-2116.

Fitzhugh, H.A. and G.E. Bradford, eds. 1983. *Hair Sheep of Western Africa and the Americas*. Westview Press, Boulder, Colorado.

Fitzhugh, H.A., H.J. Hodgson, O.J. Scoville, T.D. Nguyen, and T.C. Byerly. 1978. *The Role of Ruminants in Support of Man*. Winrock International Livestock Research and Training Center, Morrilton, Arkansas.

Hodges, J. 1984. Review of the FAO/UNEP programme on animal genetic resources conservation and management. *Animal Genetic Resource Conservation by Management: Data Banks and Training*. FAO Animal Production and Health Paper 44/1. Food and Agriculture Organization of the United Nations, Rome.

International Livestock Centre for Africa (ILCA). 1979. *Trypanotolerant Livestock in West and Central Africa*. ILCA

Monograph 2. International Livestock Centre for Africa, Addis Ababa, Ethiopia.

McDowell, R.E. 1977. *Ruminant Products: More than Meat and Milk.* Winrock International Livestock Research and Training Center, Morrilton, Arkansas.

McDowell, R.E. 1979. Role of animals in support of man. *World Food Issues.* Center for the Analysis of World Food Issues, Cornell University, Ithaca, New York.

National Academy of Sciences. 1983. *Little Known Asian Animals with a Promising Economic Future.* Board on Science and Technology for International Development, National Research Council. National Academy Press, Washington, D.C.

Peel, L. and D.E. Tribe, eds. 1983. *Domestication, Conservation and Use of Animal Resources.* Elsevier, New York.

Pino, J.A. 1986. Management of plant and animal genetic resources. *BOSTID Developments* 6(3):8-10. Available from the Office of International Affairs, National Research Council, Washington, D.C.

Renard, J.P. 1984. Methods of conserving gametes and embryos of farm mammals. *Livestock Production Science* 11(1):49-59.

Simon, D.L. 1984. Conservation of animal genetic resources–a review. *Livestock Production Science* 11(1):23-26.

Smith, C. 1984a. Economic benefits of conserving animal genetic resources. *Animal Genetic Resources Information* 3:10-14.

Smith, C. 1984b. Genetic aspects of conservation in farm livestock. *Livestock Production Science* 11(1):37-48.

Stoney, C. 1987. The day of the iguana. *VITA News* (October). Available from Volunteers in Technical Assistance, Arlington, Virginia.

Vietmeyer, N. 1982. The untapped science of animal power. *Ceres* 15(4):42-45.

PART THREE

Traditional Conservation and Use of Biodiversity

Introduction

Margery L. Oldfield and Janis B. Alcorn

> Understanding a conservation system means understanding not only the nature of what is being conserved, but also the viewpoint of the conserver. (Johannes 1978:349)

> Over the short term, the ideas of civilization A might appear vastly superior to those of civilization B. But over the long term . . . the apparently "primitive" practices of civilization B . . . incorporated deep wisdom that was unintelligible to civilization A. (Watt 1972:82)

> The presence of people may enhance the species richness of an area, rather than exert the effect that is more familiar to us. (Ehrenfeld 1989:9)

Biodiversity conservation in the United States, Europe, and other modern, industrialized nations arose primarily as an issue among educated, urban, upper middle class people concerned about the loss of aesthetically pleasing wildlands. However, older, more pragmatic roots of biodiversity conservation are found in the traditional resource management systems of developing countries. In recent years, the ecological and evolutionary contributions, ethnobiological knowledge, and resource management strategies of traditional agricultural peoples have received considerable attention from Western scientists. However, the conservation strategies of traditional societies have received comparatively little attention from the Western conservationists who are currently attempting to craft biodiversity conservation plans for the future (Lohman 1991). This bias in Western thinking persists despite the fact that most traditional cultures have had a profound ecological and evolutionary influence on wild biota. For example, long term selection pressures wrought by ancient cultures may have been primarily responsible for evolution of the cooperative (group) antipredator defense strategy of the African elephant (Douglas-Hamilton 1972) and the guiding behavior of the Greater Honeyguide (Dean et al. 1990).

Tribal and aboriginal cultures have survived in natural ecosystems for thousands of years, in large part because they are culturally well adapted to their local environment. Oral histories and traditions guide them in their use and management of biological diversity. Nabhan (1987:30) has distinguished such indigenous "subsistence farmer-hunter-gatherer communities"–cultures that have gradually accumulated knowledge about their local ecosystems within the same environment for many centuries–from present or past cultures of "rapidly developing city-state civilizations" that have been geared to accumulation of power and wealth via intensive exploitation of highly valued resource species. Today, however, the concept of employing traditional ecological and ethnobiological knowledge in conservation and development strategies is controversial. For example, Clay (1985:5) maintains that "the key to understanding sustained activities in fragile environments begins with the local residents. Their knowledge is valuable to the future of the earth's environment and peoples." He contends that First World conservationists will have few proponents in much of the Third World if they cannot develop ecologically sound land management techniques to meet the basic needs of local people. In contrast, Redford (1990:48) states that while "indigenous knowledge is tremendously important" and "offers fascinating insights of ecological value," it "only occasionally" offers methods that can be modified for use by people in the modern Neotropics.

There is, indeed, a broad gap between these differing perspectives, and the reasons and rationales behind these opposing views are probably multitudinous. The differences in opinion may emanate, in part, from evaluating indigenous cultures or traditional management systems that have been recently influenced by modern cultures (Bower 1989). One side asks: To what extent have past indigenous cultures failed to conserve wild species or adequately manage natural ecosystems in the past, opportunistically utilized any technological innovation made available to them at any time, and employed cultural traditions or belief systems that were effective only because their populations were always limited by external (natural) factors? Alternatively, one might ask: To what extent has the forced relocation of indigenous cultures onto reservations or into foreign environments, the imposition of sedentary lifestyles on previously nomadic peoples, the worldwide spread of market-driven economies, and the intentional introduction of modern weaponry and other technological innovations contributed to the destruction of traditional management systems that would otherwise have conserved biodiversity? The contributors for Part Three were asked to provide their own perspectives on the role of indigenous cultures in conserving or managing populations of wild plants or animals.

In Chapter 7, Gary Nabhan and his associates present ethnobotanical information from the U.S.-Mexico borderlands–a region of impressive cultural and biological diversity where 44 million acres are managed by indigenous people. They provide four different case studies of rare or threatened plant species that highlight the significance of traditional harvesting and management practices employed by local indigenous cultures. They conclude that indigenous people utilize a variety of *in situ* and *ex situ* conservation techniques that effectively conserve plants populations.

In Chapter 8, Victor Toledo summarizes the lessons learned from a ten year study of the biological resources of the Lake Pátzcuaro Basin in México. He argues that this region has not suffered the environmental degradation characteristic of other lake areas in Michoacán, despite thousands of years of occupation by dense human populations. He attributes this primarily to the indigenous knowledge, use, and traditional resource management practices of the P'urhépecha culture. Toledo contends that regulation of socioeconomic differentiation along with traditional land use patterns, indigenous household production strategies, and land tenure arrangements have maintained the ecological health of the Lake Pátzcuaro region.

In Chapter 9, Raymond Hames presents information from his study of wildlife hunting and conservation methods of present-day Amazonian tribal cultures. In addition, he reviews the literature on wildlife conservation in traditional societies, and offers hypotheses that should be tested to assess and guide management of indigenous hunting activities in protected areas. He also provides empirical data on the hunting efficiency of Yanomamö and Ye'kwana Indians, and concludes that these aboriginal hunters operate to maximize hunting efficiency rather than to conserve wildlife or game populations. Hames, like Redford (1990), criticizes claims that indigenous tribal societies live in harmony with their environment, observing that modern day native Amazonian cultures do not possess the social institutions necessary to enforce wildlife conservation.

Taken as a group, these three papers highlight an important conservation dilemma–a dilemma that our ancestors faced in the past and one that we still face today. On one hand, it is comparatively easy to conserve viable populations of most species of plants, fungi, microorganisms, insects and other invertebrates, and short-lived vertebrates–whether they are associated with traditional or modern resource management systems (e.g., see Cohn 1989, Dufour 1990). Nabhan and his associates demonstrate that, despite circumstances of intensive use and exploitation, even rare plants can be effectively conserved by traditional resource management techniques. Toledo's contribution provides further evidence that many different types of resource species, including native fishes (see also Chernela 1985, Dufour 1990), may be effectively conserved via traditional management techniques despite intense, long term exploitation.

Hame's contribution, on the other hand, highlights the fact that it is comparatively difficult to conserve viable populations of long-lived vertebrates. Conservation of such vertebrates remains a problem for both traditional and modern resource managers. As Hames observes, post-Pleistocene trends indicate that many different prehistoric cultures overexploited populations of large-bodied vertebrates and ocean island endemics, or caused sufficient habitat disturbance to lead to their demise. Nevertheless, the "track record" for modern, technologically advanced cultures or for Western cultures during recorded history has been no better, and appears to be even worse (Ziswiler 1967, Ehrenfeld 1972). In general, extinction rates for "higher vertebrates"–the mammals and birds–appear to have increased at least tenfold during the last three centuries as a consequence of overexploitation and habitat loss or disturbance due to agriculture, urbanization, and industrialization (Oldfield 1984). Notable examples include the American extinction (in 1914) of the Passenger Pigeon (*Ectopistes migratorius*)–once the most abundant bird species on the North American continent, and the European extinctions (in 1627 and 1851, respectively) of the aurochs *(Bos primigenius)*–the presumed wild progenitor of all *Bos* cattle breeds, and the tarpan or European wild horse (*Equus caballus ferus*)–the presumed wild progenitor of all domesticated horse breeds.

The post-Pleistocene demise of the vertebrate "megafauna" has most likely been due to a combination of both human-induced and biological factors. These interaction effects include: 1) the fact that a larger fraction of long-lived vertebrates fall into the most desirable or suitable prey size-category for human hunters in comparison to the proportions of shorter-lived vertebrates, invertebrates, microbes, or plants and fungi that fall into this size category, 2) that most large-bodied or long-lived vertebrates are killed during their reproductive years to yield unique, valuable, or extremely useful biological products that humans cannot obtain from other taxa, and 3) that long-lived vertebrates tend to be more susceptible to human-induced extinction than shorter-lived species because they exhibit a suite of life history traits, such as low reproductive rates, delayed sexual maturity, offspring and cooperative defense behaviors, and strong social ties, that have been often cited as "extinction-prone" biological attributes (e.g., see Ehrenfeld 1972, Myers 1979, Oldfield 1984, 1988). Although many of these evolved attributes are thought to be key life history traits that have contributed to the former ecological and evolutionary success of these vertebrates, and they may still serve an important adaptive function in wild populations, they also unfortunately may enhance the vulnerability of these species to human exploitation or disturbance (Oldfield 1988).

Conservation of extinction-prone species, such as large, long-lived vertebrates, is a global conservation challenge that remains a problem for both modern and traditional cultures. The dominant cultures of nearly all industrialized nations, such as the United States and many European countries, have often failed to effectively conserve populations of such extinction-prone species or to set aside large areas of natural habitat for their own resident populations of large-bodied or long-lived vertebrates. Current U.S. controversies over conservation of climax forest species–such as the Spotted Owl (a relatively long-lived carnivore) and the Red-cockaded Woodpecker (a highly social species that has unique antipredator adaptations)–are cases in point. At the same time that we are haggling about conserving comparatively small acreages of natural habitat, most of which are no longer virgin, we are requesting developing nations in the Third World to set aside vast acreages of virgin tropical rain forest. In fact, many developing countries have already established parks or reserves that dwarf the conservation efforts of developed nations like the United States.

Given the urgency of achieving effective conservation action, we should endeavor to work together to conserve viable populations of as many species as it is possible for us to conserve for present as well as future generations. This includes, of course, conservation of large, long-lived mammals and birds–often viewed as "charismatic" or "flagship" species by Western conservationists, and as "sacred" or "dangerous" species by traditional cultures (Eliade 1964, Reichel Dolmatoff 1975, Alcorn 1984). Yet, as rural, subsistence economies continue to shift to more intense participation in market-driven, cash economies (see Alcorn, Epilogue, this volume), then large-bodied taxa which still comprise the largest fraction of prey species of suitable size for human exploitation (Ehrenfeld 1972) are primarily viewed as cash in the form of individual animals or their body parts. These commodities are in great demand by consumers, e.g., for trophies, furs and skins, and animals for the live trade, by Asian consumers–for folk medicines, and by African and Latin American consumers–for meat and other products. In many instances, the future role of rural people in conserving large mammals will be largely determined by international market demand. For example, in the Zimbabwe Campfire Program, local communities are now protecting elephants because they are economic assets that generate tourist revenues (The Zimbabwe Trust 1990, B. Wyckoff-Baird, pers. comm.). Similarly, poaching dropped off dramatically in Zambia when safari concessions began to generate income for local communities (Lewis et al. 1990, see also Dasmann, Ch. 1, this volume). It is unfortunate that the devastation of African elephant populations in recent decades has required a total ban on the international ivory trade (Lewin 1989), which works against such local community conservation efforts.

Conservation of most of the world's biodiversity, will depend on empowerment of local people to make decisions based on cultural expectations concerning proper, long term land use decisions (see papers in this section), local economic opportunities (see contributions in Part Four), and modern ethical commitments to biodiversity conservation (see Alcorn's Epilogue). Callicott (1990:19) suggests that the only viable conservation philosophy is a vision of "a mutually beneficial and enhancing integration of the human economy with the economy of nature." His vision was inspired by Aldo Leopold's 1939 essay on "The Farmer as Conservationist" which describes an American farmstead "in which the wild and domesticated floral and faunal denizens of a Wisconsin farmscape are feathered into one another to create a harmonious whole" (Callicott 1990:19). The farm provided small game, fish, fruit and nuts from its fallow fields, woodlot, and wetlands, as well as songbirds, wildflowers, and "intellectual adventures" in natural history. This idyllic farmstead was achieved through skillful management of the habitat and biota. However, today, overall levels of biodiversity in modernized landscapes are much lower than those on farmsteads in developing nations that are still manipulated to maintain high levels of biodiversity. Thus, Leopold's essay more closely describes today's traditional farmsteads in Indonesia (e.g., Peluso and Padoch 1991), the Philippines (Conklin 1954), Amazonia (e.g., Posey and Balee 1989, Dufour 1990), Mexico (e.g., Alcorn 1984; Nabhan et al., Ch. 7 and Toledo, Ch. 8, this section)–and many other tropical nations.

It was this type of habitat management that elicited Ehrenfeld's comment (see above). He refers to Nabhan's (1982) comparison of two desert oases–one managed as a farmstead by Papago Indians, and the other managed as a bird sanctuary by the U.S. National Park Service. In this case, the Papago-managed landscape maintains more birds and a more heterogeneous habitat for other wild biota than the Park Service reserve (which the Papago were barred from farming when the reserve was established). Ehrenfeld's observation led Callicott (1990:20) to ask, "Can we succeed, as the Papago seem to have done, in enriching the environment as we enrich ourselves?" Callicott's question takes us beyond the contention that traditional peoples cannot conserve biodiversity if they are not successful in conserving many large or long-lived vertebrate species. Although conserving "charismatic" or "flagship" species is an important strategy that can be employed to conserve other biodiversity within the large areas of habitat required for maintenance of viable populations of long-lived or large-bodied vertebrates, most other biodiversity can be conserved by alternative means.

In sum, the ability of traditional peoples to conserve biodiversity should not be measured solely by how many tigers, rhinos, primates, or elephants they maintain in the habitats that they manage. Recognition of the biodiversity conservation successes of indigenous peoples should not be denied merely because of their failure to maintain large, long-lived mammals or birds that most modern, industrialized cultures also fail to maintain. Instead, their performance should be evaluated in terms of all of the various habitat types and taxa—plants, insects and other invertebrates, microorganisms, fungi, and other less obvious or "charismatic" vertebrates—that they maintain within their zone of influence.

REFERENCES

Alcorn, J.B. 1984. *Huastec Mayan Ethnobotany.* University of Texas Press, Austin.

________. 1989. Process as resource. Pages 63-77 *in* D.A. Posey and W. Balee (eds.), *Resource Management in Amazonia: Indigenous and Folk Strategies.* New York Botanical Garden, Bronx, New York.

Bower, B. 1989. A world that never existed. *Science News* 135:264-266.

Callicott, J.B. 1990. Whither conservation ethics? *Conservation Biology* 4:15-20.

Chernela, J. 1985. Indigenous fishing in the neotropics: The Tukanoan Uanano of the blackwater Uaupés River basin in Brazil and Colombia. *Interciencia* 10:78-86.

Clay, J.W. 1985. Parks and people. *Cultural Survival Quarterly* 9:2-5.

Cohn, J.P. 1989. Iguana conservation and economic development. *BioScience* 39(6):359-363.

Conklin, H.C. 1954. The relation of Hanunoo culture to the plant world. Ph.D. Dissertation, Yale University, New Haven, Connecticut.

Dean, W.R.J., W.R. Siegfried and I.A.W. MacDonald. 1990. The fallacy, fact, and fate of guiding behavior in the Greater Honeyguide. *Conservation Biology* 4(1):99-101.

Douglas-Hamilton, I. 1972. On the ecology and behaviour of the African elephant: The elephants of Lake Manyara. Ph.D. Thesis, Oriel College, University of Oxford, Oxford, United Kingdom.

Dufour, D.L. 1990. Use of tropical rainforests by native Amazonians. BioScience 40(9):652-659.

Ehrenfeld, D.W. 1972. *Conserving Life on Earth*. Oxford University Press, New York.

________. 1989. Life in the next millenium: Who will be left in the earth's community? *Orion Nature Quarterly* 8(2):4-13.

Eliade, M. 1964. *Shamanism*. Princeton University Press, New Jersey. (Translated from French by W.R. Trask)

Lewin, R. 1989. Globan ban sought on ivory trade. *Science* 244:1135.

Lewis, D., G.B. Kaweche and A. Mwenya. 1990. Wildlife conservation lessons outside protected areas: Lessons from an experiment in Zambia. *Conservation Biology* 4:171-180.

Lohman, L. 1991 (in press). Who defends biodiversity? Conservation strategies and the case of Thailand. *The Ecologist*.

Johannes, R.E. 1978. Traditional marine conservation methods in Oceania and their demise. *Annual Review of Ecology and Systematics* 9:349-364.

Myers, N. 1979. *The Sinking Ark*. Permagon Press, New York.

Nabhan, G.P. 1982. *The Desert Smells Like Rain*. North Point Press, San Francisco, California.

________. 1987. Ethnobiology and conservation: Valuing diversity. *Environment Southwest* 519:28-31.

Oldfield, M.L. 1984. *The Value of Conserving Genetic Resources*. U.S. Department of the Interior, Washington, D.C.

________. 1988. Threatened mammals affected by human exploitation of the female-offspring bond. *Conservation Biology* 2(3):260-274.

Peluso, N. and C. Padoch. 1991. Changing resource rights in managed forests of West Kalimantan. Paper presented at the Conference on Interactions Between People and Forests in Kalimantan, New York Botanical Garden, Bronx, New York.

Posey, D.A. and W. Balee, Eds. 1989. *Resource Management in Amazonia: Indigenous and Folk Strategies*. New York Botanical Garden, Bronx, New York.

Redford, K.H. 1990. The ecologically noble savage. *Cultural Survival Quarterly* 15:46-48.

Reichel-Dolmatoff, G. 1975. *The Shaman and the Jaguar*. Temple University Press, Philadelphia.

Watt, K.E.F. 1972. Man's efficient rush toward deadly dullness. *Natural History* 81(2):74-77, 80, 82.

Zimbabwe Trust, The. 1990. *People, Wildlife and Natural Resources-The Campfire Approach to Rural Development in Zimbabwe*. The Zimbabwe Trust, The Department of National Parks and Wild Life Management, and the Campfire Association. Belgravia, Harare, Zimbabwe.

Ziswiler, V. 1967. *Extinct and Vanishing Animals*. Springer-Verlag, New York.

7

Conservation and Use of Rare Plants by Traditional Cultures of the U.S./México Borderlands

Gary P. Nabhan, Donna House,
Humberto Suzan A., Wendy Hodgson,
Luís Hernández S. and Guadalupe Malda

INTRODUCTION

> In a country that is characterized by the cultural diversity of its rural inhabitants, it is difficult to design a conservation policy without taking into account the cultural dimension; the profound relationship that has existed since time immemorial between nature and culture.... Each species of plant, group of animals, type of soil and landscape nearly always has a corresponding linguistic expression, a category of knowledge, a practical use, a religious meaning, role in ritual, an individual or collective vitality. To safeguard the natural heritage of the country without safeguarding the cultures which have given it feeling is to reduce Nature to something beyond recognition; static, distant, nearly dead. (Toledo 1988:p. 1)

The country to which Toledo is referring is México. But a similar statement can be made for most places in the world where indigenous cultures have persisted in their attempts to produce food in their homelands while utilizing many of the same natural resources that their ancestors used hundreds or thousands of years ago. One such region where cultural and natural diversity are tightly linked is known as the *Frontera Norte* or Northern Frontier to Victor Toledo and other México City scholars. To U.S. historians, it is known as the Spanish borderlands. But to the present group of authors, this area (Figure 7.1) is known as "home" and as the study site of the FLORUTIL Conservation Project. This project endeavors to identify, study and conserve the threatened useful flora of the six Mexican and four U.S. states adjacent to the international boundary.

Over three-fourths of the U.S./México borderlands occur within arid and semi-arid climates. However, the prevailing aridity of the

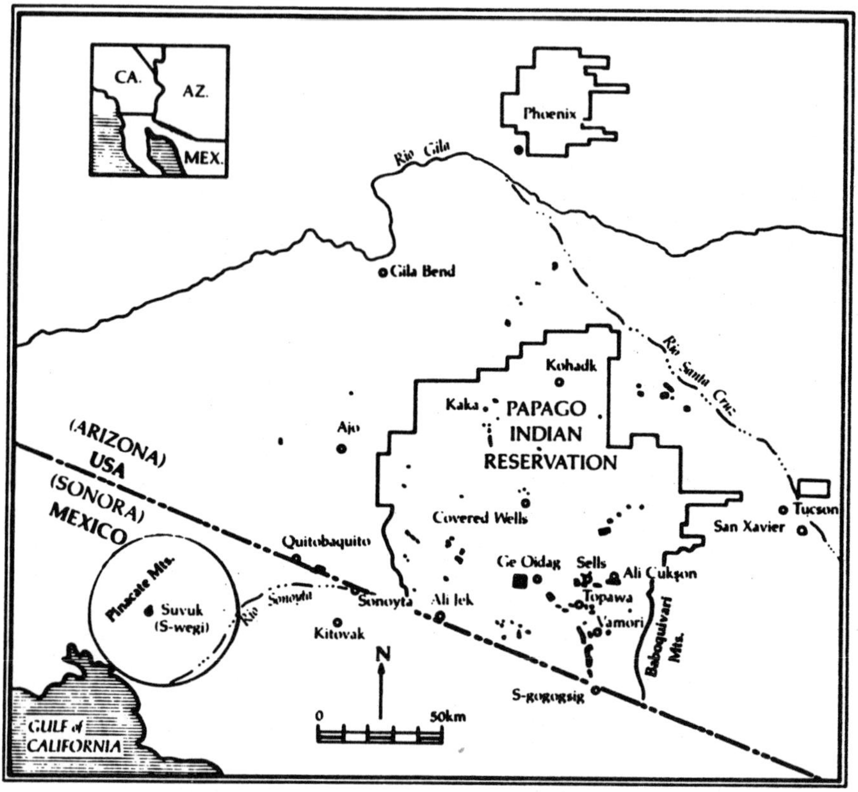

FIGURE 7.1 Map of the U.S./México borderlands region. Several plants featured in the case studies below are endemics of the binational Sonoran Desert region, while other case studies focus on the Chihuahuan Desert and adjacent Tamaulipan thornscrub.

region does not mean that it is depauperate biologically; deserts are not necessarily <u>barren</u>, nor are deserts all that this region offers. Seventeen vegetation provinces are represented within this interanational region (Bailey 1978, Bye 1983). The sizes of the floras traversing this international border are: 2500 species in the Sonoran Desert (Shreve and Wiggins 1964), 3500 species in the Chihuahuan Desert (Henrickson and Johnston, In press), and 6000-6500 species in the Tamaulipan region (Gonzalez-Medrano 1973, Toledo 1988).

Not surprisingly, the endemic plants of the U.S./México borderlands contribute a disproportionately large percentage of the threatened and endangered species lists of their respective nations. For example, more than 180 of the taxa listed as candidates or as protected species in the United States' *Federal Register* publication of threatened and endangered plants have distributional ranges which extend into México (P. Olwell, pers. comm.). Nearly 100 Sonoran Desert endemics may be considered threatened, endangered or vulnerable in México or the United States, and 37% of these occur in both countries (Nabhan et al. 1989). The overwhelming majority are narrowly distributed cacti and succulents. Hodgson et al. (1987) have estimated that 89 species in the Cactaceae and 54 species in the Agavaceae may be at risk of extinction in the U.S./México border states. Some succulent species are threatened by desertification resulting from overgrazing, groundwater depletion, fuelwood cutting, fire suppression, and land clearing and attendant local climate change in the borderlands (Humphrey 1987, Balling 1988). In addition, many showy succulents of the region are being overcollected for illicit international trade to horticultural fanciers (Anderson 1987).

This region of rich but imperiled biotic diversity is also globally famous for its persistent cultural diversity. In addition to the numerically dominant Hispanic- and Anglo-American cultures, more than 350,000 Native Americans, representing 73 linguistically distinct indigenous groups, reside today in the binational borderlands. About 75 Indian reservations occur on the United States side alone, and there are no less than 25 *reservas indígenas* or other cultural enclaves for indigenous peoples in northwestern México. The significance of these indigenous peoples in potentially conserving rare plants and their habitats can be underscored by a simple comparison. The Nature Conservancy has, up through the last decade, managed approximately 2.5 million acres of preserves in the United States (Center for Plant Conservation 1986), but this is less than the size of the second largest Indian reservation of the border states–the Tohono O'odham Reservation. Collectively, 32 million acres in the United States are held as trust lands for Native American residents (U.S. Department of Interior 1988), and more than 12 million acres in México are managed primarily by indigenous tribes native to the border states. Altogether, there may be 44 million acres under Indian domain in the U.S./México border states, compared to the 5.1 million acres under Nature Conservancy management in all of the United States, Canada, Latin America and the Caribbean.

The rate of habitat conversion on these Indian lands has been far slower than that on other private lands in either the United States or México. Given this fact, can these indigenous reserves be considered as protected zones for the plant life held within their boundaries? On one hand, we do not claim that endangered plants are offered as much capital-intensive protection and management on indigenous reserves as they receive on the Nature Conservancy preserves. On the other, indigenous peoples have detailed knowledge about particular rare plants, and traditional management practices may contribute to their local conservation. Some case studies will be presented to confirm that indigenous conservation traditions do exist, and are not merely romantic notions. At the same time, habitat degradation and plant overcollection have occurred on indigenous reserves. Nevertheless, there is much to be gained by considering native peoples as full participants in plant conservation efforts, particularly those that build on their ethnobotanical knowledge and resource management traditions, yet also offer them social, economic or environmental benefits.

MAJOR ISSUES AND CONCERNS

Regardless of the potential for building on indigenous peoples' plant traditions to further the conservation of rare species, certain ethnocentric attitudes remain among Western-trained conservation biologists which may keep this potential from being fully realized. Because many biologists are intent on analyzing so-called natural systems, they often ignore that they are really observing relationships between organisms and environments that have been influenced by humankind over thousands of years. As Rea (1987:p.3) has written, "the myth is that there is a world of nature without 'man'–there is some ideal out there without human influence." Even when they do not ignore human influences, such "natural systems" biologists typically treat human presence as a purely negative phenomenon, a nuisance or intrusion. This tendency is particularly foolhardy in México and southwestern United States, where rare plants and animals have persisted amid some of the oldest continuously inhabited settlements in North America (Nabhan and Reichhardt 1983, Nabhan 1989).

As a case in point, Diamond (1986), a prominent conservation biologist, has claimed that all preindustrial peoples–regardless of whether they have been farmers, hunter-gatherers, or city-state merchants–are equally as prone to cause deforestation and species extirpation as modern industrialized societies. While some fallacies of Diamond's arguments have been challenged elsewhere (Nabhan 1987, 1989), it remains that many conservation biologists generally fall in one of two groups–one that emphasizes the negative effects of direct collection of plants and animals, or one that is preoccupied with habitat degradation. Seldom does either "camp" differentiate the effects of traditional land and plant management practices of indigenous communities with long residency in one place, versus those of habitat conversion or specimen collecting done, often on a much larger scale, by commercial enterprises with foreign or absentee owners.

More often than not, conservation biologists are unaware that indigenous people have detailed knowledge of a particular rare plant or animal that is under study. Even when not biased against traditional peoples, these biologists often lack the skill of ethnographic interviewing and the incentives to learn from indigenous people who live in close proximity to rare species. To date, only one U.S. endangered species recovery plan, which recommended local Navajo participation in the habitat protection and plant population recovery efforts for *Carex specuicola*, has included ethnobotanical data derived from indigenous people. House (1987), a university-trained Navajo biologist, wrote this recovery plan while she was working for the Navajo Nation's Natural Heritage Program. In February 1988, she met with ethnobotanists from the United States and México who had initiated informal collaborations in 1987. They agreed to collaborate on recording traditional knowledge of indigenous cultures about rare plants in the U.S./México borderlands with the aim that this information would be more often recognized by conservationists and, therefore, integrated into conservation plans. The resulting FLORUTIL Conservation Project has the following goals: 1) to develop a shared bilingual data base on threatened useful plants of the borderlands; 2) promote *in situ* conservation of rare plants, involving local communities in these efforts; 3) conserve traditional knowledge about rare plants; and 4) investigate the effects of current and historic plant uses and land management practices on species rarity. The acronym FLORUTIL stands for "useful flora."

In this region, there are few well documented cases of plant uses by indigenous cultures that have driven a rare taxon into local extirpation or species extinction (see below). Nevertheless, one could hypothesize a variety of negative or positive effects on a plant's vegetative and sexual reproduction that might be generated by various harvesting pressures (Anderson 1987). Such hypotheses need to be carefully tested and removed from the realm of conjecture. To adequately test these hypotheses, FLORUTIL participants collected detailed field data so that the impacts of traditional gathering activities could be more objectively evaluated. The following regional surveys and case studies were initiated toward that end.

FLORUTIL METHODOLOGY

To accomplish these objectives, FLORUTIL collaborators initiated a multiple-step process to: 1) determine which plant species were considered rare, threatened or vulnerable in the region, 2) survey existing literature on the effects of various human uses and land management activities on these species, 3) collect new field data on plant rarity, uses and threats, and 4) select particular rare, useful species for follow-up field studies to confirm or cast doubt on the suggestions made in previously published literature.

Because México's SEDUE agency has not yet released an official national inventory of threatened and endangered species that parallels that of the inventory for United States published in the *Federal Register*, it was necessary for us to develop, with the help of other conservation biologists (Vovides 1981), a list of plant species and subspecies at risk in the U.S. and Mexican border states that are adjacent to the international boundary. The "plants at risk" inventory is actually much larger than that for rare plants that are legally protected or considered as current candidates for protection by the U.S. Endangered Species Act, CITES, and state laws. We wanted the list to include many vulnerable species that have been overlooked for protection in the past. For instance, we included plants with narrowly restricted ranges in the borderlands, even though they may have dense populations at known localities. Such locally common endemics are particularly vulnerable to land use changes, such as chemical control of vegetation, groundwater depletion, or dam construction—changes that might potentially affect the status of all of the remaining populations concentrated in one small area. Semicultivated plants or "wild" plants that are "encouraged" by humans, i.e., those that are strongly associated with ethnic farmers, were also considered, because they are endangered by the demise of traditional agriculture and are becoming increasingly rare today.

For each taxon on the roster, "plants at risk," a profile was developed that is maintained on-line in FLORUTIL's computerized data bases, as well as on file as "hard copy" data sheets documenting the published references, field notes and interviews from which the information was drawn. FLORUTIL database programs, written for us by Kevin Dahl in DBASE III, are accessible to project collaborators in English and Spanish. Information encompassing 38 fields is entered in the "PLANTS" database regarding taxonomy, life history, reproductive biology, biogeography, ecology, threats and forms of protection of each particular species. For each useful part of a particular plant, information is entered into the "USES" database, comprised of 18 fields for the harvesting method, specific uses, seasonality, economic value, and harvesting impacts. Further details on the FLORUTIL database programs and our methodology are described in Nabhan et al. (1987). Statistical summaries of results on threats and harvesting impacts derived from data entered through

March 1990 are presented in Tables 7.1 and 7.2. These data suggest that collecting impacts are the most frequently mentioned threats to the species surveyed to date. Commercial versus traditional exploitation are rarely distinguished in the literature. Overcollection accounts for about 27% of the threats tallied from literature reviews and fieldwork. Table 7.2 indicates that only three species surveyed thus far show indications of having been overcollected to the point of local extirpation; four have become regionally rare, and seven—locally rare. Rarity has encouraged local traditional peoples to protect five species and to cultivate ten species.

To further refine and test the hypothesis that traditional harvesting and land management practices generate qualitatively different plant population responses than other human activities, we have undertaken intensive data collection in paired field plots where different harvesting and management techniques affect the same rare species. We used a variety of field ecology methods to measure the degree of association between a rare useful plant and other components of the vegetation, assessed plant population structures for indications of decline or regeneration, and documented evidence of particular human impacts on plant populations. Finally, we interviewed local residents about the history of use of the selected plant, and history of land management in the vicinity. Summaries of each of the following case studies highlight specific methodological techniques.

CASE STUDIES

Night-Blooming Cereus Cacti

Our first case study illustrates the range of botanical knowledge still residing within native cultures of the borderlands. It involves O'odham (Pima and Papago Indian) knowledge of two night-blooming cereus cacti, *Peniocereus greggii* (Englelm.) Britton and Rose var. *transmontanus* (Engelm.) Brackenburg and *Peniocereus striatus* Brandegee. Both species are protected by CITES and Arizona state law, and have at times been listed as candidate species for federal protection in the United States. Both have wide ranges, but generally occur in densities of less than 20 plants per acre. Despite their naturally low numbers in any given locality, night-blooming cereus cactic are widely known and distinguished from other cacti in the folk taxonomies of desert peoples. In O'odham plant taxonomy, *P. greggii* is distinguished from the multiple tubers of *P. striatus* by its large taproot, and by its use as both food and medicine (as opposed to purely medicinal uses for *P. striatus*).

TABLE 7.1
Threats Cited as Pressures on Useful Plants of the U.S./México Borderlands

Type of Threat	Times Cited
Overcollection	39
Invertebrate damage	21
Vertebrate damage	7
Domestic livestock	22
Urbanization	11
Clearing for agriculture	8
Mining and oil drilling	4
Environmental contamination	1
Fire suppression	6
Climate change	1
Off-road vehicles	8
Vandalism	1
Changes in river flow	2
Flooding (especially by dams)	6
Groundwater depletion	13
Loss of pollinators or dispersers	3
Loss of nurse plants	3
Narrow endemic	13
Scarcity of numbers	9

Both of these cacti have thin, lead-colored stems that tend to be camouflaged by those of the creosote bush (*Larrea tridentata*), one of the six to eight perennial "nurse plants" beneath which these cacti commonly grow. This cryptic tendency has not kept desert dwellers from locating these cacti; they have utilized the flowers, fruits, and roots in various ways. Of particular interest are uses of the succulent tuberous roots of *P. greggii*, which weigh as much as 20 kg for plants of advanced age. Over the last 75 years, O'odham plant-gathering experts of the Sonoran Desert have consistently mentioned two major uses of the root. First, as a food, it can be eaten raw—or baked in ashes, then peeled and eaten. Second, it can be used as a medicine; in this case, the root is cut into chunks, steeped in boiled water, then it is eaten or ingested as a tea or broth. It has long been considered as an aid to digestion, and used to treat respiratory ailments and headaches.

TABLE 7.2
Impacts Cited for Useful Plants at Risk of Extinction in the U.S./México Borderlands

Type of Impact	Times Cited
Collected to local extirpation	3
Collected to regional rarity	4
Collected to local rarity	7
Collection positively affects sexual reproduction	0
Collection negatively affects sexual reproduction	18
Collection positively affects vegetative reproduction	4
Collection negatively affects vegetative reproduction	1
Collection impact related to other environmental impacts	5
Rarity in wild has encouraged local indigenous protection	5
Rarity in wild has encouraged local indigenous cultivation	10

An additional medicinal use of *P. greggii* roots–as a treatment for diabetes–emerged between 50 to 70 years ago among the O'odham, perhaps through diffusion from Mexican-American herbalists. This ethnobotanical use became exceedingly important, however, as the frequency of noninsulin-dependent diabetes mellitus increased rapidly among the O'odham, other desert Indians, and neighboring Mexican-Americans. In response to the demand for this plant from Mexican-American herbalists around 1930, the O'odham began to gather these roots for market. Oral history suggests that by the 1940s, specialty collectors had already depleted its populations near the Gila River in central Arizona, and its commercial use apparently waned. While the utility of both species as medicines is still widely recognized among the O'odham, they are harvested rather infrequently from extant populations close to their villages.

Since the 1930s, however, more lasting pressures have been exerted on night-blooming cereus populations by habitat conversion to modern agriculture and urban development, as well as by habitat degradation via overgrazing and fuelwood cutting. The severity of these pervasive pressures can be demonstrated by comparing the nurse plant ecology of *Peniocereus striatus* near the expanding Mexican and O'odham borderland community of Sonoyta, Sonora, México with that of the protected populations in nearly identical habitats across the border in Organ Pipe Cactus National Monument, Arizona. Sampling 250 m^2 plots in these habitats, we found the night-blooming cereus growing under the protective canopies of a number of different nurse plant species, listed here in order of decreasing canopy size: palo verde (*Cercidium microphyllum*); ironwood (*Olneya tesota*); wolfberry (*Lycium* sp.) and bursage (*Ambrosia deltoidea*). Compared with the random points tossed within the same sampling areas, there was a greater than expected association with the nurse plants (using Chi square, the null hypothesis was rejected the 99.5% level of confidence). The mean distance from a nurse plant for *P. striatus* was 12.2 cm, as opposed to 35.4 cm for the random tosses. By inference, we have hypothesized that *P. striatus* (and, from our preliminary observations, *P. greggii* as well) depends upon the perennial cover to provide buffered microenvironments and prey refugia (from livestock and rabbits). However, on the unprotected, Sonoyta side of the border, 60% of the trees showed evidence of coppicing for fuelwood, which exposes the night-blooming cacti beneath them to damage by mammals. Just outside of the sample plots, we located one dead *P. striatus* that had been totally exposed by the cutting of all live branches off an ironwood trunk. In the protected area, no trees within the plots exhibited cut branches, shade was denser, and rabbit damage to plants, though still evident, appeared to be reduced.

It is probable that direct medicinal uses of these cacti by indigenous people are having less effect on the viability of their plant populations than are the increasing demands for fuelwood that have followed the demographic explosion in Sonoran Desert border towns, where a sevenfold increase in human population has occurred since 1950 (Jamail and Ullery 1979). By December 1990, ironwood firewood cut from the desert near these study sites was being sold in Sonoyta, Sonora for 500,000 pesos per pickup truck load, or roughly $160 (US) per cord of low quality fuelwood. Palo verde and mesquite (*Prosopis*) fuelwood considered to be of lower quality have already been depleted from large acreages around many urban areas in Sonora and Baja, California, within the ranges of these two *Peniocereus* species.

As a response to the increasing local rarity of these cacti and their nurse trees around urban areas, several O'odham families we know have brought these species into cultivation in their dooryard gardens. The utilitarian value of these night blooming cacti, as well as the fragrance of their blossoms are cited as reasons for the protection of these plants. More significantly, perhaps, is the protection of these species and other desert plants by the Gila River Indian Community's Ordinance Number 04-85. This is the first native plant law of its kind initiated by a reservation community. The Akimel O'odham (Gila River Pima), Tohono O'odham (Papago) and Maricopa Indian members of this community now prohibit the destruction or removal of these plants by outsiders. A Natural Resources Standing Committee must approve any collection of these protected plants.

In summary, indigenous cultures of the Sonoran Desert retain detailed knowledge of two rare night-blooming cereus cacti, continue to use them on a modest scale, and have initiated both formal and informal means to protect and encourage the continued availability of these cultural resources. Although indigenous collecting can be considered an impact, it is far less a threat to these species than fuelwood collection from their protective nurse plants, overgrazing, land conversion, and groundwater depletion. Although these two cacti species are protected within the Organ Pipe Cactus National Monument biosphere reserve and on Indian reservations, they remain vulnerable over a large portion of the range along the U.S./México border.

Red Barrel Cactus

The following case study compares the impact of traditional food uses with those of commercial cactus collecting for ornamental trade. *Ferocactus pilosus* (Salm-Dyck) Wenderm., is also known as *viznaga colorada* or "the red barrel cactus" because of its brilliant reddish-pink spines. Its stems can be found solitary, but are often found in massive clumps with 6-15 stems (ramets) forming one genetic individual (genet). *Viznaga colorada* is found in the Chihuahuan Desert almost as far north as the Rio Grande near Big Bend and as far south as Hidalgo in the semi-arid Mexican highlands. The majority of our data for this widespread but highly vulnerable cactus comes from the vicinity of Miquihuana, Tamaulipas, in the transition zone between rain-shadow Tamaulipan scrub and Chihuahuan Desert vegetation.

Viznaga colorada has two distinct sets of uses: 1) traditional food uses and 2) ornamental uses. First, its lemony-tasting, acrid fruits are highly-favored locally in the Miquihuana Valley, where they are commonly used as the flavoring for a lemonade. In addition, flower buds are pickled or cooked in various ways throughout its range. Barrel cacti stems are widely used to make *dulces*, a crystallized sugar candy. This species is not normally favored for candy-making, but when the preferred, sympatric *Echinocactus platyacanthus* has been overexploited, *viznaga colorada* can be depleted as the second choice for this consumptive use. Nevertheless, these traditional uses have not had as drastic an impact on *F. pilosus* populations as has the extraction of juvenile sizes for the ornamental cactus trade. Single-stemmed individuals from 30 cm to 1 m in height are collected, and sold in clay pots. There are gaps in the size-class distribution within one population we have studied, indicating plant removal.

Although these gaps may be due to sporadic recruitment and establishment of this species in an unpredictable climate, we hypothesize that they are due to recent collection episodes by commercial nurserymen. If extraction continues, few individuals may arrive at reproductive age over the next generation. Other impacts on the population ecology of *viznaga colorada* include *Neotoma* (packrat) midden-building in multi-stemmed clumps, and goat trampling of seedlings and browsing of their nurses. Although we have yet to quantitatively assess these factors, they could impact more than one life stage of this cactus species. This case study, like that of the night-blooming cereus cacti, points to the need for threat assessments not only of direct plant collection, but of the more indirect effects of land management as well.

Thornber's Fishhook Cactus

The third case study illustrates the conservation status of a rare plant on Indian reservations where it is still used in comparison with its status in and near a U.S. National Park Service (NPS) reserve in an urban area. Thornber's fishhook cactus, *Mammillaria thornberi* Orcutt, occurs in patches over a 4,000 square mile range centered on the Tohono O'odham reservation along the Arizona-Sonora border. The northernmost locality of this species was recently discovered by a young Pima (Akimel O'odham) plant enthusiast on Gila River Indian Community lands. In U.S. Fish and Wildlife Service listings, it is a Category 3C species, and it is also protected by CITES (App. II), Arizona state law, and the Gila River Indian Community's Ordinance No. 04-85 (noted above). Current threats include construction of the Central Arizona Project aqueduct, urbanization, groundwater depletion, overgrazing, and overcollection.

The O'odham call this plant *ban mauppa* "coyote's paws" or *ban ha-i:swig* "coyote's hedgehog cactus," (*Echinocereus engelmannii* [Parry] Lemaire). Like the folk names of the more common pincushion cactus, *Mammillaria grahamii* Engelm., reference to coyote for *M. thornberi* is rooted in O'odham lore that coyotes spoiled their fruits, making them sour (Nabhan 1982). The strawberry-like fruit of the hedgehog cactus have been favored over the fruit of other cylindroid cacti by both the contemporary and historic O'odham. Judging from the frequency of *Echinocereus* versus *Mammillaria* macrofossils in prehistoric Hohokam remains, the same may have held true in pre-Columbian times. Pima Indian children still use cut *Mammillaria* stems as a source of moisture during extended periods of play in the desert, but the stems have no major food or beverage status now. Historically, the Pima rubbed *Mammillaria* fruit on arrow shafts to color them blood-red, and sliced boiled, dethorned stems to use as a poultice for earaches (Curtin 1949); such uses are uncommon today. Although they are no longer of great utilitarian import, coyote plants are humorously tolerated and occasionally protected or cultivated by O'odham families. In one case, a clump seen by a Tohono O'odham family near their saguaro cactus fruit gathering camp was carefully transplanted back to their desert homestead, on the edge of a backyard religious shrine. In another case, it persists in many microsites within an historic O'odham village that is known to have been farmed and managed by Indian people for well over 280 years.

While this species' status is relatively stable on tens of thousands of acres of Indian lands, its populations may be declining in or near two National Monuments managed by NPS. At Organ Pipe Cactus National Monument, NPS has documented local depletion of Thornber's fishhook cactus by both collection and trampling. Bennett et al. (1987) recorded that the degree of illicit poaching of small cacti by tourists is higher within NPS boundaries than on other federal, state or private lands nearby. With respect to *Mammillaria thornberi* alone, NPS biologists documented the loss of 44% of the original number of plants sampled at the onset of their study. Apparently, tourists fail to heed NPS's strong interpretive message and regulations promoting conservation, and in their quest for memorabilia from their desert visit, they illegally take living elements of the monument home with them.

With regard to Saguaro National Monument, the depletion of Thornber's fishhook cactus has occurred immediately outside park boundaries, due to uncontrolled suburban development by residents attracted to the proximity of "pristine desert." In a 64,000 acre study area in and around Saguaro National Monument West, approximately 10,470 acres formerly found in desert vegetation in 1937 had been converted to urban, industrial and agricultural uses by 1987. Ironically, only 4,420 acres of *M. thornberi* habitat occurs inside the 18,650 acre Monument, and the habitat loss rate outside the Monument increased 67% between 1967 and 1979, as Tucson's urban growth reached the edges of the newly established West unit. More difficult to assess is the impact of habitat fragmentation on this cactus, a predominantly outcrossing species which may presumably suffer inbreeding depression if gene pool sizes drop below a certain level.

The accelerated conversion of natural habitat immediately outside of the Saguaro National Monument is not a dilemma unique to this one protected area. Wherever there is an economy external to a park which may gain economic benefits from the its proximity, there will inevitably be costs to the conservation goals of the park itself. In this case, the Monument boundaries were not established to accommodate minimal viable populations of the rarest plant species within its reach; in fact, the scenic values of desert mountains were the predominant guides to boundary establishment. Unfortunately, few of the Thornber's fishhook cacti occur on the steep-sloped, coarse soils of the mountains within the Monument. Instead, they are more frequently found on the fine soils of the valley below, which has low scenic value, but is suitable for housing developments and irrigated fields. As long as parks can not legally zone the buffer areas around them or design their boundaries to accommodate the rarest species within their reach, such "protected areas" will not be enough to save plant species from extinction.

Hohokam Agave

Nearly every case study of *in situ* conservation of useful plants which we have seen published has as its focus a plant still used by contemporary peoples, presumably protected to continue this use. The story of the Hohokam agave, *Agave murpheyi* Gibson, departs from this trend because it describes a feral domesticated crop which has persisted without human care on prehistoric Hohokam terraces last tended more than 500 years ago.

The story of the presumed "wild" Hohokam agave seemed relatively simple until nine years ago, for no prehistoric or historic cultivation of agaves had ever been recorded north of México (Fish et al. 1985). This species was described in 1935 from what was thought to be a wild population near Tonto Basin of central Arizona–an area now known for its impressive cliff dwellings, extensive prehistoric fields, and rich ethnobotanical remains. In addition to two nearby populations (now obliterated) in central Arizona, a few others were later found more than 200 miles away, at a truck stop in Sonora, México and in gardens on the Tohono O'odham reservation. This geographic disjunction was not the only anomaly associated with the Hohokam agave. Its inflorescence begins elongating in late winter, much earlier than other agave species in Arizona. Flowers, often deformed, appear in spring. These are followed by vegetative bulbils which have the potential to root if placed in loose soil. The bulbils, however, are parasitic on the mother plant, and usually fail to establish themselves when they fall to the desert ground surface and are unaided by humans. Instead, the principal natural means of reproduction of untended plants are clonal offshoots arising from rhizomes at the base of the mother plants. Yet, many thousand bulbils have been planted over the last half century because the type clone was considered so attractive as an ornamental that a large horticultural trade developed based solely on this one agave species.

Recently, our perspective on this species has radically shifted as a result of ethnobotanical inquiries. Virtually every one of the 54 clones of the Hohokam agave has been found in proximity to prehistoric, historic or contemporary Indian gardens and fields. In Sonora, all known clones are within close proximity to historic or current Tohono O'odham villages, where the plants have been used for the production of food, fiber or alcohol within this century. The Tohono O'odham name for this agave means "it has progeny," apparently referring to the prolific production of plantlets by the mother plant. Elderly O'odham claim that during their grandparents' era (1860-1910), this agave grew above winter campsites where it was harvested and roasted as a major foodstuff for the winter months. They now maintain in their ornamental gardens remnant clones which were formerly extensively utilized by their ancestors.

In 1982, archaeobotanist Charles Miksicek identified agave fibers in flotation samples from prehistoric Hohokam roasting pits. Shortly thereafter, extensive prehistoric agave cultivation in south-central Arizona was deduced from the strong association of roasting pits and mescal knives with rock pile fields, terraces and check dams covering tens of thousands of hectares of desert bajada slopes. Since then, evidence of prehistoric cultivation of two agave species has been found over a 1242 square mile stretch of Arizona formerly occupied by Hohokam and Salado cultures. Unfortunately, none of the archaeobotanical remains of fiber and carbonized leaf fragments have been definitively identified to the species level. However in February 1988, the late Rick DeLamater, a Desert Botanical Garden research assistant, brought to our attention *Agave murpheyi* clones near a prehistoric village site in the New River watershed. His own botanical reconnaissance there had established the presence of more than twenty clones of this species, four times as many as had been cumulatively documented elsewhere to date. When the entire FLORUTIL team went to make vouchers of these clones, it discovered that these clones did not exist in natural vegetation *per se*, but that they were growing on manmade terraces, on linear cobblestone borders and in rock piles that had been mapped by archaeologists a decade before without any reference to agave presence (Gummerman et al. 1975). Recently, agave remains have been found in association with roasting pits, mescal knives or runoff fields in the adjacent Agua Fria watershed (Green 1986, A. Rankin, pers. comm.). In many recently excavated Hohokam sites, agave remains are as frequent as those of corn. Unfortunately, tens of thousands of acres of probable habitat for prehistoric agave fields have been cleared in Arizona for modern irrigated agriculture and urban expansion of Phoenix and Tucson. Of the 50 extant clones of Hohokam agave, some are now threatened by planned urbanization, expansion of flood control reservoirs, road construction, overgrazing, and recreational development. An incredible story of prehistoric agriculture could conceivably be reconstructed if this plant and its cultural context could be legally protected.

To date, FLORUTIL collaborators have had success in initial steps to gain adequate protection for the Hohokam agave. In August 1988, the U.S. Fish and Wildlife Service (USFWS) was petitioned to protect this plant as a threatened species. On March 6, 1990, the FWS determined that protection was warranted due to the species' limited distribution and abundance and present threats to its survival (Finn 1990). However, it precluded immediate listing as a threatened species due to "listing actions of higher priority" and considers Hohokam agave a candidate for protection in Category 2. We have also initiated steps to nominate agave fields in the New River watershed for the National Register of Historic Places and to describe a second Hohokam-Salado agave species (Hodgson and Ecker, in press). In addition, the State Historic Preservation Office has agreed to enforce the Antiquities Act and Historic Preservation Act with respect to agave field sites where cultural artifacts remain part of the plant's context. The Desert Botanical Garden has convened and now chairs an interagency task force on Prehistoric Agave Resources" which includes representatives from the Bureau of Reclamation, U.S. Forest Service, Arizona Game and Fish, USFWS, Arizona State Historic Preservation Office, and the Arizona State Museum. This may well be the first example of *in situ* conservation within U.S. boundaries for a plant that is both a cultural and a natural resource.

CONCLUSIONS

From these case studies and others accomplished by FLORUTIL team members (Nabhan et al. 1986, Suzan Azpiri et al. 1989), it is clear that indigenous cultures have their own reasons for conserving rare, useful plant resources, which are put into practice in the following ways:

- Recurrent regeneration and consistent protection of plant populations over many human generations.
- "Seed-banking" in hermetically-sealed pots, caves, or granaries for back-up supplies.
- Translocation of clones to botanical gardens, sacred mountains, shrine sites and land reserves where rare or valuable plants can be offered additional protection.
- Taboos, codes, or other social constraints to discourage depletion or overexploitation.

- Promotion of these plants as cultural identity symbols, ethnic foods, or objects of aesthetic and historical appreciation, so that their relationships with cultural communities will persist.

However successful these strategies have been in the past, they are now vulnerable to the impacts of tremendous social and economic forces working against their perseverance in the future. Genetic erosion and plant extinction cannot be isolated from either socio-economic change or habitat alteration, and both of these processes are accelerating. It is clear to us that plant genetic erosion has been triggered by impacts on indigenous cultures far more pervasive and lasting than recent Green Revolution agricultural developments and tropical deforestation. To save these rare plants for future generations, the indigenous cultures of the U.S./México borderlands must be encouraged to participate in the conservation process in new ways as well as through traditional modes. Although such requisites mandate cohesive conservation strategies far more complex than that of *ex situ* seed banks, it is desirable to conserve these species as cultural resources, not merely as germplasm. To paraphrase the famous American geographer, Carl Sauer, whether or not a natural resource is considered worth saving is ultimately a cultural appraisal.

To encourage this kind of "cultural bonding" with rare plants, FLORUTIL team members in Tamaulipas have recently initiated a nursery for threatened succulents, including some recovered from confiscations at the border, and from plants eroding out of degraded lands nearby. The nursery is situated very close to one of the localities most frequently visited by foreign collectors, and has been developed on land held in common by the local community, with *campesinos* from the community being provided part-time employment there. Some of them had formerly assisted foreign collectors for extremely poor wages compared to what the collectors gain for selling the rare cacti overseas. Recently, when they were informed how collection was contributing to extirpation of these species locally, and were told of the large monetary returns to others as a result of their work, they decided not to collect wild plants further, but to assist with the nursery and reintroduction efforts instead.

One of the local nursery workers, Juan García, told us about foreign collectors who recently requested his help in locating the endemic cactus, *Mammillaria carmenae* (M. Castaneda), which is near extinction in his area. Juan agreed to drive them into the field because he needed the money, but he took them long distances in his vehicle before letting them get out and hike to sites where no cacti could be found. When the collectors expressed their frustrations, Juan replied, "Sorry, it looks like there are no cacti left anymore."

The truth was that Juan knew where *Mammillaria carmenae* grew, but he took them to other sites far from there, where none could be found. With the retelling of this story in the local community, pride in protecting rare species has become transmitted to others, and conservation has become reinforced as a community value.

REFERENCES

Anderson, E. 1987. Black magic in cactus country. *Garden* 11:2-5, 32.

Anderson, M.K. 1988. Southern Sierra Miwok plant resource use and management of the Yosemite region. M.S. Thesis, Department of Forestry and Resource Management, University of California, Berkeley.

Bailey, R.G. 1978. *Description of the Eco-Regions of the United States*. U.S. Forest Service, Odgen, Utah.

Balling, R.C. 1988. The climatic impact of a Sonoran vegetation discontinuity. *Climatic Change* 13:99-109.

Bennett, P., R.R. Johnson and M.R. Kunzman. 1987. Cactus collection factors of interest to resource managers. Pages 215-224 *in* T.S. Elias (ed.), *Conservation and Management of Rare and Endangered Plants*. California Native Plant Society, Sacramento.

Bye, R.A. 1983. Vegetation and soils. Pages 98-104 *in* E.R. Stoddard, R.L. Nostrand and J.P. West (eds.), *Borderlands Sourcebook*. University of Oklahoma Press, Norman, Oklahoma.

Center for Plant Conservation. 1986. *A Garland of Generations*. (Video). Available from the Center for Plant Conservation, Jamaica Plains, New Jersey.

Curtin, L.S.M. 1949. *By the Prophet of the Earth*. San Vicente Foundation, Santa Fe, New Mexico.

Diamond, J. 1986. The environmentalist myth: Archaeology. *Nature* 324:19-20.

Finn, V.M. 1990. Endangered and threatened wildlife and plants; notice of finding on petition to list *Agave murpheyi*. *Federal Register* 55(44):7920-7921.

Fish, S.K., P.R. Fish, C. Miksicek and J. Madsen. 1985. Prehistoric agave cultivation in southern Arizona. *Desert Plants* 7:107-112.

Gonzalez-Medrano, F. 1973. *La Flora y Vegetación de Nordeste Tamaulipas*. Annales del Instituto de Biología, UNAM, México City.

Green, M. 1986. Settlement, subsistence, and specialization in the northern periphery: Research design for mitigative recovery at sites in the new Waddell Dam borrow areas. *Archaeological Consulting Services Report* 40:1-54.

Gummerman, G.J., J.A. Hanson, D. Brew, K. Tomoff and C.S. Weed. 1975. *The Hydrology of Prehistoric Farming Systems in a Central*

Arizona Ecotone. Lyndon B. Johnson Space Center, Houston, Texas.

Henrickson, J. and M. Johnston. In Press. *The Flora of the Chihuahuan Desert Region*. VPT Publication, Austin, Texas.

Hodgson, W., A. Zimmerman, G. Malda Barrera, L. Hernandez Sandoval and G.P. Nabhan. 1987. *FLORUTIL/Cactaceae & Agavaceae*. Desert Botanical Garden Pages 1-6, Phoenix, Arizona.

Hodgson, W. and L. Echer. In Press. A new Agave in Arizona. *Madroño*.

House, D. 1987. Navajo sedge (*Carex specuicola*) recovery plan. U.S. Fish and Wildlife Service, Albuquerque, New Mexico.

Humphrey, R.R. 1987. *90 Years and 535 Miles: Vegetation Changes Along the Mexican Border*. University of New Mexico Press, Albuquerque, New Mexico.

Jamail, M.H. and S.J. Ullery. 1979. International water use relations along the Sonoran desert borderlands. Arid Lands Resource Information Paper No. 14. Office of Arid Lands Studies, University of Arizona, Tucson, Arizona.

Nabhan, G.P. 1982. *The Desert Smells Like Rain: Naturalist in Papago Indian Country*. North Point Press, San Francisco, California.

Nabhan, G.P. 1989. *Enduring Seeds: Native American Agriculture and Wild Plant Conservation*. North Point Press, San Francisco, California.

Nabhan, G.P. and K.L. Reichhardt. 1983. Hopi protection of a rare sunflower, *Helianthus anomalus. Southwestern Naturalist* 28:231-235.

Nabhan, G.P. and M.J. de Wet. 1984. *Panicum sonorum* in Sonoran Desert agriculture. *Economic Botany* 38:65-81.

Nabhan, G.P., W. Hodgson and R. Greenhouse. 1986. At the edge of extinction: Useful plants of the border states of the United States and Mexico. *Arnoldia* 46(3):36-46.

Nabhan, G.P., W. Hodgson and L. Hernandez. 1987. Plant genetic resources threatened in North American deserts: The FLORUTIL conservation project. Pages 59-65 *in Strategies for Classification and Management of Native Vegetation for Food Production in Arid Zones*. U.S. Forest Service, Fort Collins, Colorado.

Nabhan, G.P., E. Saucedo Monarque, P. Olwell, P. Warren, W. Hodgson, C. Gallindo-Duarte, R. Bittman and S. Anderson. 1989. A preliminary list of plants at risk in the Sonoran Desert of the U.S. and Mexico. *Agave* 3(3):19.

Rea, A.M. 1987. Introducing ethnobiology. *Environment Southwest* 519:3-4.

Shreve, F. and I. Wiggins. 1964. *Vegetation and Flora of the Sonoran Desert*. Stanford University Press, Stanford, California.

Suzan Azpiri, H., Malda Barrera, J.L. Jimenez Perez, L. Hernández Sandoval and M. Martínez. 1989. Evaluación de plantas amenazadas y en peligro de extinción en el Estado de Tamaulipas. *Biotam* 1(1):20-27.

Toledo, V.M. 1988. La diversidad biológica de México. *Ciencía y Desarollo*. Conacyt, México City.

U.S. Department of Interior. 1988. *Indian Agriculture/Range Program*. Bureau of Indian Affairs, Washington, D.C.

Vovides, A.P. 1981. Relación de plantas Mexicanas raras o en peligro de extinción. *Biótica* 6:222-228.

8

Pátzcuaro's Lesson: Nature, Production, and Culture in an Indigenous Region of México

Victor M. Toledo

INTRODUCTION

Lakes, oak-pine forests, and woodlands were the natural habitats that allowed the development of prehispanic societies in the central region of Mesoamerica. In fact, the lake area of central México was not only significant in the development of large sedentary populations, but also coincides with the most important locale of highland civilizations in Mesoamerica, represented by the core of the Aztec and Tarascan states. Not only were fish and other aquatic animals (ducks, frogs, shrimps, insects) important as food resources, but the marshy lake shores supplied a dense growth of tule reeds used throughout Mesoamerica for making mats, baskets and other materials. In addition, the oak-pine forests and woodlands around the lakes provided deer, rabbits, quail, doves, and honey for food, fuelwood, medicines, and housing materials. Finally, between the lakes and the terrestrial vegetation, the flatlands and gentle slopes were utilized for different agricultural systems, such as the *chinampas* (see also Gliessmann, Ch. 10, and Chapin, Ch. 11, this volume) and crops, principally maize, beans, squash and amaranth.

Today after 450 years, most of the 25 lakes originally distributed on the volcanic plateau of central México (Tamayo and West 1964) are either greatly deteriorated or have disappeared. Lake Pátzcuaro is, however, one remarkable exception. The largest lakes of the neo-volcanic axis of central México were artificially drained (Valley of México), reduced by diverting water to irrigation (Lake Cuitzeo), or polluted with industrial and urban wastes (Lake Chapala). But the Pátzcuaro Basin remains a relatively well conserved habitat. In fact, although there are several processes provoking environmental impacts (deforestation, slope soil erosion, and eutrophication) in some portions of the basin, the region has relative ecological stability (Barrera-Bassols 1986). The lake itself is only moderately polluted, despite being close to the town of Pátzcuaro (Mazari 1981, Saavedra 1982, Velasco 1982). Both the aquatic as well as the terrestrial habitats of the Pátzcuaro Basin are in a certain environmental balance, resulting in a high degree of regional biological richness. The natural ecosystems of the lake, for instance, include four communities of hydrophytes (Lot and Novelo 1988) that provide the habitat for a rich flora and fauna comprising 48 aquatic plant species, 14 species of fishes, 13 species of aquatic birds (10 migratory ducks and others) and some endemics such as the bizarre Mexican salamander (*Ambystoma dumerilii*) (Argueta 1988). The terrestrial ecosystems are represented by the oak and pine forests (covering one third of the area) and intervening shrub and grasslands with a estimated flora of over 1000 species (J. Rzedowski pers. comm.). These ecological conditions of the Pátzcuaro Basin seem even more striking considering two facts: 1) there is evidence of agricultural activity (presence of maize pollen in the lake sediments see Watts and Bradbury 1982), in the basin at approximately 3500 BP which denotes a very long human presence in the area and, 2) the population of the Basin in the early 16th century, during the florescent Tarascan state, has been estimated between 60,000 and 100,000 inhabitants distributed in 90-95 settlements (Gorenstein and Pollard 1983). In addition, the region is at present part of the most populated, urbanized and industrial area of México, and is located scarcely 400 km away from México City, the second largest urban concentration in the world.

What are the cultural, economic, ecological, or historical mechanisms or processes responsible for conservation of the natural habitats of the Lake Pátzcuaro Basin? I contend that the relative environmental well-being of the Lake Pátzcuaro Basin is directly due to the presence of a strong traditional culture in the area, the P'urhépecha (or Tarascan) Indians, heirs of the ancient Tarascan civilization who are still the principal consumers of the region's natural resources. The ability of the P'urhépecha Indians to maintain a certain ecological balance with the aquatic and terrestrial ecosystems of the region must be directly linked with the capacity of the indigenous communities to overcome, resist, neutralize and/or assimilate the external disruptive forces brought about by a rapidly changing world. In fact, unlike other cases where the geographical and economic isolation of indigenous peoples allow a relatively unchanged situation, the P'urhépecha peasant communities of Pátzcuaro are very dynamic social entities. The P'urhépecha have retained their essential cultural traits even though they have established themselves within the national and international economic spheres. To support the previous thesis, information obtained from a long term research project carried out in the Pátzcuaro region since 1977 will be utilized. The aim of this project[1], which is based on an interdisciplinary, human ecology approach, has been to study indigenous knowledge, use, and management of the natural resources of the Lake Pátzcuaro Basin.

THE STUDY AREA

The basin is one of four regions that make up the modern day geographic area of the P'urhépecha culture (West 1948). In prehispanic times, it was the center of the Tarascan Empire. Despite social and cultural changes, it continues to be one of the most extensive sites of indigenous cultures in México, and, possibly, the only relict of the ancient lacustrine culture of Mesoamerica. It is located in the transverse neovolcanic belt in the northern part of the state of Michoacán in western México (Figure 8.1). The area is approximately 1,000 km^2 and ranges from 2,043m (at the lake level) to 3,200 m. There are five obvious physiographic zones: the lake islands, shoreline, hillsides, intermountain valleys, and mountains. The lake itself occupies about 100 km^2, and is 8m-12m deep.

Forests are often dominated by oaks (*Quercus* spp.), pines (*Pinus* spp.), and fir (*Abies religiosa*) with intervening shrub and grasslands. Agricultural lands, situated predominantly on the shoreline and hillsides, cover about 40% of the basin's terrestrial landscape. The region shows evidence of recent volcanic activity, including basalt flows and cinder cones, and is a complex edaphological mosaic with ten main soil types and more than 20 soil units (Barrera-Bassols 1986). Although the climate is temperate (mean monthly temperature is 16°C) with mild winters, temperatures below freezing occur often during December and January. A well-marked dry season extends from November to May, and the rainy season is from June to October. Annual precipitation is about 1,000 mm.

SETTLEMENTS AND POPULATION

Without exception, all the main present day settlements of the Lake Pátzcuaro region were already in existence during the prehispanic era (1450-1520) (Gorenstein and Pollard 1983). Today, however, the region is a culturally complex mosaic formed by indigenous and mestizo communities, and by transitional settlements in between these two extremes. In addition, there are a few settlements of recent foundation. As a result of acculturation processes during the Colonial period and national integration during more recent times, the Lake Pátzcuaro Basin is primarily dominated by three large, nonindigenous

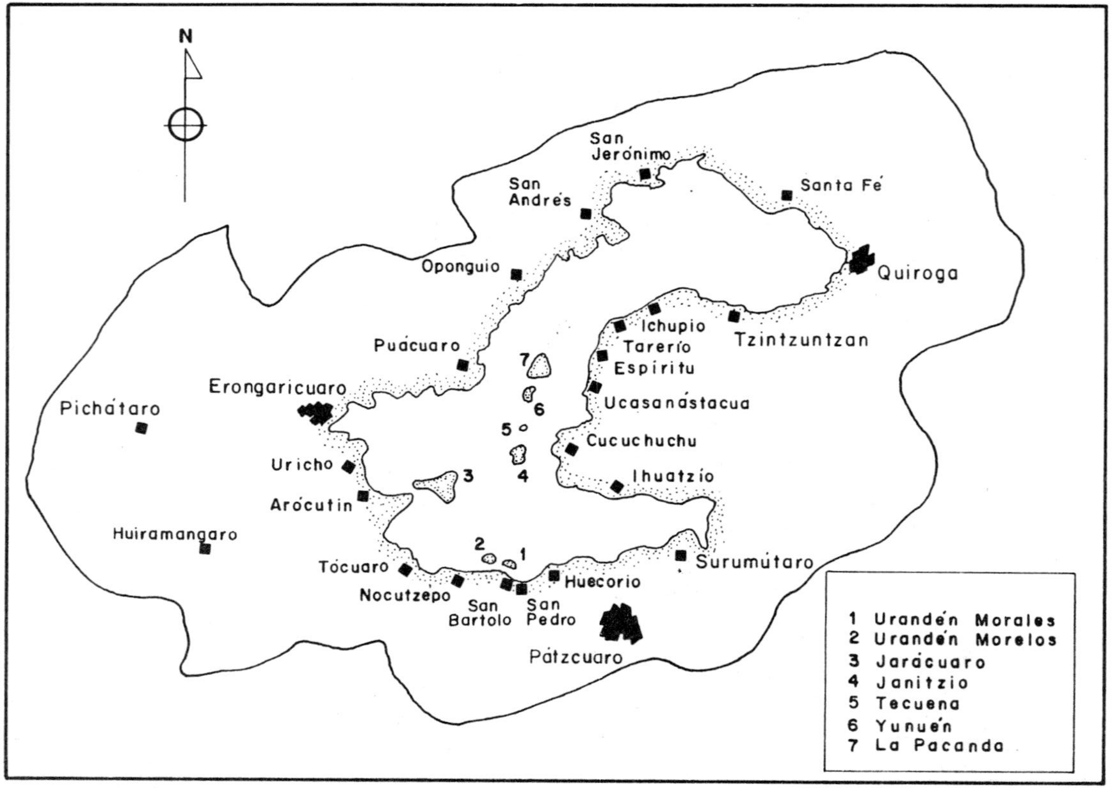

FIGURE 8.1 Major indigenous and mestizo settlements of the Lake Pátzcuaro Basin.

settlements (Pátzcuaro, Erongarícuaro, and Quiroga) that are strategically located around the lake (Figure 8.1). Despite this situation, the 25 remaining indigenous communities are still the principal users of the region's terrestrial and the aquatic natural resources. The west and north shores of the lake are dotted with villages still largely indigenous in speech, except for Erongarícuaro, which was transformed into a mestizo settlement as early as the 16th century. The eastern and southern shores, although still occupied by indigenous communities (Ihuatzio, Cucuchuchu, Ucasanástacua, Espíritu, and Tarerio), have been more exposed to change. There acculturation is more advanced, including Tzintzuntzan, the former capital of the Tarascan Empire (see Foster 1948). Finally, all the lake islands, the mountains, and intermountain valleys on the west are still settled by indigenous villagers (Figure 8.1).

According to the 1980 Mexican National Census, a population of 103,216 individuals lived in the five municipios that form the Lake Pátzcuaro region. These people were distributed in 17,553 households within 112 settlements. This figure corresponds to a population density of 103 inhabitants per km^2. Only two of the 112 settlements can be considered urban: Pátzcuaro (32,902 inhabitants), the regional market center, and Quiroga (10,127). The rural population of 60,187 individuals, is distributed among six major villages that have between 2,500 and 5,000 inhabitants (including three indigenous communities), eleven medium-size settlements with populations of between 1,000 and 2,500, and 93 settlements below 1,000 inhabitants.

The indigenous population that was listed in the 1980 Census as P'urhépecha speakers totaled 19,868 individuals or 22.5% of the total population and approximately 33% of the rural population. However the above figures underestimate the indigenous presence in the area, because several settlements considered as mestizo closely resemble indigenous settlements in various traits other than language (for example social and political organization, land tenure, religious activities, and modes of appropriation of nature). Thus, many of the so called "nationalized indigenous communities" (Moone 1973) are, in practice, indigenous villages.

ECOLOGICAL IMPORTANCE OF THE INDIGENOUS COMMUNITIES OF PÁTZCUARO

The conservationist attitudes of traditional peoples have not been carefully investigated, but it is clear that the use of natural resources is regulated through cultural, social, economic, political and technological means (Nietschmann 1982, McCay and Acheson 1987). In the case of Pátzcuaro, research efforts to understand this topic should be concentrated on how the Indian community works. The Indian community is, in practice, the social unit through which the human populations of the region appropriate nature. Therefore, the enormous adaptive capacity of these age-old social systems provides a possible explanation for the persistence of the basin's natural resources. Four major characteristics of the P'urhépecha indigenous communities of Pátzcuaro apparently operate as four social keystones to achieve resource conservation: 1) collective but limited access to natural resources; 2) the nature of the indigenous economy; 3) the traditional strategy of appropriation of nature, and 4) the holistic orientation of indigenous knowledge to the natural environment.

The Collective Appropriation of Nature

A principal goal of the Mexican social revolution, proclaimed in 1910 and legalized in the 1917 Constitution, was the redistribution and restitution of lands to peasant communities. As a consequence of this land reform, by 1970, 50% of Mexican territory was managed collectively by groups of peasants under two specific forms of land tenure: *ejidos* and Indian communities. In both systems, each producer (the representative of each recognized household in the community) has the right to the exploitation of a piece of nature (limited access to resources). The right, however, is not equivalent to private property since it is restricted to direct exploitation. Legally, the producer cannot sell, rent, or mortgage his plot of land (or other natural resources). Inheritance is also regulated by certain legal stipulations, which are promulgated by the community assembly or by the authorities formally elected by the producers. In both forms of social property, the producers and their families are, in reality, no more than the individual user-managers of a set of natural resources that belongs to everyone in the community. Thus, they exercise their rights and responsibilities on behalf of the collectivity. In addition, the natural resources of a given *ejido* or Indian community tend to be allocated under an implicit rule of equitable distribution. Thus in theory, each household is engaged in the exploitation of equivalent natural resources. This practice tends to minimize class differentiation in peasant communities. Finally, although these social properties are collectively oriented, exploitation of natural resources is largely individually based. Thus, each household exploits limited amounts of natural resources under the regulation of the entire community.

In the Lake Pátzcuaro Basin, the communal system is the predominant form of land tenure among the indigenous settlements (Moone 1973). The majority of the P'urhépecha communities regulate access to resources through collective mechanisms (the Indian community);

some villages (e.g., Napízaro and Puácuaro) operate under the *ejido* system. On the other hand in mestizo settlements, use of resources tends to occur privately. According to the National Agricultural Census, 72% of the region's territory was exploited by social or non-private units by 1970, and only 28% by private property owners. This pattern of common property resource use was even more pronounced in the three *municipios* (county-like units) where indigenous settlements clearly predominate: Erongarícuaro (91%), Quiroga (85%), and Tingambato (91%).

Since it is possible to distinguish three main landscapes in the region (forests and woodlands, pastures, agricultural lands, and the lake) it is interesting to review the modalities for exploitation adopted by the indigenous peoples according to these environmental units. Although all the natural resources are owned communally, the agricultural lands tend to be considered, in practice, as private property in that continued use of the land makes it inalienable. However, this is not the case for forested lands, pastures or the lake. In some cases (e.g., Pichátaro; see Alvarez-Icaza 1988) the producer rents, gives custody for, or mortgages portions or all of his agricultural lands to other members of his community. In most cases, these unusual transactions are conducted with the permission of the community authorities, but there are many examples of illegal transactions. Where the indigenous political organization has been, or is, less strong, the individualistic and private-oriented concept of land use tends to transform and replace the original agrarian structure of the communities (e.g., Uricho, Huiramangaro, San Pedro, and San Bartolo).

The forests, pasture, shrublands, and the lake are more strictly visualized as collective resources by the P'urhépecha communities. The forests are still considered communal resources. In the past, perhaps as a consequence of population size and economic isolation of the villages, the forests, pastures and shrublands were free for the use of all community members, even though no one could claim special proprietary rights. Today, with higher pressure, all nonagricultural, terrestrial resources have been divided for individual exploitation (for resin, wood, firewood, and certain foods). An excellent example is the case described by Alvarez-Icaza (1988) in Pichátaro, where almost 4,000 ha of pine-oak forests held by this indigenous community have been evenly distributed among the eight *barrios* (wards) of the village (see Figure 8.2). Thus, each of the 559 households makes use of a similar amount of forest resources.

Similarly, the lake is legally considered national territory, and the lake resources are used by over 700 fishermen of 21 settlements (Table 8.1). Fundamentally, the lake is an indigenous managed resource. All but two of the settlements (Pátzcuaro and Erongarícuaro)

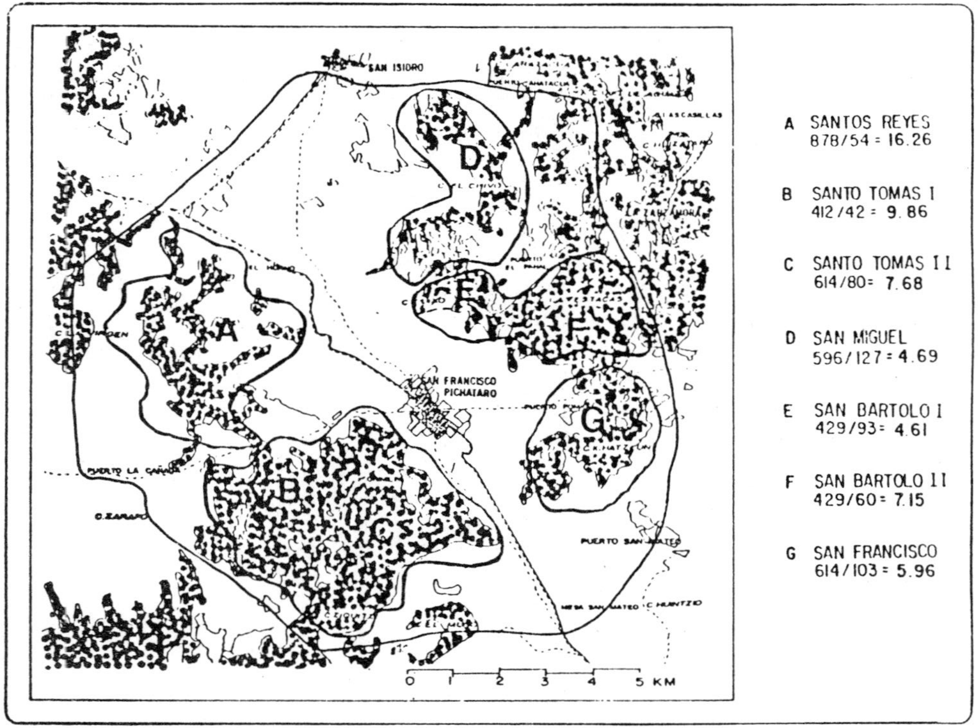

FIGURE 8.2 Distribution of pine-oak forests in Pichátaro wards; the average forest plot per household is measured in ha² per total households for each ward. (After Alvarez-Icaza 1988).
(Average values range between 4.5-8 ha² per household, with exception of Santo Tomas I and Santo Reyes; the latter ward has a very high average value of 16 ha² per household due to its association with a sparsely forested area.)

TABLE 8.1
Number of Fishermen (F), Canoes (C), and Large Nets (N) in Lake Pátzcuaro, 1980

Communities	F	C	N
Janitzio	85	109	1476
Tecuena	27	27	756
Yunuén	23	23	94
La Pacanda	43	52	387
Jarácuaro	17	---	----
Uranden-Morelos	59	52	1375
Napízaro	30	20	110
Arócutin	27	16	97
Pátzcuaro	20	15	272
Uricho	31	25	108
Erongarícuaro	19	19	435
Puácuaro	59	19	435
Oponguio	22	---	----
Ihuatzio	24	39	692
Cucuchuchu	23	23	138
Ucasanástacua	37	---	----
Ichupio	30	---	----
Tarerio	32	---	----
San Andrés	43	---	----
San Jerónimo	40	43	217
Santa Fé	19	25	132
TOTAL	710	701	6976

Source: Toledo and Barrera-Bassols 1984.

that engage in the exploitation of lake resources are indigenous communities. As in the case of forests, the lake body has been distributed among the communities, and then among the fishermen of each settlement, through certain agreements and collectively established rules. Therefore, each community and each fisherman have exclusive access to well defined portions of the lake border (the primary source of tule reeds, fishes, ducks and other aquatic animals). These "aquatic plots" are marked on the lake surface through the creation of channels using tule plants (Argueta 1988).

The Prestige Economy: Non-Accumulative Production

As in other traditional Mesoamerican villages, the implicit objective of the production process of Pátzcuaro Indian communities is the simple reproduction of the peasant household and consequently, the entire community. In the context of this social imperative, the peasant household behaves not as an isolated entity, but as the cell of an organism, that is to say, as members of a community. The purpose of production is, therefore, not the creation of value, even though peasants may perform surplus labor in exchange for foreign goods. Instead, the purpose is the maintenance and reproduction of the producer and his family as well as the communal body as a whole. In this perspective, household production is regulated through several mechanisms that inhibit internal class differentiation. Particularly important is the role played by the fiesta or cargo system which is organized on the principle of *mayordomia* or sponsorship, and serves as a social and collective control over household (or individual) accumulation (Smith 1977). By means of these social mechanisms, it is customary for heads of households to spend considerable amounts of time and resources in communal activities, for instance to contribute considerable sums of money, food, offerings and so forth to finance village rituals (generally the cult of the community saints). Since these responsibilities are circulated periodically among those able to pay, the community simultaneously obtains a way to demonstrate and enhance its solidarity as well as to level wealth distinctions within its membership. Unlike a market economy, in which the objective of production is capital accumulation, such "nonproductive consumption" is part of the prestige economy of the Indian community because the only means to achieve local authority and reputation is continuous disaccumulation by individuals (Varesse 1986).

As a result of acculturation processe, in Pátzcuaro and in other P'urhépecha regions, all these mechanisms tend to have been replaced by more individualistic forms of consumption. However in the majority of indigenous communities, vigorous internal prestige systems represented by various forms of social, religious, and political participation still predominate. To date, there are several studies, that document prestige systems among the P'urhépecha communities of both the Lake Pátzcuaro Basin (Ihuatzio, Santa Fé, Pichátaro, San Gerónimo) and other regions, e.g., Cherán (Beals 1946, van Zantwijk 1967, Gortaire 1971, Young 1981). In Ihuatzio, for instance, each head of household is obligated to participate in activities of five different social institutions of the community: the nuclear and extended families, the kinship group, the wapánekwa (a communal work organization), and the *barrio* (ward) (van Zantwijk 1967). Since individuals and households must expend a great deal of effort and goods to sponsor ceremonials and community activities that serve to undergird and exemplify the solidarity of their Indian communities, production processes are not oriented toward capital accumulation but rather toward sustaining the village or community.

The Multiple Use Strategy of Natural Resources Use

Traditional subsistence patterns of use of natural resources are based on multiple use of ecosystems (Toledo et al. 1976, Toledo 1980, 1990). As members of a subsistence non-commodity economy, indigenous people carry out production based more on the principle of resources and production diversity than on specialization. By utilizing more than one ecosystem, indigenous people achieve the integration of different practices, multidimensionality of human activities, and diversification of natural products obtained from each ecosystem. This strategy may operate as much on the level of the household as on that of the community or even an entire region. This pattern is carried out through time and space. Spatially, maximum utilization of all available ecosystems is sought. Temporally, the goal is to obtain a maximum number of essential products that each ecosystem offers throughout the year. Although subsistence production for livelihood does not exclude surplus production, the commodities exchange sphere remains subordinate to the aim of self-sufficiency. In this sense, production is based more on ecological exchanges (with nature) than on economic exchanges (with markets). Consequently, indigenous producers are obliged to adopt mechanisms that guarantee an uninterrupted flow of goods, materials, and energy from the natural environment. This multiple use strategy allows indigenous peoples to manage different ecogeographical units and different biotic and physical components, thus avoiding specialization in their natural spaces and productive activities. This explains why indigenous producers, in general, are not solely agriculturalists, even though agriculture is usually the central productive activity of indigenous households. Instead, it is always complemented (or replaced) by such practices as plant-gathering, forest product extraction, fishing, hunting, cattle raising, and handicrafts. The combination of these practices buffers indigenous households and communities against both market fluctuations and environmental changes and hazards. The same diversified arrangement tends to be reproduced at a microlevel with multi-species and multi-story crops (polyculture) favored over monoculture. In sum, variety in geographical, biological, ecological, and genetic terms is therefore the main feature of indigenous production, because variety itself is a risk-reducing mechanism.

In the Lake Pátzcuaro Basin where topographical, vegetational, and soil complexity result in a mosaic of terrestrial and aquatic resources, this diversified use of nature can be observed at the regional, community, and household levels. The P'urhépecha economy is based on several activities that include: seed-, vegetable- and tree-agriculture, fishing, aquatic and terrestrial hunting, gathering, cattle raising, aquatic and forestry extraction, and derived handicrafts (pottery, weaving based on aquatic plants, wheat and palm products; textiles; bakery and wooden products) (Toledo et al. 1980). Based on ecological, technological and socioeconomic criteria, it is possible to distinguish fourteen agricultural landscapes or systems that include three rainfed, one humidity, and six irrigated agricultural systems, as well as two silvicultural systems, and two home gardens (Mapes et al. 1991). In addition, it is possible to distinguish ten different lake fishing systems based on the instruments used for capture (Argueta et al. 1985). Finally, there is an enormous variety of products obtained through all these practices. For instance, during an ethnobiological study conducted in the Indian peasant market of Pátzcuaro, a total of 222 products corresponding to 138 species of plants, animals, and mushrooms were recorded (Mapes and Toledo 1991). All these products were obtained from species living or growing in four natural and four transformed ecogeographical units (Table 8.2).

There is a relative predominance of some production activities at the community level. However, the indigenous (as well as the mestizo) villages of the region tend to maintain a variety of biological resource use activities (Table 8.3). The exceptions are the fishing villages established on the three smallest islands of the lake (Janitzio, Tecuena and Yunuén) communities. A similar situation is found at the household level. In his detailed survey on the production activities of the mountainous village of Pichátaro, Alvarez-Icaza (1988) reported that 88% of the 559 families carry out a combination of different practices and only 12% are specialized units. This pattern tends to be repeated with more or less emphasis in other indigenous communities of the region, for example, La Pacanda (Toledo and Barrera-Bassols 1984), Ihuatzio (van Zantwijk 1967), Jarácuaro (Dinnerman 1974) and Santa Fé (Gortaire 1971).

Holistic Knowledge of Natural Resources

Figure 8.3 summarizes updated information about the P'urhépecha knowledge of natural resources of the Lake Pátzcuaro Basin (Mapes et al. 1981a, 1981b, Caballero and Franco 1983, Cortes and Mapes 1985, Argueta et al. 1985, Argueta 1988, Barrera-Bassols 1988). This figure shows that the indigenous cognitive corpus integrates a global

TABLE 8.2
Number of Plant, Mushroom, and Animal Species by Use Category and Ecosystem Type in Indigenous Markets of Pátzcuaro

	Natural Ecosystems				Modified Ecosystems			
	Lake	Forests	Shrub-lands	Grass-lands	Cultiv. Fields	Fallow Fields	Kitchen Gardens	TOTAL
Foods	19	9	6	5	15	9	26	86
Medicines	1	13	3	0	0	1	13	31
Ornamentals	0	2	0	0	1	0	2	5
Forages	0	0	0	0	2	0	0	2
Fuelwoods	0	4	0	0	0	0	0	4
Domestics	2	1	0	0	0	0	0	3
Work/Tools	0	4	0	0	0	0	0	4
TOTAL	**22**	**33**	**9**	**5**	**18**	**10**	**41**	**138**

Source: Mapes and Toledo 1991.

body of knowledge about nature, which is intrinsically related to the set of productive activities (praxis). This knowledge covers practically all the main environmental components including plants, animals, mushrooms, soils, vegetational units, and meteorological events (such as sky forms and moon phases). The high number of labeled species, soil types, and vegetational units (Table 8.4) is an indicator of the accuracy of indigenous knowledge.

Besides the nomenclatural dimension, the P'urhépecha mind houses a complex set of taxonomical systems which cover each one of the main groups of natural elements. Especially noteworthy is the case of soils (Figure 8.4) where, based on morphogenetic, physical, and biological criteria, the P'urhépecha distinguish 18 soil types and are capable of recognizing, naming, and describing each horizon in the soil profile (Barrera-Bassols 1988). Also notable is P'urhépecha classification of pines and oaks (J. Caballero, pers. comm.), maize (Figure 8.5) and other domesticated plants (Mapes 1987, Mapes et al. 1991), mushrooms (Figure 8.6), aquatic plants (Cortes and Franco 1983) fishes (Figure 8.7), and other animals (Argueta 1988).

As theoretically expected, indigenous knowledge of nature is not circumscribed by structural aspects of landscapes (biological and physical components). On the contrary, there is a rich repertoire of

TABLE 8.3
Productive Activities in the Indigenous Communities of the Pátzcuaro Basin*

Communities:	A	B	C	D	E	F	G	H	I	J	K	L	M	N	O	P	Q	Tot
Urandén Morales	-	2	3	1	4	-	-	-	-	-	-	-	-	-	-	-	-	4
Urandén Morelos	-	-	-	1	2	-	-	-	-	-	-	-	-	-	-	-	-	2
San Bartolo Pareo	2	1	4	-	5	-	-	-	-	-	-	-	-	-	-	-	3	5
San Pedro Pareo	2	1	4	-	5	-	-	-	-	-	-	-	-	-	-	-	3	5
Arócutin	1	-	-	3	-	-	4	-	-	-	-	-	-	2	-	-	-	4
Uricho	1	-	-	-	-	-	4	3	-	-	-	-	-	2	-	2	-	5
Puácuaro	1	-	-	2	-	-	-	-	-	-	-	3	-	-	-	-	-	6
Tziróndaro	1	-	-	2	-	-	4	-	3	-	-	2	-	-	-	-	4	6
Purenchépuaro	1	-	-	2	-	-	4	-	3	-	-	2	2	-	-	-	-	6
Santa Fé	2	-	4	5	-	-	3	-	-	-	1	-	-	-	-	-	-	5
Ichupio	3	-	4	1	-	-	4	-	-	5	-	2	2	-	-	-	-	7
Tarerio	3	-	4	1	-	-	4	-	-	5	-	2	2	-	-	-	-	7
Ucasanáscua	3	1	2	3	-	-	4	-	-	-	-	2	2	-	-	-	-	7
Espíritu	3	2	4	1	-	-	-	-	-	-	-	3	3	-	-	-	-	6
Cucuchuchu	2	2	3	3	-	-	-	-	-	-	-	1	-	-	-	-	-	5
Ihuatzio	1	-	3	2	-	-	-	-	-	-	-	1	-	-	-	-	-	4
Jarácuaro	2	-	4	2	2	2	-	-	3	-	-	-	-	-	1	-	-	6
Janitzio	-	-	3	2	2	-	-	-	-	-	-	-	-	-	-	-	-	4
Tecuena	-	-	-	1	-	-	-	-	-	-	-	-	-	-	-	-	-	1
Yunuén	-	-	-	1	-	-	-	-	-	-	-	-	-	-	-	-	-	1
La Pacanda	2	-	-	1	-	-	-	-	-	-	-	-	-	-	-	-	-	2
Pichátaro	1	4	3	-	-	7	4	-	2	-	-	-	-	5	-	6	3	9

*Numbers indicate the level of importance of each practice in the community: A: seed culture, B: vegetable culture, C: tree culture, D: fishing, E: aquatic hunting, F: terrestrial hunting, G: gathering, H: aquatic extraction, I: terrestrial extraction, J: bee culture, K: pottery, L: aquatic plant handicrafts, M: wheat products, N: wood products, O: palm products, P: textile products, Q: bakery products, Tot: total activities.

knowledge in P'urhépecha culture that refers to relationships among various biological species or physical elements as well as to geographical, ecological, or other biotic processes. The indigenous perception of nature includes knowledge of the habitats of plants and animals, food webs of aquatic and terrestrial animals, flowering periods, and life cycles of plants, fungi or animals (for example, metamorphosis), periodic changes of the lake level and ecological succession of forests (Argueta 1988, J. Caballero, pers. comm.).

The last salient feature of P'urhépecha indigenous knowledge is its essentially utilitarian character. Nearly all environmental knowledge is linked with one or more practical activities. Thus, ecological

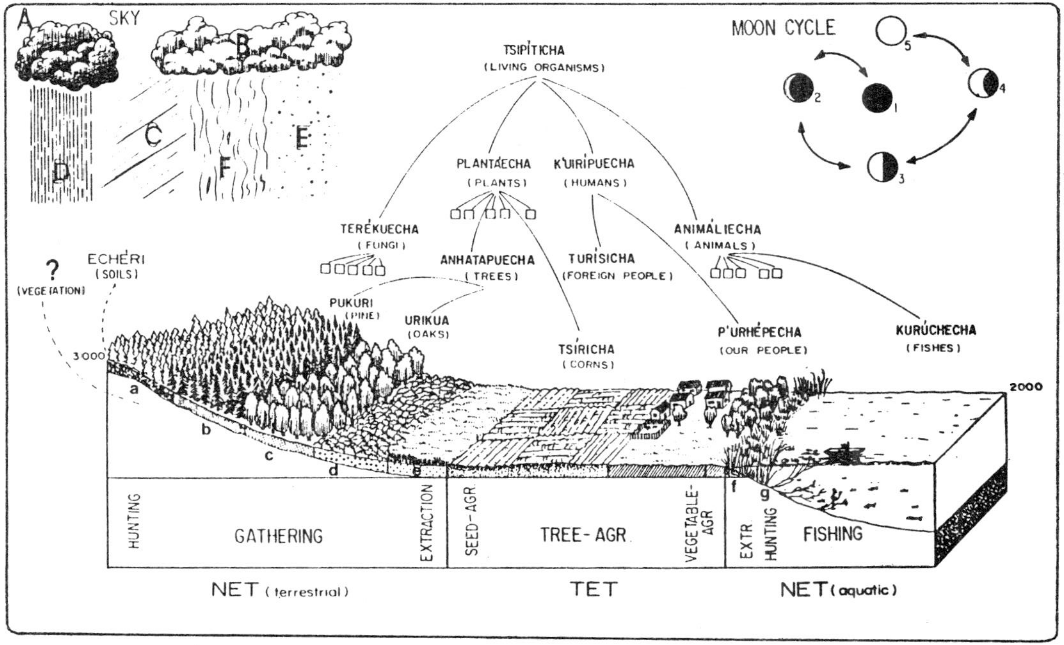

FIGURE 8.3 P'urhépecha view of the natural resources of the Pátzcuaro Basin. (Based on Argueta 1988, Barrera-Bassols 1988, Mapes et al. 1981a, Toledo et al. 1980)
(NETt: terrestrial natural environment; TET: transformed environment; NETa: Aquatic natural environment. Moon phases: 1: *Kutsi k'amarastia* (new moon); 2: *Kusti uentania sapichu*; 3: *Kutsi k'epekhuristi*; 4: *Kutsi k'epekumisatia*; 5: *Kuusi uiripiti* (full moon). Sky: A: *Xumu turipiti* (black cloud); B: *Xumu urapiti* (white cloud); C: *Tariat* (wind); D: *Janikua* (rain); E: *Shanuata* (hail); F: *Xuanda* (vapour). Vegetation: a: *Kumchkaricharu* (fir forests); b: *Pukuricharu* (pine forests); *Urikuacharu* (oak forests); d: *Anatapuicharu sapicharu* (shrublands); e: *Ambakuri* (grasslands); f: *Tupataru ka patzimuru* (tule reed marsh); g: *Uitsakuicha* (floating hydrophytes))

TABLE 8.4
P'urhépecha Knowledge of Natural Elements as Indicated by Known or Estimated Taxa and Labeled Species

Natural Elements	Number of Taxa (Known or Estimated)	Number of Labeled Species
Taxonomic Category		
Plants	1000	>400
Mushrooms	134	53
Mammals	49	31
Reptiles	25	13
Amphibians	8	8
Birds	136	75
Fishes	14	11
Ecosystem Component		
Vegetation units	8	8
Soil units	18	18

Sources: Mapes et al. 1981, Argueta 1988, Barrera-Bassols 1988, C. Mapes, pers. comm., J. Rzedowski, pers. comm.

morphological, and behavioral knowledge about mammals, aquatic birds, and fishes serves as the intellectual means to realize the operational steps of terrestrial hunting and aquatic fishing. In the same way, pottery and weaving activities are based on the repertoire of knowledge about soils and aquatic plants. Finally, the accurate knowledge of 355 species of wild and cultivated plants and mushrooms is linked to agricultural and gathering activities. A total of 224 plant and mushroom species are utilized in the Lake Pátzcuaro region as medicines (90 spp.), foods (30 spp.), household utensils (20 spp.), firewood (16 spp.), and/or ornamentals (12 spp.) (Caballero and Mapes 1985).

PÁTZCUARO'S LESSON: WHO TEACHES WHOM?

The conservation success of a human group inhabiting a given area appears to be directly linked to its ability to use and manage natural (biotic and non-biotic) resources correctly. Natural resources are utilized as the material basis of the process of production. The goal of correct natural resource management is to resolve the conflict

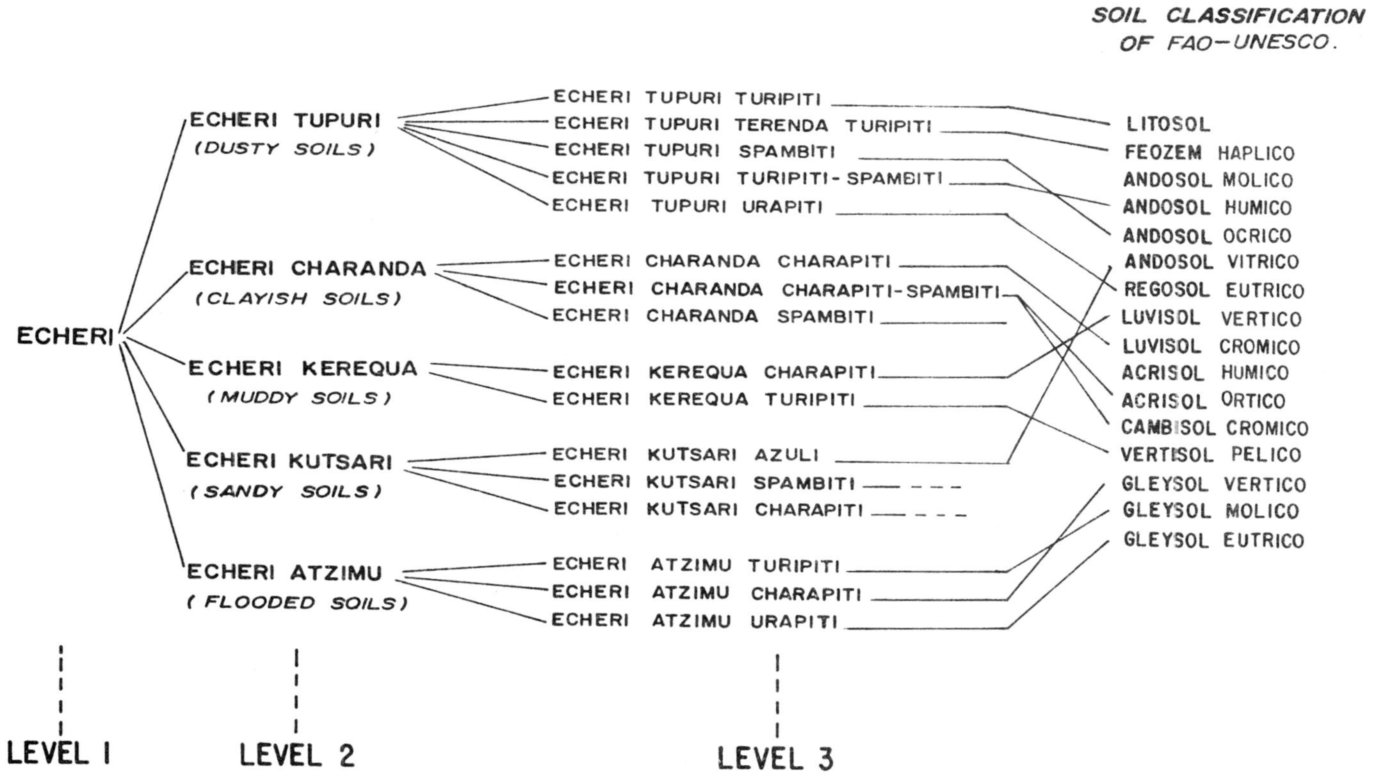

FIGURE 8.4 Soil classification in the P'urhépecha culture. (After Barrera-Bassols 1988)
(1st level: Soil taxon; 2nd level: Primary lexemes showing textual criteria; 3rd level: Secondary lexemes indicating color criteria)

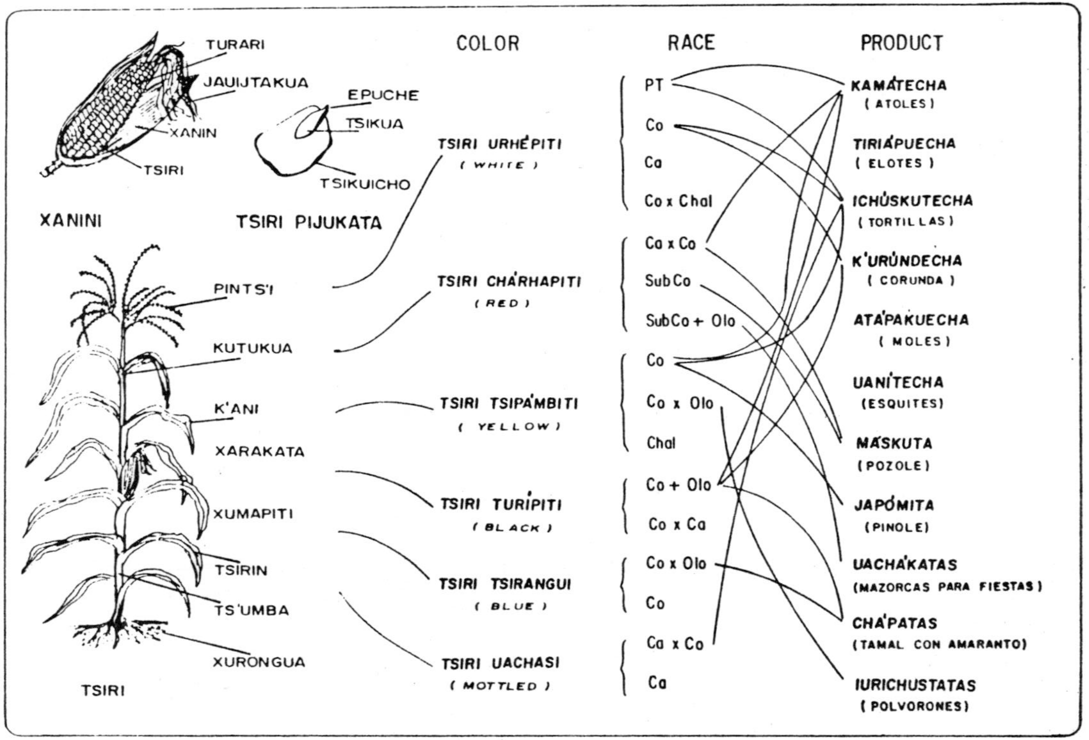

FIGURE 8.5 Morphology, classification, and uses of maize (*'Tsiri'*) in the P'urhépecha culture. (After Mapes 1987)
(Races shown are: PT: *Palomero toluqueño*; Co: *Conico*; Ca: *Cacahuacintle*; Chal: *Chalqueño*; Sub-Co: *Conico* sub-race)

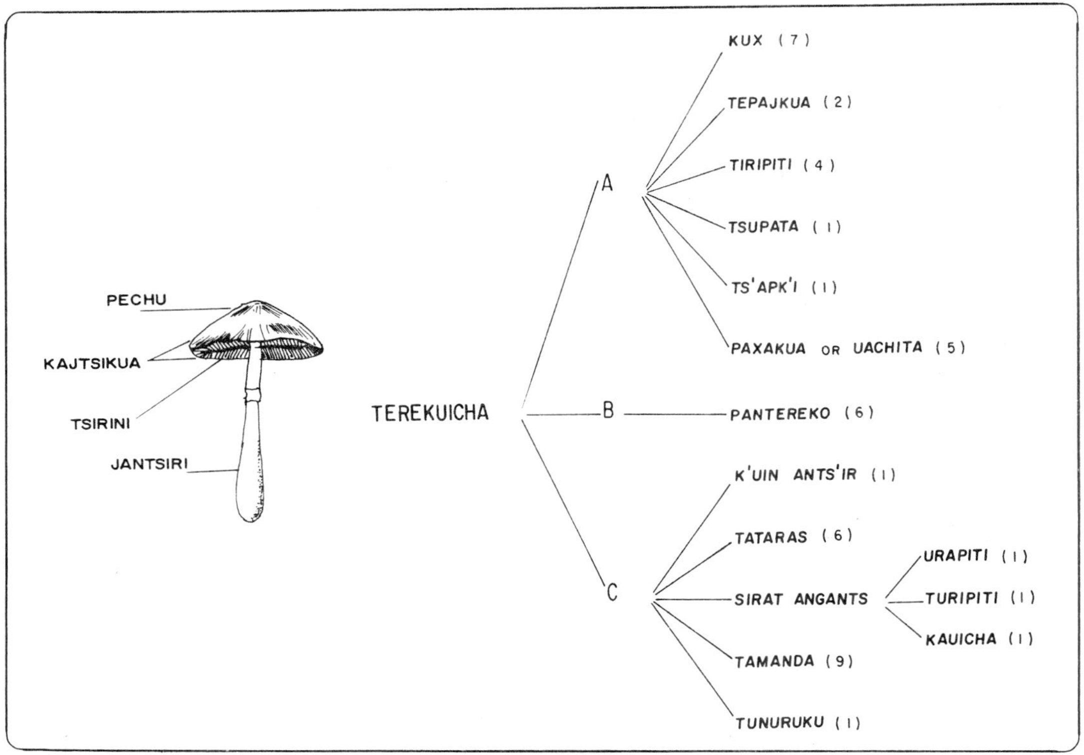

FIGURE 8.6 Classification of mushroom (*'Terekuicha'*) species in the P'urhépecha culture. (After Mapes et al. 1981b)

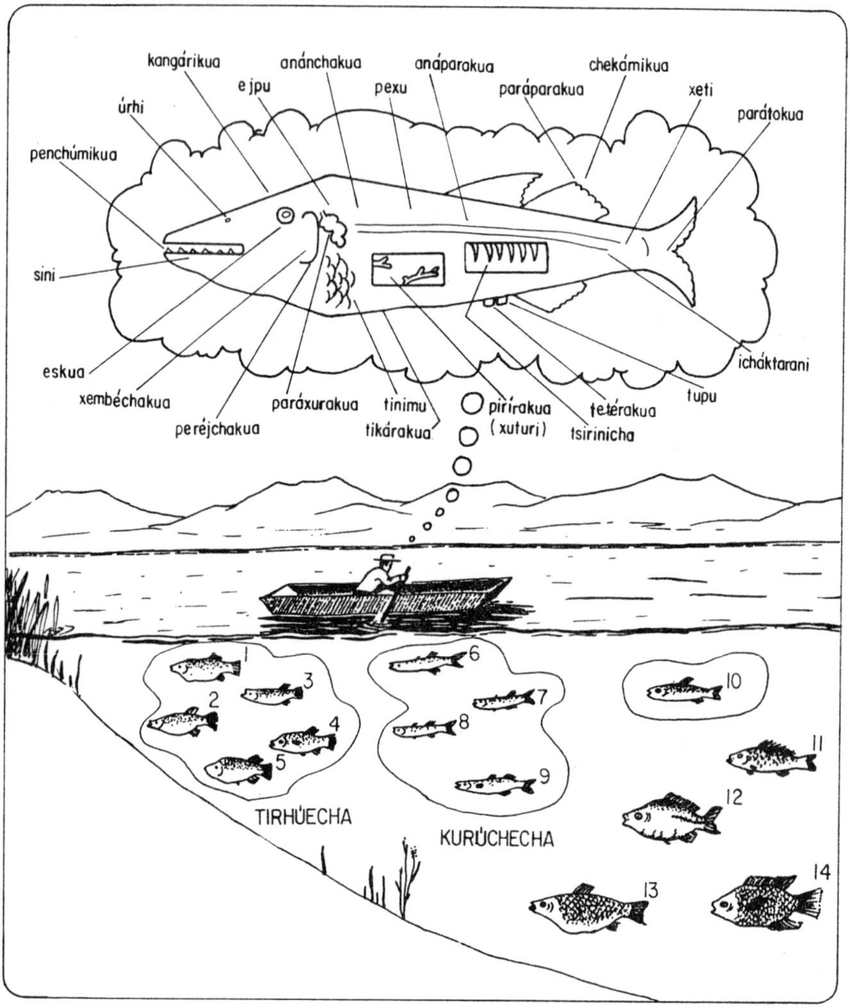

FIGURE 8.7 Classification and morphology of fishes (*'Kuruchecha'*) in the P'urhépecha culture. (After Argueta et al. 1985)

'Tirhuecha' (family Godeidae): (1) *'Choromu'* (*Noophorus diazi*), (2) *'Chehua'* (*Allophorus robustus*), (3) *'Tirhu'* (*Skiffia lermae*), (4) *'Tirhu sapichu'* (*Allotoca vivipara*), (5) *'Tirhu pitsupiti'* (*Goodea atripinnis*). *'Kuruchecha'* (family Atherinidae): (6) *'Chakuami'* (*Chirostoma grandocule*), (7) *'Kuerpo'* (*C. patzcuaro*), (8) *'Kuerepo turipiti'* (*C. bartoni*), (9) *'Kurucha urapiti'* (*C. estor*). Other species: (10) *'Akumara'* (*Algansea lacustris*), (11) *'Trucha'* (*Micropteris salmoides*), (12) (*Cyprinus carpio specularis*), (13) (*Ctenopharyngodon idellus*), (14) (*Tilapia melanopleura*).

between maintenance and exploitation. Sustainable long-term production must therefore avoid both extremes of uncontrolled exploitation (which often leads to disappearance of species, populations, soils and other resources) and conservative management (which leads to inefficient use). According to ecological principles, efficient management of a natural resource depends on the producer's capacity to recognize the potentialities and limits (thresholds) of the ecosystem(s) in the landscapes of his concrete territory. Thus, when the producer ignores the previous principle, production processes will provoke ecological constraints. As a result, the producer's action will be penalized by a decline in production or increased production costs ultimately. In summary, only production in concert–not in conflict–with the particular characteristics of ecosystems will allow persistence of the resource base over time.

The indigenous communities of the Lake Pátzcuaro Basin possess a set of intrinsic features that have favored conservative uses of nature, thus allowing continuous maintenance of production processes. Collective access to natural resources produces a collective concept of biological elements, such as species which are visualized as patrimonial goods that must be used prudently by individual producers, rather than as mere commodities. The tendency toward equitable distribution of natural resources also prevents overuse of the resource base by a few producers. In a similar way that the prestige economy induces continuous capital disaccumulation, the entire community executes control over each producer, thus avoiding economic overexploitation of natural resources and inhibiting detrimental environmental activities. These collective or communally based mechanisms also serve to control exploitative intrusions by people or enterprises external to the indigenous community. In summary, by exercising control over its own members, the community executes a set of regulations over the natural resources which serve as the basis for continuous satisfaction of its own material needs.

If collective appropriation and distribution of natural resources within indigenous economies serve as regulatory mechanisms to avoid resource abuse and environmental destruction, then the strategy and knowledge used to manage natural resources should tend to diversify their exploitation. As multiuse strategists, indigenous producers are obliged to decode the variety of local microenvironments, understand the potential resources encapsulated in these microhabitats in great detail, and recognize the limits of the natural resources utilized during production processes. The richness and accuracy of indigenous knowledge of natural landscapes serves as the intellectual means to conduct these operations. In contrast to the specialized mode of biological resource use in modern economies, where comparatively few resources are perceived and recognized as having production value, under the indigenous view all of the components of natural landscapes are directly or indirectly useful or usable resources. This aspect alone tends to favor a conservationist attitude toward the environment. Moreover, it seems clear that the multiple use strategy of indigenous communities is an effective mechanism for preserving and even increasing biological diversity by increasing habitat heterogeneity.

Perhaps the most difficult problem faced today in the effort to accomplish biological conservation in the Third World is the inability of educated people such as scholars, conservationists, and policy-makers to recognize and comprehend the enormous ecological value of the modes of subsistence of traditional cultures that inhabit rural areas. Most recent evidence suggests that indigenous peoples play a key role in the conservation of biological diversity. The case of Pátzcuaro Basin, where indigenous settlements and ecosystems have persisted despite 35 centuries of continuous human use, demonstrates this principle. That is the lesson!

ACKNOWLEDGMENTS

This paper has been based on discussions with the following colleagues: Pedro Alvarez-Icaza, Arturo Argueta, Narciso Barrera-Bassols, Javier Caballero, and Cristina Mapes, all of whom have made important contributions to our knowledge of indigenous perception and use of Pátzcuaro natural resources. I thank Celia de Ita, N. Barrera-Bassols, and especially Ronald Nigh and Margery L. Oldfield for correcting the English or improving the form and content of earlier versions of this manuscript. I am indebted to Ana Irene Batis and Juan Almazan for their illustrations for the figures.

This research was sponsored by the Instituto de Biología of Universidad Nacional Autónoma de México (UNAM), with the financial support of the Mexican Ministery of Education (Dirección General de Culturas Populares) and the Ford Foundation.

REFERENCES

Alvarez-Icaza, P. 1988. Economía campesina y agricultura indígena tradicional en la Región P'urhépecha. Thesis Maestría en Desarrollo Rural. UNAM-Xochimilco, México.

Argueta, A. 1988. Etnozoología P'urhépecha. Historia, utilización y nomenclatura P'urhépecha de los animales. Thesis Biólogo, Facultad de Ciencias, Universidad Nacional Autónoma de México, México City.

Argueta, A., M. Merino, T. Zubieta, S. Campos and J.L. Chavez. 1985. Japondarhu Anapu, o de la pesca en los lagos Michoacános. *Cuadernos de la Casa Chata* 122:1-197. CIESAS, México.

Barrera-Bassols, N. 1986. La Cuenca del Lago Pátzcuaro, Michoacán: aproximación al analisis de una región natural. Thesis Geografía, Facultad de Filosofía y Letras, Universidad Nacional Autónoma de México, México City.

__________. 1988. Etnoedafología P'urhépecha: Conocimiento y uso de los suelos en la cuenca de Pátzcuaro. *México Indígena* 24:47-52.

Beals, R.L. 1946. *Cherán: A Sierra Tarascan Village*. Smithsonian Institution, Institute of Anthropology, Publication No. 2, Washington, D.C.

Caballero, J. and C. Mapes. 1985. Gathering and subsistence patterns among the P'urhépecha Indians of México. *Journal of Ethnobiology* 5:31-47.

Cortes, M.E. and V. Franco. 1983. *Artesanía de plantas acuáticas en la región del Lago de Pátzcuaro, Michoacán*. Cuadernos de Etnobiología 3. Dirección General de Culturas Populares, SEP, México.

Dinnerman, I. 1974. *Los Tarascos, campesinos y artesanos de Michoacán*. Colección Sep-Setentas 129, Secretaría de Educación Publica, México.

Foster, G.M. 1948. *Empire's children: The people of Tzintzuntzan*. Smithsonian Institution, Institute of Social Anthropology, Publication No. 6. Washington, D.C.

Gorenstein, S. and H.P. Pollard. 1983. *The Tarascan civilization: A late prehispanic cultural system*. Vanderbilt University Publications in Anthropology 28. Nashville, Tennessee.

Gortaire, A. 1971. *Santa Fé: Presencia etnológica de un pueblo-hospital*. Universidad Iberoamericana, México.

Lot, A. and A. Novelo. 1988. Vegetación y flora acuatica del Lago de Pátzcuaro, Michoacán, México. *The Southwestern Naturalist* 33:167-175.

Mapes, C. 1987. El maíz entre los Purepecha de la cuenca del lago de Pátzcuaro. *America Indígena* 47:345-379.

Mapes, C., G. Guzman and J. Caballero. 1981a. *Etnomicología P'urhépecha: El conocimiento y uso de los hongos en la cuenca del lago de Pátzcuaro*. Cuadernos de Etnobiología 2. Dirección General de Culturas Populares, Secretaría de Educación Publica, México.

__________. 1981b. Elements of the Purepecha mycological classification. *Journal of Ethnobiology* 1:231-237.

Mapes, C., V.M. Toledo, N. Barrera-Bassols and J. Caballero. 1991. La agricultura en una región indígena: La cuenca del lago de

Pátzcuaro. *In*: T. Rojas (ed.), *La Agricultura Indigena: Pasado y Presente*. CIESAS, México.
Mapes, C. and V. M. Toledo. 1991. The indigenous market of Pátzcuaro (México): An ethnobiological study. *Journal of Ethnobiology*.
Mazari, M. 1981. Evaluación de la calidad del agua desde el punto de vista bacteriológico y fisicoquimico en el lago de Pátzcuaro, Michoacán. Thesis Biólogo, Facultad de Ciencias, Universidad Nacional Autónoma de México, México City.
McCay, B.J. and J.M. Acheson (eds.). 1987. *The Question of the Commons: The Culture and Ecology of Communal Resources*. University of Arizona Press, Tucson.
Moone, J.R. 1973. *Desarrollo Tarasco: Integración nacional en el occidente de México*. Instituto Indigenista Interamericano, México.
Nietschmann, B. 1982. Indigenous island peoples, living resources and protected areas. Paper presented at the International Union of Conservation of Nature. World National Parks Congress, Bali, Indonesía.
Saavedra, J. 1982. Macroinvertebrados bentónicos como indicadores de la calidad del agua en el lago de Pátzcuaro, Michoacán. Thesis Biólogo, Facultad de Ciencias, Universidad Nacional Autónoma de México, México City.
Smith, W.R. 1977. *The Fiesta System and Economic Change*. Columbia University Press, New York.
Tamayo, J.L. and R.C. West. 1964. The hydrography of middle America. Pages 84-121 *in* R.C. West (ed.), *Handbook of Middle American Indians*. University of Texas Press, Austin.
Toledo, V.M. 1980. La ecología del modo campesino de producción. *Antropología y Marxismo* 3:35-55.
__________. 1990. The ecological rationality of peasant production. *In*: M. Altieri and S. Hecht (eds.), *Agroecology and Small Farm Development*. CRC Press, Boca Raton, Florida.
Toledo, V.M. and N. Barrera-Bassols. 1984. *Ecología y Desarrollo Rural en Pátzcuaro*. Instituto de Biología, Universidad Naciónal Autónoma de México, México City.
Toledo, V.M., A. Argueta, P. Rojas, C. Mapes and J. Caballero. 1976. Uso multiple del ecosistema, estrategias de ecodesarrollo. *Ciencia y Desarrollo* 11:33-39.
Toledo, V.M., J. Caballero, C. Mapes, A. Argueta, N. Barrera and M. A. Nunez. 1980. Los purepechas de Pátzcuaro: Una aproximación ecologica. *America Indígena* 40:17-37.
van Zantwijk, R.A.M. 1967. *The Servants of the Saints: The Social and Cultural Identity of a Tarascan Community in México*. Royal Van Gorcum, Assen, Netherlands.

Varesse, S. 1986. Multietnicidad y construcción hegemonica: Proyectos índios y futuro global. *La Palabra y el Hombre* 57:23-40.

Velasco, A. 1982. Evaluación de la calidad del agua con base en el fitoplancton del lago de Pátzcuaro, Michoacán. Thesis Biólogo, Facultad de Ciencias, Universidad Nacional Autónoma de México, México City.

Watts, W.A. and J.P.Bradbury. 1982. Paleoecological studies at Lake Pátzcuaro on the west central Mexican plateau and at Chalco in the Basin of México. *Quaternary Research Review* 17:56-70.

West, R.C. 1948. *Cultural geography of the modern Tarascan area.* Smithsonian Institution. Institute of Social Anthropology Publication 7. Washington, D.C.

Young, J.C. 1981. *Medical Choice in a Mexican Village.* Rutgers University Press, New Brunswick, New Jersey.

9

Wildlife Conservation in Tribal Societies

Raymond Hames

INTRODUCTION

It is widely alleged (e.g., Bodley 1976, Martin 1978, Clad 1984) that unacculturated tribal peoples lived in harmony with the environment. This very general concept is most frequently used to imply that aboriginal use of the environment approached a steady state such that demands for renewable resources did not exceed environmental replenishment. This claim is most forcefully made in reference to exploitation of wild game, fish, and plant resources (Sahlins 1968) by hunter-gatherers or low density shifting cultivators (e.g., native Amazonians) with high dependence on wild species. Three lines of independent evidence are used to support this claim. First, there is little contemporary evidence (until recently) that relatively unacculturated native tribal populations have caused local extinctions of fish and game species upon which they depended for their subsistence.

Second, the population densities of many such tribal groups appear to be in equilibrium, i.e., are not substantially increasing or decreasing. For example, Birdsell (1958, 1968) has argued that an equilibrium density of Amazonian aboriginal hunter-gatherers is achieved by cultural means such as infanticide, abortion, and sexual regulations that prevent individual groups from placing heavy demands on the ecosystem. Sahlins (1968, 1972) has similarly argued that, in addition to cultural population regulation mechanisms, culturally determined desires or wants in tribal societies are moderate and thus prevent tribal people from overtaxing their resource base. These twin processes of population regulation and moderate desires led to the maintenance of tribal populations at approximately 50% of the carrying capacity of the environment (Sahlins 1972).

Third, native ideological systems have been adduced as evidence for wildlife conservation. Martin (1978) argued that reverence of game animals in many North American native societies was designed specifically for conservation. Reichel-Dolmatoff (1976) suggested for the Colombian Tukano and many other native Amazonian groups that religion promotes a balance between humans and wildlife by preventing humans from overexploiting game resources. McDonald (1977) and to some extent Ross (1978, but see below) suggested that among many Amazonian groups, animal taboos represent an effective mechanism designed to prevent overhunting.

It is easy to conclude that native peoples were or are conservationists given the paucity of contemporary evidence that native peoples caused local extinction of game species, the fact that such societies apparently exist below the carrying capacity of their environment and that the veneration of game animals is a common ideological element in their societies. Even negative evidence for conservation among native people has been used to demonstrate aboriginal conservation. There is clear evidence that the introduction of guns to Plains Indians led, in part, to local extinctions of bison (Ewers 1955). However, since the gun was non-native, native peoples did not have adequate time to adjust to this remarkable advance in hunting technology. Among subarctic Indians, the steel trap and a commerical market for furs led to the local decimation of valuable furbearers (Leacock 1954). In this case, there was not only a technological introduction, but more importantly, Indians were forced into a non-native economic system that stressed individualistic aggrandizement. In a convoluted thesis, Martin (1978) argued that North American native peoples began slaughtering the species they formerly conserved because of the introduction of European diseases. In a sense, these exceptions prove the rule: instances of non-conservation are the result of loss or disruption of aboriginal culture as a result of Western acculturation.

The aim of this chapter is to put the issue of native wildlife conservation on a firm empirical and theoretical foundation. A number of serious methodological, empirical, and theoretical problems with claims about native systems of conservation are discussed and illustrated with recent data on Amazonian groups—contemporary peoples who tend to have had the least contact with the Western world. The goal is not to generalize about the existence of native conservation, but rather to develop ideas about the conditions under which a conservation ethic will evolve, the kinds of evidence required to determine whether conservation is practiced in a particular society, and the degree to which native peoples may be able to successfully exploit game resources if their land becomes a park or reservation (Redford and Robinson 1985).

METHODOLOGICAL AND THEORETICAL PROBLEMS

Deducing conservation from ideology alone is dubious at best. Such analyses rest on no sure scientific foundation; this can lead to highly divergent interpretations. For example, the exchange between two of the best known anthropologists of recent times, Levi-Strauss (1973) and Harris (1976), on their radically different interpretations of the meaning of a particular Bella Bella myth clearly indicates the lack of consensus in this area of anthropological research. Harris suggested that the myth provides ecologically adaptive information on clam exploitation, whereas Levi-Strauss claimed it defines critical social relationships. Even if consensus is reached about the proper way to interpret the meaning of native ideology, one still must demonstrate how ideology provides an effective code for behavior. To bring this point home, what can one conclude about Christian North America by reading the Bible? White (1967) suggested that Christianity presents an anti-conservation ideology, while Berry (1979) claimed that the Bible has a conservationist program. The White-Berry and Harris-Levi-Strauss debates are clear demonstrations that diametrically opposite interpretations regarding conservation can be derived from a common ideological source. Tuan (1968), in a similar vein, points out that Confuscian ideology stresses harmony with the environment, yet this has not prevented widespread environmental degradation in China during premodern times. Finally, Kay (1985: 133) in a cross-cultural survey of environmental ethics notes "The historical record does not yet indicate a consistent or necessary relationship between ecological beliefs, practices and impacts."

Callicott (1990) recently reviewed a number of symposium papers dealing with the ideological bases of conservation among a variety of tribal peoples. He cogently shows that researchers' analyses of similar ideological systems claim a variety of motives behind conservation ranging from utilitarian (designed for optimum sustained yield), religious reverence (with conservation as a side-effect), ecological awareness (understanding environmental dynamics and wanting to preserve all regardless of economic importance), and environmental ethics (animals must be treated as if they were people, therefore one should follow the golden rule when dealing with game). Martin's (1978) widely cited thesis argues that precontact North American native peoples were conservationists and only began indiscriminate slaughter of the species they formerly conserved because of the introduction of European diseases.

A serious blow to Martin's 1978 thesis was provided by Brightman (1987). Using subArctic ethnohistoric data, he directly counters three of Martin's most fundamental claims about native North American conservation. While there is limited, early eyewitness evidence for conservation and management among Canadian Indians, Brightman says, "The above findings must be placed in the context of a far greater body of evidence for the lack of conservation and intentional management and for a proclivity to kill animals indiscriminately in numbers beyond what were needed for exchange or domestic use" (p. 123). Like the North American bison hunters, the Canadian hunters stripped delicacies (tongues, livers) from large game, leaving most of the carcass to rot. Ideologically, Cree and other Algonkian Indians believed that the more animals they killed, the more that would be available, or that animals immediately reincarnated themselves after proper disposal of blood and bones (Brightman 1987). Brightman argued that conservation did emerge in the early 1800s as a result of fur trading. For similar critiques of Martin's thesis as it relates to subArctic foragers, the edited collection of Krech (1981) should be consulted.

There are several lessons one can draw from attempts by researchers to use ideological data to make claims about the existence of conservation. First, there is no generally accepted way to ascertain the meaning of ideological information contained in myth and ritual. This is especially difficult when one is attempting to say something about the relationship between the Western concept of conservation and native concepts. Second, there should be some correspondence between a belief system and the behavioral system it allegedly determines, but examination of this relationship has low priority for those who analyze ideology. Most importantly, if the behavior of the people in question satisfies stipulated theoretical expectations, this is sufficient to establish the reality of conservation. If a people have a conservation ideology but do not act as conservationists, then they are not conservationists. Conservation deals with how humans treat the resources in their environment. How a hunter feels about the deer he hunts and how those feelings may have been engendered by his religion are interesting psychological questions. But the hunter's actual treatment of deer is the only thing that matters.

Brightman (1987) and Berkes (1987) suggested that ideological and behavioral evidence must be combined to determine whether conservation or management of wild-derived resources occurs. They argued that for conservation to occur, hunters must understand the relationship between hunting activities and game depletion (see also Nelson [1979] on the Koyukon) and must be motivated to regulate their activity in order to maintain ecological balance. That is, there must be some sort of conscious intent. Although conscious intent based on some understanding of game population dynamics may enhance the probability of successful conservation or management, at the same time, it is neither sufficient nor necessary for conservation. Conservation is a matter of performance–not intent.

According to Brightman (1987:131), modern Cree who engage in conservation believe "that ritual procedures for disposing of animal bones and blood prefigure and influence animal regeneration and reincarnation." But such rituals need not have any affect on game populations. Suppose, for example, that the sparing of a mated pair of beavers was done by Cree to avoid supernatural punishment (Brightman 1987:123). This sort of belief is presumably adaptive because it will serve to effectively maintain the breeding stock, and thus the beaver population. The relationship between supernatural retribution and killing of a mated pair of beavers is false. The ideological rationale for engaging in a practice, need not correspond to ecological reality. It is merely sufficient that the belief leads to behavior that has the effect of conservation or management. Brightman (1987) and Berkes (1987) appeared to be troubled by a lack of agreement between scientific and local native explanations of native behavior. Brightman, for example, noted that "The Cree are not practitioners of scientific reductionism" (p. 89). The evolutionary-ecological explanations pursued here are not models of how natives think any more than optimal foraging theory is a model how a praying mantis perceives its prey (Charnov and Orians 1973).

Ideological approaches to the problem of conservation in tribal society have suffered from a number of theoretical shortcomings because they have not adequately dealt with conditions, maintenance, design, and alternative hypotheses. Almost invariably, the conditions under which conservation should evolve have not been specified. It is never made clear why conservation is practiced by one society, but not another. The cause of conservation is often seen as a group's ethos or ideology, but the conditions that favor the development of such belief systems are never made specific. How conservation is maintained by insuring that people do not cheat is often not examined. Many studies simply have assumed that people are slaves to ideology. Other studies (see below) pointed out, for example, that the consumption or killing of a taboo animal will lead to sickness. However, these same studies indicate that not everyone follows the rules, and in most tribes (as opposed to chiefdoms and states [Service 1975] and below), legal enforcement of such rules is impossible.

Finally, most of the ideological studies assumed implicitly or explicitly that conservation is adaptive for the group. If a behavior is adaptive, then one must show that it is well designed for the task and it is not the result of some other process (Williams 1966). This failure has already been noted by Callicott (1990). For the last 20 years, modern evolutionary biology has generally recognized that behavioral or other traits do not evolve in a population because they enhance the well-being of the group (i.e, as a consequence of group selection). Instead they are hypothesized to persist primarily because they are advantageous for individuals within populations. Anthropologists Lees and Bates (1979) have convincingly shown that population regulation models such as those proposed by Birdsell and Sahlins (cited above), are based on a group selection model. The following sections present my own recommendations about proper method, data evaluation, and theory in dealing with the complex problem of conservation.

COMPARATIVE CASE STUDIES OF TRIBAL FORAGING STRATEGIES IN AMAZONIA

The best way to determine whether aboriginal humans practiced deliberate conservation, as opposed to conservation as a side-effect of poor technology or low population density, is to examin archaeological, paleontological, and ethnological behavioral data on the relationship between human predation and demography of game populations. The archaeological and paleontological data are sparse and controversial, while the ethnographic data are extensive. As a result, the prehistoric data will be quickly reviewed, and there I will focus on recent Amazonian tribal data. The Amazonian research data were chosen for several reasons. First, most groups studied have had weak to no contact with external markets such that their hunting and fishing activities are largely aimed at domestic consumption. Second, low population density Amazonian populations have achieved little in the way of landscape modification that, in itself, has serious consequences for game populations. And third, Amazonian researchers have collected some of the finest behavioral data available on human predation (Hames and Vickers 1983a).

The idea that humans were in large part responsible for the extinction of Pleistocene megafauna in North America was first put forward in a series of publications by Martin (1967a, 1967b, 1973, 1984) in what has become known as the "Pleistocene Overkill" hypothesis. This hypothesis was based on a correlation between human entry and expansion into the New World and the timing of megafaunal extinctions. Martin forcefully argued that North American megafauna succumbed because they had not evolved adaptations to cope with the sudden appearance of this new "superpredator." Initially, this hypothesis raised a great deal of controversy, but more recently, a large number of paleontologists appeared to agree that humans may have played at least a partial role in these Pleistocene extinctions (see Diamond 1982, 1984, 1986, and contributions in

Martin and Klein 1984). However, recent archaeological evidence on hu-man antiquity in the New World suggests that Martin will have to modify the 'blitzkrieg' portion of his hypothesis (Guidon and Delibrias 1986).

Perhaps the best archaeological and paleontological data on human-caused extinctions of animal species comes from islands. Data from New Zealand on the extinction of moas (Anderson 1984, Trotter and McCulloch 1984), Madagascar–on prosimians (Richard and Sussman 1975, Dewar 1984), and insular Mediterranean–on pygmoid relatives of mainland animals (Davis 1987) suggest that human hunting (as opposed to landscape modification from agriculture) has led directly to extinction. The balance of the island studies indicates that hunting, coupled with landscape modification through agriculture plus the introduction of foreign or exotic species (e.g., pigs and rats), caused the extinction of numerous Hawaiian bird species (McGleone 1983, Olson and James 1984), birds and marsupials in Melanesia (Cassels 1984), and a variety of vertebrates in the West Indies (Steadman et al. 1984). That insular extinctions seem to be more common than their mainland counterparts may result from several factors, some of which may interact. First, islands have lower species diversity than mainland areas of comparable latitudes (MacArthur and Wilson 1967), forcing hunters to take a more restricted range of game. Second, since islands are small, delimited ecosystems, local extinction is tantamount to total extinction because recruitment from neighboring islands is difficult over the short term. And third, human population density is often high in island ecosystems, leading to extreme pressure on local resources. Because of their small size and limited biological diversity, island ecosystems intensify human ecological and evolutionary constraints and enhance probabilities of extinction.

The best available way to determine whether modern day tribal populations practice conservation is to observe their behavior in the face of declining resource yields (Redford and Robinson 1985). Unfortunately, today there are no tribal populations that have not been affected by nation states. Thus, it is impossible to claim that such populations operate under the same conditions that have characterized humanity for most of its history as a species. Nevertheless, tribal populations that engage in little or no cash cropping and manage their own political affairs provide the clearest view of how social and economic life were conducted prior to the advent of the nation state. One could argue that technological introductions, e.g., use of shotguns instead of bows (Hames 1979) or the introduction of Christianity, radically alters tribal environmental relations. But the most dramatic kind of environmental change occurs when resource procurement ceases to be aimed at subsistence and is instead market-driven, especially when accompanied by loss of a traditional authority that regulated resource exploitation. That is, demand for resources is not based on local nutritional or technological needs, but rather on the opportunities presented by an external market.

The data presented below on hunting and fishing efficiency and time allocation come from Amazonian tribal populations that are fundamentally subsistence-based economies, i.e, they directly appropriate resources to satisfy local needs rather than the needs of an external market. All things being equal, resource depletion is governed by time allocated to resource acquisition. If human foragers are interested in decreasing resource depletion and thus conserving resources, then time allocated to acquisition should decline at some point as resources are depleted to allow fish and game resources to rebound. If, on the other hand, foragers have no conservation strategy and instead are interested in maintaining a relatively stable level of fish and game consumption, then time allocated to fishing and hunting should increase as depletion increases. While there are many studies that have documented time allocated to hunting and fishing, there are none that have coupled such observations with direct measures of fish and game densities. However, if hunting and fishing efficiency (kg of game or fish harvested per hour of foraging effort) is a fair index of protein density, then comparative data are available. A positive correlation between hunting/fishing time and hunting/fishing efficiency should provide evidence for conservation, while a negative correlation between the two should indicate lack of a conservation strategy. In a previous study (Hames 1989:66), I demonstrated a significant negative correlation between foraging time and foraging efficiency in a sample of 15 native Amazonian groups. Thus, generally, Amazonians react to depletion by increasing time allocated to hunting or fishing. While this correlation provides evidence against a conservation ethic, it is by no means conclusive. One could argue that levels of game depletion have not yet reached sufficiently critical levels for native hunters to reduce their predatory impact on fish and game populations. Such crosscultural comparisons do not have the inherent controls found in studies of neighboring village populations of the same culture, or in longitudinal comparisons of foraging activities of a single village over time. While local tests are less general than crosscultural tests, they are likely to be more reliable. Below, a number of local level studies are analyzed to evaluate the relationship between time allocation and foraging efficiency.

A number of Yanomamö villages in Brazil experienced fish and game depletion resulting from the building of a spur of the Trans-Amazon highway. Fish and game were taken by road crews and were driven out by road building. Streams were blocked, diverted, and muddied by run-off (Saffirio and Hames 1983). Although the crews have departed, the damage remains. Saffirio collected data on time allocation and hunting and fishing returns of Yanomamö hunters and fishers who lived adjacent to the road after departure of the road crews. At the same time, he collected the same kind of data on hunters and fishers from a village unaffected by the road building. For the highway villagers, hunting and fishing time was greater than for forest villagers who had greater hunting/fishing efficiency. Therefore, the Yanomamö responded to resource depletion (albeit not of their own making) by increasing their exploitation pressure. A similar comparison can be made between different Yanomamö villages studied by Lizot (1978) and Colchester (1984). Those studied by Lizot gained game at a rate of 2.8:1 (kcal output per kcal input) and hunted 74 minutes per day, while those studied by Colchester gained game at about half Lizot's rate (1.8:1) but hunted about twice as intensively–at 139 minutes per day. Again, there is clear evidence of a negative correlation between hunting time and efficiency which indicates that the Yanomamö respond to game depletion by increasing their hunting efforts.

Baksh's (1985) longitudinal study of Machiguenga fishing clearly showed an increase in fishing time as yields declined. After the Machiguenga set up a new village, fishing returns were high because they exploited undepleted streams near the village. Over time, fish became depleted, and the villagers were forced to fish further from the village. In addition, villagers were forced to turn to labor-intensive poison (*barbasco*) fishing as yields declined. But even labor-intensive techniques produced diminishing yields. For example, one stream yielded 768 kg when it was first poisoned in the wet season (when poison is minimally effective). It was poisoned again during the dry season (when poison is maximally effective), but yielded less than 100 kg. Baksh concluded that the Machiguenga coped with declining yields by working longer hours. Other Amazonian groups such as the Ye'kwana also engage in frequent barbasco fishing. However, they will not poison the same stream twice in one year–not because they fear depleting resources, according to informants, but because the low yield will not be worth the effort (Hames 1978).

Vickers' (1980, 1988) ten years of longitudinal research on the Siona-Secoya of Ecuador provide the best available data on the process of game depletion over time. His data (1980) revealed a consistent negative correlation between hunting efficiency and time allocated to hunting. Furthermore, hunting led to near local extinction of woolly monkeys, curassows, and trumpeters in heavily exploited areas (Vickers 1988). Most importantly, he showed that game populations have reached an equilibrium (with the exception of the three species mentioned above) such that foraging efficiency and effort appear to have leveled. This important study suggests that the Siona-Secoya are attempting to maintain a stable game intake, and have not caused the extinction of basic game resources even though they have no apparent environmental ethic or conservation policy. One might conclude this is because of the low demand the Siona-Secoya make on game resources (which may be a function of Siona-Secoya population density).

Qualitative and quantitative data on the Ye'kwana and Yanomamö of the Venezuelan Amazon may provide some insight into the attitudes and behaviors these Indians manifest toward game species that supply them with 80% of their high quality protein intake (Chagnon and Hames 1979). The Yanomamö, like many Amazonian groups, have taboos against the consumption and/or killing of certain wild animals. For example, in most areas of their tribal distribution, capybara, deer, and otter are taboo. However, this does not prevent the killing of these animals should the situation present itself. Once, while traveling in a canoe with two Yanomamö companions, we spotted a herd of capybara on the shore. I was entreated to shoot one, which I did successfully. Upon disembarking, both of my companions began poking the dead animal with sticks. I asked who was going to butcher the animal, and was told that no Yanomamö would touch such a "filthy" animal, much less prepare it for consumption. I asked why they had asked me kill it? One said he wanted to see a big one up close, while the other said he wanted to see how it died.

On another occasion, I was hunting in the forest with a Ye'kwana. His dogs cornered a giant anteater–an animal tabooed as a source of food. After failing in his attempt to divert the dogs from their quarry in order to move on to proper game, he killed the anteater. He even prevented the dogs from eating the animal, claiming this would only encourage them to hunt it again and that it would cause them to become very ill just like a human. Many similar examples of treatment of taboo animals could be cited. Kracke, in reference to the Tupian Kagwahiv, stated "but their exclusion from the diet does not prevent Kagwahiv from killing them for their skins whenever they are encountered" (1981:109). The point is that at least among the Ye'kwana and Yanomamö, tabooing is not associated with conservation attempts (McDonald 1977, Ross 1978) or any general reverence for animal life.

Finally, Rambo's (1985) monographic account of the Malaysian Semai, shifting cultivators with a high reliance on foraging of wild animals, directed attention to the environmental impact a tribal population can have on its environment. Although the Semai are not Amazonian, they live in a similar environment and have similar ecological adaptations. Rambo documented widespread habitat destruction and befoulment, and noted that the switch from blowpipe hunting to the blunderbuss led to the local decimation of many game species (Rambo 1978).

A preponderance of the evidence indicates that Amazonian tribal populations make no active or concerted effort to conserve fish and game resources. At the same time, it is clear in most cases there may be no need for a conservation policy, because current local subsistence demands on resources have not led to severe resource shortages. At this point, it would be useful to formulate empirical hypotheses about how foragers should behave if their goal were one of conservation rather than maximization of foraging success, and specify the conditions under which a conservation ethic is likely to evolve.

DIFFICULTIES IN THE DEVELOPMENT OF EMPIRICAL HYPOTHESES

An important goal in ecological anthropology should be the formulation of hypotheses that will allow determination of the foraging practices of tribal peoples designed (*sensu* Williams 1966) to accomplish conservation as opposed to those that will maximize foraging efficiency or some other goal (Hunn 1982, Hames 1987). A potential problem is that maximization of hunting efficiency may lead to conservation. The importance of determining whether tribal practices are designed for conservation can be illustrated by the World Conservation Strategy definition endorsed by Berkes (1987:84): "Conservation is the maintenance of essential ecological processes and life support systems, the preservation of genetic diversity, and the sustainable utilization of species and ecosystems." Should environmental conditions change or resource demand increase, hunters whose practices are designed to maximize efficiency may cause environmental destruction while hunters whose practices are designed for sustainable yields will not upset the ecological balance. Below a series of paired hypotheses are described that may help to determine whether foraging practices are aimed at conservation or the opportunistic maximization of efficiency. As will become clear, the task is not simple because many of the paired predictions are identical. This leaves us with the possibility that maximization of foraging efficiency and conservation are not mutually exclusive goals in some environments (Berkes 1987:89).

In central place foraging theory, differential allocation of time in hunting patches leads to patterns of radial game depletion. After a site is settled, hunters hunt close to the village since game is equally dense in all areas. Hunting near the village brings the highest rate of return because travel time is minimized. Through time, areas close to the village become depleted of game, and hunters extend their range because the added travel to distant zones is repaid by higher rates of return in those zones. This process leads to a depletion gradient, with game populations densest in the most distant hunting zones and most sparse in the closest zones. The result of this process is shown in Figures 9.1 and 9.2 for the Yanomamö and Ye'kwana, respectively. Hunting efficiency was used as an index to game depletion, and these figures clearly demonstrate that game densities tend to increase with increasing distance from the village. Given these conditions, conservationists should allocate more time to distant areas with the greatest density of game, thereby allowing the closest zones to recover by releasing hunting pressures. However, hunters who wish to maximize their rate of return in hunting would follow the same strategy since this would yield the highest rate of return. Figures 9.3 and 9.4 indicate that the Yanomamö and Ye'kwana appear to be following this pattern since both groups of hunters allocate more time to rich than poor hunting zones.

If a hunter wishes to hunt in a game-rich distant area, he must first travel through depleted hunting areas adjacent to the village. This leads to another pair of predictions. Hunters interested in conservation should not take game in depleted areas while they are in transit to more distant, undepleted areas. On the other hand, hunters interested in maximizing their efficiency should take game wherever encountered, even in depleted areas. Unfortunately, there are no quantitative data to evaluate this hypothesis. However, in numerous hunts with both Ye'kwana and Yanomamö hunters, I always observed them to pursue game in depleted areas while they were en route to more distant hunting areas.

Tabooing by its very nature should aid in conserving species. Ross (1978) suggested that variability in game taboos among Amzonian tribal populations presents an important insight into the dynamics of tropical forest adaptations. He seemed to simultaneously maintain that tabooing is designed to allow hunters to maintain the demographic integrity of game animal populations (particularly those that reproduce slowly) and to also hunt at maximal efficiency (Hames 1987). Following Slobodkin's (1974) "prudent predator" model, tabooing should be designed to facilitate long-term maximization. According to Ross, tabooing would be most likely to occur among those groups with stable settlement patterns (e.g., the Achuara Jivaro of the Peruvian Montaña), but least likely among expanding populations, e.g., the Yanomamö.

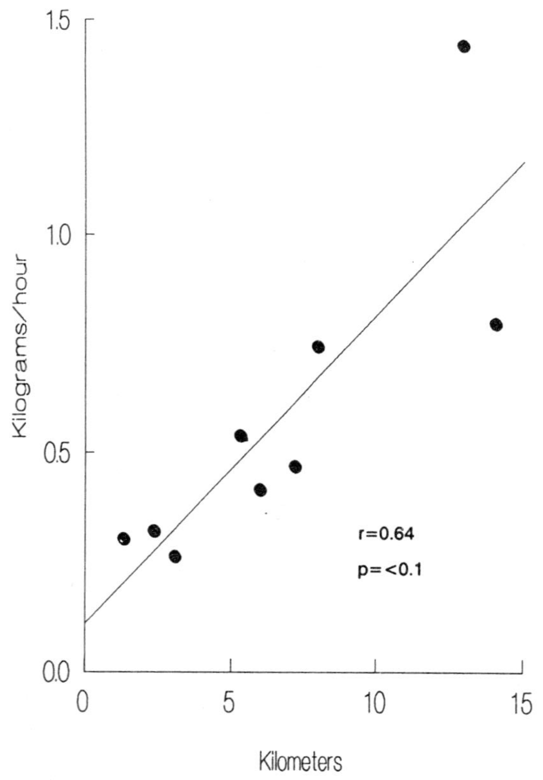

FIGURE 9.1 Yanomamö hunting efficiency and hunting zone distance.

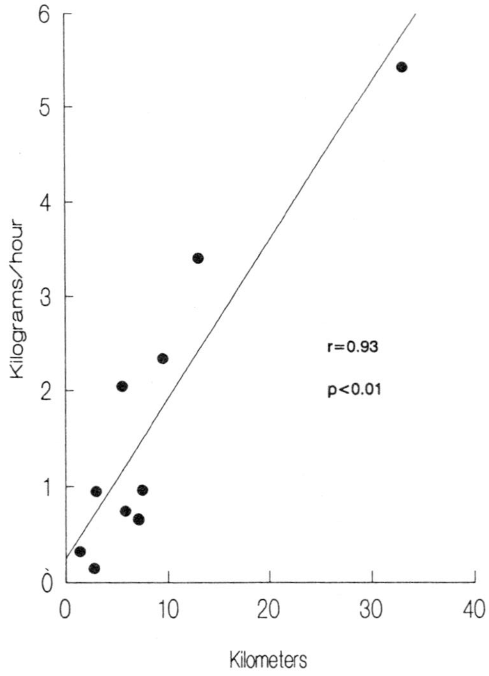

FIGURE 9.2 Ye'kwana hunting efficiency and hunting zone distance.

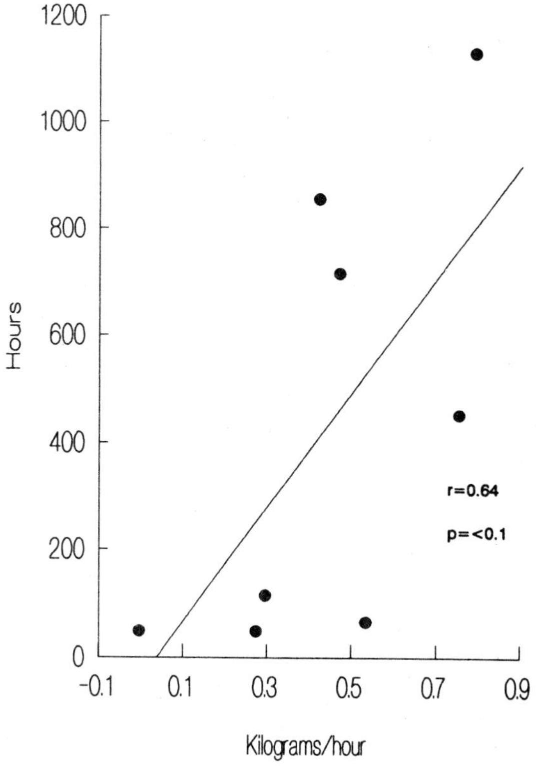

FIGURE 9.3 Yanomamö time allocated to hunting and hunting zone efficiency.

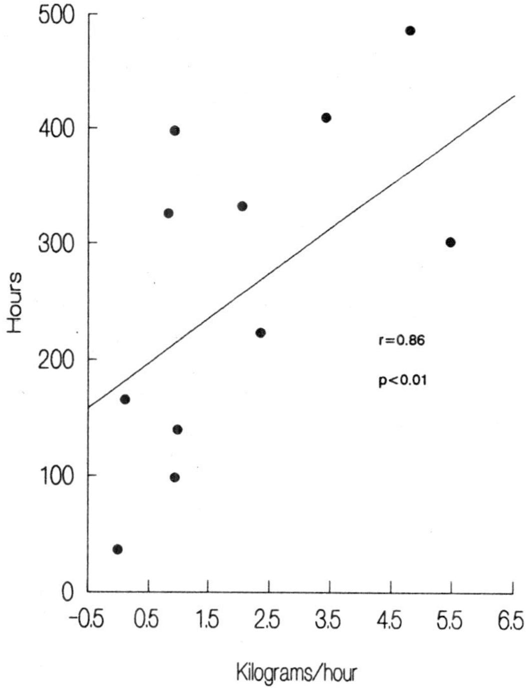

FIGURE 9.4 Ye'kwana time allocated to hunting and hunting zone efficiency.

If hunters instituted a taboo on a species to conserve its population size(s), then one would expect the taboo to work in a density-dependent manner. Over time, tabooing would be switched on or off–depending on a species' level of depletion. But hunters out to maximize their hunting efficiency might also institute a taboo on depleted species, not because they are depleted but because their rate of return upon encounter was declined. In this case, tabooing is designed to maximize efficiency, and conservation is a side-effect of efficient hunting. Species that have been subjected to heavy predation frequently exhibit behavioral depression (Charnov et al. 1976). That is, they become more wary, nocturnal, or refuge themselves in habitats (e.g., thickets) that make hunting more difficult. As a result, their net rate of return knocks them out of a hunter's optimal diet breadth (Charnov and Orians 1973). With a conservation strategy, such animals should lose their tabooed status when they lose those attributes that made their cost of acquisition high. Maximizers differ in response from conservationists, because for the former it is the cost of acquisition that is important, while in the latter it is game density.

The only evidence for taboos being switched on and off comes from Kensinger's (1981) work on the Cashinahua. Inedible animals become edible "when a village is old and the preferred game supply has been depleted by years of hunting to the point that a dependable supply of the kuin [edible] animals is unavailable" (1981:161). No data other than the above statement was supplied. Unfortunately, Kensinger's observation is not detailed enough to understand why certain animals were tabooed initially and whether taboos are lifted because of changing rates of return or depletion.

Kent Redford (pers. comm.) argues that tabooing could not lead to the maximization of rates of return in hunting. He points out that if one, for example, stumbled onto a sleeping tapir (Ye'kwana and Yanomamö hunters have recounted instances to me) it should be taken since the likely rate of return upon encounter is extremely high. There are two ways of dealing with this criticism. First, calculations of the optimal diet breadth (the game species one should always go after upon encounter) are based on average rates of return, which means that an animal is either in or out of the diet breadth independent of circumstances of encounter. However, this response exposes a weakness in the classic simple optimal diet breadth model: it does not deal with variance in rates of return upon encounter that are reliably correlated with environmental cues (Stephens and Charnov 1982). Nevertheless, optimal diet breadth models can be easily modified to take this condition into account. Second, and most important, the claim that one strategy is more adaptive than another is not the same as saying that it is the best strategy conceivable (Williams 1966).

For the taboo to be adaptive (never take the animal upon encounter), it must merely be superior to the alternative (always take the animal upon encounter).

While Ross (1978) and others (e.g., McDonald 1977) are concerned with general or blanket taboos, or those that extend to all members of a population, many Amazonian groups have taboos on consumption of game species limited to specific population segments (e.g., see the contributors to Kensinger and Kracke 1981). For example, pregnant women, children, and men undergoing certain long term initiation rites may be prohibited from consuming certain kinds of fish or game that others may consume freely. Furthermore, whether the taboo is general or restricted, tabooed animals may be killed, although they may not be consumed by certain people. The obvious effect of restricted taboos may be to depress (as opposed to reducing to zero under blanket prohibitions) the rate at which certain game species are taken. Thus, they may function like a quota system or 'bag limit' in Western hunting systems. These sumptuary regulations are extremely common throughout lowland Amazonia, and there is no simple reason why a hunter out to maximize his net rate of return would want to be subjected to such regulations. However, a hunter prohibited from consuming a particular game species could be motivated to hunt that species, even if his wife and family were not able to consume it if he could gain prestige by giving it to another family. In most Amazonian groups, hunters are expected to share large game, e.g., among the Yanomamö, approximately 60% of all game consumed by a family was acquired by a hunter who is not a family member (Hames, unpubl. data). This figure is consistent with other quantitative examinations of exchange in Amazonian societies (Aspelin 1979, Kaplan and Hill 1987).

A satisfactory, materialistic explanation of restricted versus generalized hunting and fishing taboos may prove extremely difficult. Ross's hypothesis (1978), as mentioned previously, is the only rigorous ecologically based model available; but it does not deal with restricted taboos. It may be that fish and game taboos, whether general or restricted, have more to do with psychology than anything else. Both the Ye'kwana and Yanomamö appear to be extremely reluctant to eat foods with which they have no familiarity. Once a Yanomamö explained to me that he was reluctant to eat a mata-mata turtle even though many in his village consumed it because he had not "learned to eat it yet"; he then implied it might cause him to become sick. In addition, the Yanomamö sometimes associate illnesses with things they have recently consumed, especially non-garden foods (see Kracke 1981:102 for a similar observation on the Brazilian Kagwahiv). Conservatism in eating habits may be adaptive in an environment where food spoils easily and where plants are loaded with toxic or unpalatable secondary compounds that probably serve as a defense against herbivory, i.e., it may be that many food taboos have nothing to do with conservation or hunting efficiency.

A final way that conservation might be obtained is by sparing females. As is well known, the demographic viability of a population is dependent on the number of reproductive females. Game laws in most states regulate the taking of females in species as diverse as pheasants and deer. Therefore, tribal hunters interested in maintaining game populations at a sustained level of harvest should spare mature females. Hunters out to maximize their rates of return, on the other hand, should take game independent of the sex or reproductive status of the species. Ye'kwana and Yanomamö hunters have no compunction in taking females. In fact, a common deer hunting tactic comes into play when a hunter encounters a fawn frozen in undergrowth. A hunter strangles the fawn, and then waits for the mother to return so she may be killed. Perusal of the literature on Amazonian hunting tactics fails to supply any evidence that females are spared.

Determining whether conservation occurs is not as theoretically important as specifying the conditions under which conservation will evolve and be maintained. This has obvious implications for the development of effective conservation policies, and it would immensely aid in the interpretation of the design of tribal foraging patterns. Two mechanisms that are dependent on an ecological variable must be in place to allow conservation to develop. From an ecological-evolutionary perspective, humans, like all other organisms, should be genetically selfish. That is they should maximize their inclusive fitness (Hamilton 1964). As a result, they should forego attempts to maximize their net rate of return while foraging only if it is in their net reproductive interest to do so. This assumes there is a positive correlation between foraging efficiency and reproductive success (see Kaplan and Hill 1987). They can be either "prudent predators" (Slobodkin 1974), interested in maximizing hunting returns over a long period of time, or opportunistic predators interested in maximizing their instantaneous rates of return.

The problem with prudence is that a co-resident may cheat and take selfish advantage of another hunter's restraint (Hunn 1982). As a result, the first requirement for conservation is a social mechanism that punishes or threatens punishment for those who fail to practice prudent predation. In states, and to some extent chiefdom level societies, there are legal bureaucracies that can enforce such regulations through supernatural means or physical coercion (Service 1975). However, all the Amazonian societies reviewed above are tribes that lack such institutions at the present time. Leaders in tribes are headmen who must rely primarily on verbal persuasion to induce people to follow rules. Among the Yanomamö and other Amazonian groups (Txicão [Menget 1981], Shipibo [Abelove and Campos 1981:174], Kagwahiv [Kracke 1981:106]) some people adhere to taboos while others do not. The only consistent punishment appears to be supernatural and personal. For example, for a northern Yanomamö dialect group, Taylor (1981:43) noted "For the Sanuma, as is the case for so many different societies, breaking a food taboo can result in a state of illness or bodily disorder" (see also Langdon 1981:58). In other cases the consumer of a tabooed animal is considered odd or disgusting. On the other hand, according to my Ye'kwana informants, the killing of an anaconda becomes a social threat since it is believed that this will lead to disastrous flooding of the entire village. This last sort of mechanism, threat to the social group, may be more effective than the threat of individual supernatural punishment because the one who did not follow the rule endangers everyone.

The second requirement for effective conservation is some sort of social mechanism to prevent outsiders from poaching. Hardin (1968) was one of the first to stress that some means of community control of resources is necessary for effective conservation (see Chapman 1985 and Hames 1987). Further, it is obvious that mechanisms of control must not only extend legally to community members through some system of tenure, but also to neighbors through political means. Although members of a village may behave in ways that prevent overexploitation, their efforts can be nullified if a neighboring group takes endangered game in their customary hunting areas. To succeed the social group would have to be territorial.

According to ecological-evolutionary theory, territories tend to develop in areas where resources are relatively dense and predictable (Brown 1964, Davies and Huston 1984). Brown's economic model of territorial defendibility posits that the costs of defending a resource (patrolling, fighting, and advertisement) must be less than the benefits of having unrestricted access to resources in a given area. This model has been evaluated on human populations with promising results (Dyson-Hudson and Smith 1978). Areas in which resources are relatively dense and predictable are less costly to defend because less travel time is required to maintain a perimeter by detecting poachers than attempting to defend a sparse resource distributed over a larger area. Brown's model is a cost-benefit model, and costs and benefits may vary in different ecological circumstances. Thus, density and predictability are indices–not direct measures–of costs and benefits.

In sum, it seems unlikely that many Amazonian tribal societies meet the specific conditions that should lead to conservation of wild game and fish populations. Amazonian game resources tend to be sparsely rather than densely distributed and, with the exception of the white-lipped peccary (Kiltie and Terborgh 1983), game tends to be spatially and temporally predictable, with most species as year-round residents. The issue of territoriality among Amazonian groups has not been systematically addressed although some (e.g., Ross 1978) have related it to warfare and protection of hunting areas. Others (Chagnon and Hames 1979, Hames 1983) have strongly denied that war is a matter of competition over scarce game resources through territorial defense. In addition, modern day native Amazonians do not possess the social institutions to enforce conservation among covillagers, although there are indications that chiefdoms may have once existed in Amazonia prior to European conquest (Roosevelt 1989). The model developed here suggests a more careful look at chiefdom level societies that depend heavily on resources that are economically worth defending.

CONSERVATION IN OCEANIC CHIEFDOMS

Conditions necessary for the development of native conservation systems; i.e., institutionalized enforcement of rules and territoriality seem to be met among chiefdom level societies in Micronesia and Polynesia. Chiefdoms differ from tribes, such as the egalitarian Amazonian tribal societies described above, in at least two fundamental ways. First, social status is based on heredity; only those having the proper status can rise to become leaders. And second, the most powerful position of leadership is the chief, instead of the headman in tribal egalitarian society. Chiefs, unlike headmen, have the ability to enforce their decisions and punish transgressors through a variety of social mechanisms unavailable to headmen (Service 1975). Territoriality, as manifested socially by a variety of resource tenure practices, extends not only to agricultural land but also to fishing areas.

According to Chapman (1985), chiefs and their councils in Micronesia and Polynesia habitually regulated the exploitation of communally owned fishing areas through a variety of mechanisms such as rotation of closed and open fishing areas, restriction of effective fishing technology, and limitations on the number of fishermen in particular coastal fisheries. In a comparative study of Micronesian tenure, Sudo (1984:226), observed "it is common to all societies that some persons (chiefs) or organizations (councils) exist to control and conserve marine resources. Those [sic] have right to regulate the use of particular sea sections, and have responsibilities to protect against the exhaustion of food resources." Importantly, Chapman (1985) pointed out that when open access to fisheries occurs, depletion follows because chiefs can no longer manage resources for the benefit of the local community. The importance of existence of adequate social mechanisms to enforce conservation was made more compelling when Chapman (1985) compared the absence of conservation and management in Melanesian fisheries. She pointed out that Melanesian societies are less politically complex than chiefdoms. Instead of chiefs, Melanesian groups are led by "big men", a kind of glorified headman. Like headmen, big men lack the coercive abilities of a chief. As a result, they do not have the ability to conserve or manage resources, even if they so desired.

It is highly possible that conservation of resources occurred among the hunting, fishing, and gathering chiefdoms of the Northwest Coast of North America–a coastal area stretching from northern California to southern Alaska. In nearly all of the chiefdoms, economically important fishing, hunting, and gathering areas were owned. Chiefs and other high status individuals determined who, when, and how intensively individuals could exploit resource areas (Drucker 1965, Oberg 1973:88, Richardson 1982). Unfortunately, it has proved difficult to uncover precise information on the relationship between resource depletion and harvesting restraint, although it is frequently implied (e.g., Richardson 1982). Of interest is Forde's observation on the Nootka (1963:78), "If a run in a particular stream began to fail they actually restocked it, obtaining spawn from another river at breeding season and carrying it back in moss-lined boxes to start a new generation in the depleted stream." Clearly, the Northwest Coast is worthy of further investigation to determine the extent of native systems of conservation.

CAN NATIVE PEOPLES COEXIST WITH PARKS?

Throughout much of human history, and especially in the last five hundred years of European expansion, tribal peoples have had their land and lives taken from them by state peoples. Contact has led to reduction of their areal base, introduction of new technology and market exchange, and settlement concentration. The strength of these variables and the ways in which they interact are the main determinants of a tribal people's ability to maintain an equilibrium with their environment. The balance of this chapter is devoted to a description of these factors and how they effect tribal adaptation. The aim here is to inform those responsible for setting up parks or reservations who seek to maintain cultural and/or biological diversity.

Native bows are much less efficient in taking game than shotguns or rifles (Rambo 1978, Hames 1979). The common effect of general purpose weapons, such as the shotgun, is reduction of hunting time (Hames 1979), which suggests that demand for game is relatively constant. This parallels the introduction of steel axes in New Guinea: the amount of forest cleared for gardens remained the same, but the time required to do the work decreased (Salisbury 1962). Although game depletion occurs sometimes (Binford and Chasko 1976:107, Rambo 1978), it may be that it is conditioned by the interaction of other factors (see below). Specific purpose weapons, especially those designed to exploit a narrow range of species, can lead to severe game depletion. The Ye'kwana of the Padamo (Hames 1979, 1980) and Yanomamö who live around the missions of the upper Orinoco have heavily depleted riverine game such game as capybara, herons, egrets, otters, and caiman by using outboards on their dugouts and headlamps. Headlamps and outboards, coupled with shotguns, make hunting of riverine game so efficient that hunters specialize. It may be that technology that reduces hunting costs over a wide range of game does not lead to game depletion, whereas technology that permits specialized cost reductions for a small range of game does lead to depletion.

Entrance into an external market economy changes the value of game animals. Aboriginally, game value is based on local subsistence demand. External markets increase game value depending on their utility for nonaboriginals as biological specimens, pets, and their products, e.g., skins, feathers, and furs. The increase in value leads to increased hunting and depletion (Nietschmann 1972), not infrequently among species that had no previous local value. Entrance into a market economy is undoubtedly the most potentially devastating change in aboriginal environmental relations.

Agents of contact such as missionaries and government officials routinely install trading posts and other sites such as medical clinics and schools that are by their very nature designed to attract and concentrate large numbers of natives. For example, among the Yanomamö, a mission was established in 1968 at the mouth of the Mavaca River. This mission initially attracted a village of approximately 100 Yanomamö (Chagnon 1985), but today there are seven villages, with a population of approximately 550, in the same immediate area. As a result, hunting and fishing pressures probably have increased more than five-fold (missionaries, their staff, and visitors are frequently supplied with local fish and game). Having worked at the mouth of the Mavaca for about three months during 1985-1987, it is my impression from observations of food exchange and casual discussion with long-time hunters of the area that hunting there is less rewarding than in non-mission areas. Just as importantly, populations such as the Yanomamö tend to have a mobile settlement patterns; villages are commonly relocated every five years or so (Hames 1983). Mobile settlement patterns reduce game depletion (although they are probably not specifically designed to do so). But Western presence leads to a less mobile settlement pattern (Chagnon 1985), thus intensifying the problem of game and resource depletion. These twin processes result in unrelieved, high hunting and fishing pressure.

Nevertheless, simultaneous conservation of ethnic and biological diversity is possible. Native peoples can exist in equilibrium with game populations if the technology they use is regulated, if hunting and fishing are aimed solely at meeting local nutritional demands, and if settlement patterns remain dispersed, mobile, and at low population densities. The simplest way to achieve such a goal is to regulate the activities of non-natives. Recently, it has been suggested that natives be barred from biological reserves toward the end of maintaining the natural functioning of neotropical ecosystems. This policy is erroreous for two reasons. First, native peoples have been a part of neotropical ecosystems for thousands of years in the role as top predators. They are therefore an integral part of it as much as natural predators such as the endangered jaguar and puma. Second, Indians have been systematically deprived of their land for the sake of "economic development" ever since the entrance of whites in the neotropics. Now other non-natives are asking them to give up their land for the sake of "biological conservation".

Given strong competing interests among missionaries, anthropologists, governments, conservationists, and commercial interests for native peoples, non-intervention is unlikely. Added to this are the desires of native peoples for contact, change, and the material accoutrements the Western world can give them. Technically, the maintenance of cultural and biological diversity is a simple problem. Politically, it is, and will probably continue to be, enormously difficult.

REFERENCES

Abelove, J. and R. Campos. 1981. Infancy related food taboos among the Shipibo. Pages 174-177 *in* K. Kensinger and W. Kracke (eds.), *Food Taboos in Lowland Amazonia*. Working Papers on South American Indians. Bennington College, Bennington, Vermont.

Anderson, A. 1984. The extinction of moas in Southern New Zealand. Pages 728-740 *in* P. Martin and R. Klein (eds.), *Quaternary Extinctions: A Prehistoric Revolution*. University of Arizona Press, Tucson.

Aspelin, L. 1979. Food distribution and social bonding among the Mamainde of Mato Gross, Brazil. *Journal of Anthropological Research* 35:309-327.

Baksh, M. 1985. Faunal food as a "limiting factor" on Amazonian cultural behavior: A Machiguenga example. *Research in Economic Anthropology* 7:145-177.

Berkes, F. 1987. Common-property resource management and Cree Indian fisheries in subArctic Canada. Pages 66-91 *in* B. McCay and J. Acheson (eds.), *The Question of the Commons*. University of Arizona Press, Tuscon.

Berry, W. 1979. The gift of good land. *Sierra Club Newsletter* (Nov./ Dec.):20-26.

Binford, L. and W. Chasko. 1976. Nunamiut demographic history. Pages 63-143 *in* E. Zubrow (ed.), *Demographic Anthropology*. University of New Mexico Press, Albuquerque.

Birdsell, J. 1958. On population structure in generalized hunting and collecting populations. *Evolution* 12:189-205.

Birdsell, J. 1968. Some predictions for the Pleistocene based on equilibrium systems among recent hunter-gatherers. Pages 229-240 *in* R. Lee and I. DeVore (eds.), *Man the Hunter*. Aldine, Chicago, Illinois.

Bodley, J. 1976. *Anthropology and Contemporary Human Problems*. Benjamin Cummings Publishing, Menlo Park, California.

Brightman, R. 1987. Conservation and resource depletion. Pages 121-141 *in* B. McCay and J. Acheson (eds.), *The Question of the Commons*. University of Arizona Press, Tuscon.

Brown, J. 1964. The evolution of diversity in avian territorial systems. *Wilson Bulletin* 6:160-169.

Callicott, J. 1990. American Indian land wisdom? Sorting out the issues. Pages 44-69 *in* P. Olson (ed.), *The Struggle for the Land*. University of Nebraska Press, Lincoln.

Cassels, R. 1984. The role of prehistoric man in the faunal extinctions of New Zealand and other Pacific islands. Pages 741-767 *in* P. Martin and R. Klein (eds.), *Quaternary Extinctions: A Prehistoric Revolution*. University of Arizona Press, Tucson.

Chagnon, N. 1985. *Yanomamo: The Fierce People*, 3rd Edition. Holt Rinehart & Winston, New York.

Chagnon, N. and R. Hames. 1979. Protein deficiency and tribal warfare in Amazonia: New data. *Science* 203:910-913.

Chapman, M. 1985. Environmental influences on the development of traditional conservation in the South Pacific region. *Environmental Conservation* 12:217-230.

Charnov, E. and G. Orians. 1973. Optimal foraging: Some theoretical explanations. (Mimeo, available from E. Charnov), University of Utah, Salt Lake City.

Charnov, E., G. Orians, and K. Hyatt. 1976. Ecological implications of resource depression. *American Naturalist* 110:247-259.

Clad, J. 1984. Conservation and indigenous peoples: A study of convergent interests. *Cultural Survival Quarterly* 8(4):68-73.

Colchester, M. 1984. Rethinking Stone Age economics: Some speculations concerning Pre-Columbian Yanoama economy. *Human Ecology* 12:291-314.

Davies, N.B. and A.I. Huston. 1984. Territory economics. Pages 148-169 *in* J.R. Krebs and N.B. Davies (eds.), *Behavioral Ecology: An Evolutionary Approach*. Sinauer, Sunderland, Massachusetts.

Davis, S. 1987. *The Archaeology of Animals*. Yale University Press, New Haven, Connecticut.

Dewar, R. 1984. Extinctions in Madagascar. Pages 574-591 *in* P. Martin and R. Klein (eds.), *Quaternary Extinctions: A Prehistoric Revolution*. University of Arizona Press, Tucson.

Diamond, J. 1982. News and views. *Nature* 298:787.

_______. 1984. Historic extinctions: A rosetta stone for understanding prehistoric extinctions. Pages 824-862 *in* P. Martin and R. Klein (eds.), *Quaternary Extinctions: A Prehistoric Revolution*. University of Arizona Press, Tucson.

_______. 1986. The environmentalist myth. *Nature* 324:19-20.

Drucker, P. 1965. *Cultures of the Pacific Northwest*. Chandler Publishing Company, San Francisco, California.

Dyson-Hudson, R. and E. Smith. 1978. Human territoriality: An ecological reassessment. *American Anthropologist* 80:21-41.

Ewers, J. 1955. *The Horse in Blackfoot Indian Culture*. Bureau of American Ethnology Publ. No. 23. Smithsonian Institution, Washington D.C.

Forde, D. 1963. *Habitat, Economy, and Society*. Dutton, New York.

Guidon, N. and G. Delibrias. 1986. Carbon-14 dates point to man in the Americas 32,000 years ago. *Nature* 321:769-771.

Hames, R. 1978. *A behavioral account of the division of labor among the Ye'kwana*. Ph.D Thesis, University of California at Santa Barbara.

_______. 1979. A comparison of the shotgun and the bow in neotropical forest hunting. *Human Ecology* 7:219-252.

_______. 1980. Game depletion and hunting zone rotation among the Ye'kwana and Yanomamö of Amazonas, Venezuela. Pages 31-66 *in* R. Hames (ed.), *Studies in Hunting and Fishing in the Neotropics*, Bennington College, Vermont.

_______. 1983. The settlement pattern of a Yanomam population bloc. Pages 192-229 *in* R. Hames and W. Vickers (eds.), *Adaptive Responses of Native Amazonians*. Academic Press, New York.

_______. 1987. Game conservation or efficient hunting? Pages 192-207 *in* B. McCay and J. Acheson (eds.), *The Question of the Commons*. University of Arizona Press, Tucson.

Hames, R. and W. Vickers. 1983a. *Adaptive Responses of Native Amazonians*. Academic Press, New York.

Hames, R. and W. Vickers. 1983b. Optimal foraging theory as a model to explain variability in Amazonian hunting. *American Ethnologist* 9:358-378.

Hamilton, W. 1964. The genetical evolution of social behavior. *Journal of Theoretical Biology* 7:1-52.

Hardin, G. 1968. The tragedy of the commons. *Science* 162:1243-1248.

Harris, M. 1976. Levi-Strauss et La Parlourde. *L'Homme* XVI:5-22.

Hunn, E. 1982. Mobility as a factor limiting resource use in the Columbian Plateau of North America. Pages 17-43 *in* N. Williams and E. Hunn (eds.), *Resource Managers: North American and Australian Hunter-Gatherers*. Westview Press, Boulder, Colorado.

Kaplan, H. and K. Hill. 1985. Hunting ability and reproductive success among male Ache foragers: Preliminary results. *Current Anthropology* 26:131-133.

Kensinger, K. 1981. Food taboos as markers of age categories in Cashinahua. Pages 157-185 *in* K. Kensinger and W. Kracke (eds.), *Food Taboos in Lowland Amazonia*. Bennington College, Vermont.

Kensinger, K. and W. Kracke 1981. *Food Taboos in Lowland South America*. Bennington College, Vermont.

Kiltie, R. and J. Terborgh. 1983. Observations on the behavior of rain forest peccaries in Peru: Why do some white-lipped peccaries form herds? *Zietschrift für Tierpsychologie* 62:241-255.

Kracke, K. 1981. Don't let the piranha bite your liver: A psychoanalytic approach to Kagwahiv (Tupi) food taboos. Pages 91-143 *in* K. Kensinger and K. Kracke (eds.), *Food Taboos in Lowland South America*. Bennington College, Vermont.

Krech, S., ed. 1981. *Indians, Animals, and the Fur Trade*. University of Georgia Press, Athens, Georgia.

Langdon, T. 1983. Food taboos and the balance of opposition among the Barasana and Taiwana. Pages 55-75 *in* K. Kensinger and W. Dracke (eds.), Working Paperson South American Indians. Bennington College, Vermont.

Leacock, E. 1954. *The Montagais "Hunting Territory" and the Fur Trade*. American Anthropological Association Memoir No. 18. American Anthropologist, Washington, D.C.

Levi-Staruss, C. 1973. Structuralism and ecology. *Social Science Information* 12(1):7-24.

Lees, S. and D. Bates. 1979. The myth of population regualtion. Pages 273-289 *in* N. Chagnon and W. Irons (eds.), *Evolutionary Biology and Human Social Behavior*. Duxbury, North Scituate, Maryland.

Lizot, J. 1978. L'economie primitive. *Libre* 4:69-113.

MacArthur, R. and E.O. Wilson. 1967. *The Theory of Island Biogeography*. Princeton University Press, Princeton, New Jersey.

Martin, C. 1978. *Keepers of the Game*. University of California Press, Berkeley.

Martin, P.S. 1967a. Africa and Pleistocene overkill. *Nature* 212:339-342.

______. 1967b. Prehistoric overkill. Pages 75-120 *in* P.S. Martin and H. Wright (eds.), *Pleistocene Extinctions*. Yale University Press, New Haven, Connecticut.

______. 1973. The discovery of America. *Science* 179:969-974.

______. 1984. Prehistoric overkill: The global model. Pages 354-403 *in* P.S. Martin and R.G. Klein (eds.), *Quaternary Extinctions: A Prehistoric Revolution*. University of Arizona Press, Tucson.

Martin, P.S. and R.G. Klein (eds.) 1984. *Quaternary Extinctions: A Prehistoric Revolution*. University of Arizona Press, Tucson.

McDonald, D. 1977. Food taboos: A primitive environmental protection agency. *Anthropos* 72:734-48.

McGleone, M. 1983. Archaeology of Oceania 18:11.

Menget, P. 1981. From forest to mouth: Reflections on the Txicao-theory of substance. Pages 1-17 *in* K. Kensinger and W. Kracke (eds.), *Food Taboos in Lowland South America*. Bennington College, Vermont.

Neitschmann, B. 1972. The hunting and fishing focus among the Miskito Indians of eastern Nicaragua. *Human Ecology* 1:41-67.

Nelson, R. 1979. Athapaskan subsistence adaptation in Alaska. Pages 205-232 *in* Y. Kotani and W. Workman (eds.), *Alaska National Museum of Ethnology*. Osaka, Alaska.

Oberg, K. 1973. *The Social Economy of the Tlingit Indians*. AES Monograph No. 22. University of Washington Press, Seattle, Washington.

Olson, S.L. and H. James. 1984. The role of Polynesians in the extinctions of the avifauna of the Hawaiian Islands. Pages 768-780 *in* P.S. Martin and R.G. Klein (eds.), *Quaternary Extinctions: A Prehistoric Revolution*. University of Arizona Press, Tucson.

Rambo, A. 1978. Bows, blowpipes, and blunderbusses: Ecological implications of weapons change among the Malaysian Negritos. *Malayan Nature Journal* 32(2):209-216.

Rambo, A. 1985. *Primitive Polluters: Semang Impact on the Malaysian Tropical Rain Forest Ecosystem*. Museum of Anthropology. Anthropological Papers No. 76. University of Michigan, Ann Arbor.

Redford, K. and J. Robinson. 1985. Hunting by indigenous peoples and conservation of game species. *Cultural Survival Quarterly* 9:41-44.

Reichel-Dolmatoff, G. 1976. Cosmology as ecological analysis: A view from the rain forest. *Man* 11:307-318.

Richard, A. and R. Sussman. 1975. Future of the Malagasy lemurs: Conservation or extinction? Pages 86-102 *in* I. Tattersall and R. Sussman (eds.), *Lemur Biology*. Plenum Press, New York.

Richardson, A. 1982. The control of productive resources on the northwest coast of North America. Pages 93-112 *in* N. Williams and E. Hunn (eds.), *Resource Managers: North American and Australian Hunter-Gatherers*. Westview Press, Boulder, Colorado.

Roosevelt, A.C. 1989. Resource management in Amazonia before the conquest: Beyond ethnographic projection. Pages 30-62 *in* D. Posey and W. Balee (eds.), *Resource Management in Amazonia: Indigenous and Folk Strategies*. The New York Botanical Garden, New York.

Ross, E. 1978. Food taboos, diet, and hunting strategy: The adaptation to animals in Amazonian cultural ecology. *Current Anthropology* 19:1-36.

Saffirio, G. and R. Hames. 1983. The forest and the highway. Working Papers on South American Indians No. 6 and Cultural Survival Occasional Paper No. 11. Cultural Survival, Inc., Cambridge, Massachusetts.

Sahlins, M. 1968. Notes on the original affluent society. Pages 85-89 *in* R.B. Lee and I. DeVore (eds.), *Man the Hunter*. Aldine, Chicago, Illinois.

Sahlins, M. 1972. *Stone Age Economics*. Aldine, Chicago.

Salisbury, R. 1962. *From Stone to Steel*. Cambridge University Press, Cambridge, Massachusetts.

Service, E. 1975. *Origins of the State and Civilization*. W.W. Norton and Company, New York.

Slobodkin, L. 1974. Prudent predation does not require group selection. *American Naturalist* 108:665-78.

Steadman, D.W., G.K. Pregill and S.L. Olson. 1984. Fossil vertebrates from Antigua, Lesser Antilles: Evidence for Lake Holocene human-caused extinctions in the West Indies. *Proceedings of the National Academy of Sciences* 81:4448-4451.

Stephens, D. and E. Charnov. 1982. Optimal foraging: Some simple stochastic models. *Behavioral Ecology and Sociobiology* 10:251-263.

Sudo, K-I. 1984. Social organization and types of sea tenure in Micronesia. Pages 203-230 *in* K. Ruddle and T. Akimichi (eds.), *Maritime Institutions in the Western Pacific*. National Museum of Ethnology, Osaka.

Taylor, K. 1981. Knowlege and praxis in Sanuma food prohibitions. Pages 24-54 *in* K. Kensinger and K. Kracke (eds.), *Food Taboos in Lowland South America*. Bennington College, Vermont.

Trotter, M. and B. McCulloch. 1984. Moas, men, and middens. Pages 221-239 *in* P.S. Martin and R.G. Klein (eds.), *Quaternary Extinctions: A Prehistoric Revolution*. University of Arizona Press, Tucson.

Tuan, Y.-F. 1968. Discrepancies between environmental attitude and behaviour: Examples from Europe and China. *Canadian Geographer* 12(3):176-191.

Vickers, W. 1980. An analysis of Amazonian hunting yields as a function of settlement age. Pages 7-29 *in* R. Hames (ed.), *Studies on Hunting and Fishing in Amazonia*. Bennington College, Vermont.

Vickers, W. 1988. Game depletion hypothesis of Amazonian adaptation: Data from a native community. *Science* 239:1521-1522.

White, L. 1967. The historic roots of our ecologic crisis. *Science* 155:1203-1207.

Williams, G. 1966. *Adaptation and Natural Selection*. Princeton University Press, Princeton, New Jersey.

PART FOUR

Biodiversity Conservation and Ecodevelopment

Introduction

Janis B. Alcorn

> The basic question thus becomes: how to restructure industrial society or guide its evolution in ways which will be more sustainable and regenerative? . . . (R)estructuring agriculture . . . is the key task. (Dahlberg 1987:378)

> Development must enhance ecological sustainability.. . . It must maintain and encourage biotic diversity as an insurance for the future.. . . This is where the conservation element enters ecodevelopment. (Dasmann 1984:20)

> Projects attempting to link protected area management with . . . development face considerable challenges.. . . (But) we must support and expand the Integrated Conservation Development Project (ICDP) approaches because we have no other choice. (Wells et al. 1990:4)

The definition of "development" is changing. The preanalytic vision of modern macroeconomics conceives of a circular flow of abstract exchange values with no dependence on environment or natural resources–rather like an animal with no digestive tract, only a circulatory system–so that resource flows and waste output flows are not monitored (Daly 1990). In the past, therefore, development projects have been chosen and promoted on the basis of economic analyses conducted without concern for ecological sustainability. But the new ecological economics is challenging development planners to redefine development so that it must maintain local and global ecological processes–thus ecodevelopment is receiving new attention.

Ecodevelopment projects offer opportunities to stem the trends of natural resource degradation and biodiversity losses. The discussion here will focus on the more narrow subset of ecodevelopment options that enhance biological diversity–not on "sustainable agriculture" per se. Retention of natural biodiversity has not been a goal of most sustainable agriculture efforts (e.g., MacDonald 1989, Reganold et al. 1990, Jiang 1990) which still tend to rely on homogenization of natural habitats, a few crop species, and external genetic reservoirs which provide the genetic resources in new varieties planted on the farm.

New ecodevelopment efforts are sometimes based on traditional resource management systems. For example, the new extractive in the Latin American tropics build on the rubber-tappers (*seringueros*) traditional land use systems (Fearnside 1989, Allegretti 1990). Proponents of traditional resource management systems have documented tradtional systems and indigenous knowledge that should be incorporated into development projects (e.g., see Richards 1985, Wilken 1987, Clay 1988, Browder 1989, Warren et al. 1989, Altieri and Hecht 1990, Anderson 1990, Alcorn 1991), and ecologists have investigated the ecological bases of traditional agroecosystems (e.g., see Gliessman 1990). Yet, thus far, there is no widespread effort to build on traditional resource management systems. The first two papers in Part Four offer different perspectives on the outcome of efforts to transfer an indigenous technology–*chinampa* wetland management–from the Mexican highlands to the lowlands. *Chinampa* technology transfer is one of the most frequently cited examples of development based on indigenous management systems.

In Chapter 10, biologist Stephen Gliessman explores the ecological basis of modern and prehispanic traditional wetland management in southeastern México. He describes the history of experimental wetland agroecosystems established during the 1970s, including the *chinampa* transfer effort. Although analysts from modern cultures may expect results quickly and promoters may be quick to claim success, agricultural systems that rely on biological processes must be established and adapted slowly. Gliessman stresses that these kinds of agricultural systems (such as wetland systems that rely on the slow accumulation of flood-mediated benefits) cannot be expected to yield immediate benefits. Since natural processes slowly improve the agricultural quality of soils, one must wait several years before observing increased yields. Farm families slowly work out cropping systems that fit their work schedules and needs, as well as market demands. Both of these processes proceed at their own pace. Rather than label the *chinampa* transfer models as failures because they have not led to benefits for large numbers of farmers, Gliessmann feels that there has not been sufficient time to judge their success. Significantly, however, he shows that the mature, raised fields produce higher yields than capital-intensive agriculture on drained fields in the same geographical area.

In Chapter 11, Mac Chapin, an anthropologist, points out how discussion of sustainable development has outstripped concrete action in farmers' fields. He analyzes two ecodevelopment projects in México: the Chontal *chinampa* transfer project in Tabasco (also discussed by Gliessman), and an agricultural improvement project among the Mixtec in Oaxaca. Based on these two case studies, Chapin advances a set of guidelines to enhance agroecosystem development success, and argues that ecodevelopment must incorporate the agenda of small farmers–not just the agenda of biologists, environmentalists, or policy makers. Finally, he suggests readings for biologists unfamiliar with applied anthropological literature. Chapin's contribution emphasizes the depth of multidisciplinary cooperation needed among scientists, applied development workers, policy makers, and local people if rural development is to be linked with conservation.

Both Gliessman and Chapin emphasize that it is particularly important for those of us concerned with biodiversity conservation keep a cold eye on reality as we measure the success of ecodevelopment projects. We must remember that "success" means success for centuries–not merely glimmerings of success in the last days of the 20th century. Our "ruler" for measuring success must span time periods greater than one's own lifespan; it must also be free of distortions introduced by the career objectives of any one generation or academic discipline. Although grassroots development specialists have debated various criteria for measuring success of development projects (e.g., Forte and Maguire 1988), the ruler for measuring success must be based on social, physical, and ecological criteria. One social criterion–to develop projects, programs, and policies not as "ends" per se, but rather as the "means through which people may build greater belief in themselves and greater confidence in their abilities to improve their own living conditions in their own way"–was raised by Salmen (1988:41); he observed that "This quality of reaching into a person to inspire faith in and commitment to the improvement of self and neighbors is the keynote to success in development." Such social criteria are necessary, but not sufficient, for determining the ecological sustainability of a development intervention. To evaluate the success of ecodevelopment efforts, both social and ecological parameters must be measured by outside experts and local people.

Achieving successful ecodevelopment is not an easy task, but achieving "sustainable development" along traditional lines is no easy task either. For example, an internal World Bank evaluation judged only half of over 500 of its sponsored projects as "sustainable." Unsatisfactory performance was found to be highest in agriculture. Ecologically sound rural development projects are vulnerable to the same constraints that face any development project. It is probably more difficult to attain short term success in ecodevelopment projects due to their very nature. Therefore, the second group of papers in Part Four explore a broad range of international efforts to achieve conservation and sustainable development.

In Chapter 12, Jason Clay provides an overview of lessons from projects sponsored by Cultural Survival, Inc. He asks, "Which conditions will encourage indigenous people living in fragile environments to conserve resources?" His answer is that strong, indigenous political organizations are necessary in order for indigenous people are to seek ecodevelopment on their own, or to respond to pressure from conservationists who feel that people do not belong in "protected areas." Clay notes that the first step toward achieving conservation is protection of indigenous land and forest tenure rights. A strong relationship between tenure and traditional stewardship can form the basis of future, conservative land use. Careful interpretation of the tenure-stewardship theorem is prudent; it does not mean that title or deed per se will lead to wise land use. Individual titling has been linked to commoditization of nature and its conversion to capital for short term gain. Locally-adapted systems of communal titling, such as those in Mexican *comunidades* and *ejidos*, hold great promise to guarantee local decision rights while at the same time preventing wholesale loss of lands through individual sale of land or timber rights (Lynch and Talbott 1988, Lynch and Alcorn 1991).

In Chapter 13, Bill Gregg provides a global inventory and evaluation of UNESCO-MAB biosphere reserves in terms of their potential for linkages between conservation and rural development. He describes efforts in several countries where active local participation was instrumental in achieving rural development in association with MAB biosphere reserves. Nevertheless, it is clear that the potential for these reserves to function as catalysts for regional ecodevelopment remains largely unrealized. The MAB biosphere reserve model per se should not be blamed for this shortfall. The expected linkages between conservation and development in biosphere reserves are the same kinds of linkages now being made by "Integrated Conservation Development Projects" (ICDPs) and "Buffer Zone Projects" around many types of reserves. Nowhere are these kinds of projects finding easy success (Wells et al. 1990). For example, Santana et al. (1989: 220) identify the most critical element for success of Manantlan Biosphere Reserve as "establishment of the institutional coordinating mechanism" that internalizes biosphere reserve's objectives, thus giving full representation to the reserve's inhabitants. They further observe, "(T)his is a generalized problem in many Latin American projects that must be resolved." In 1984, Western described successful integration of Masai pastoralists into park management of Amboseli. Yet, new adaptations in institutional relationships will continue to be necessary to accomodate ecodevelopment of pastoralists' communities in 1991 (C. Jensen, pers. comm.).

A global evaluation of ICDPs commissioned by the World Bank (Wells et al. 1990, see introductory quote) concluded that there are few success stories for several reasons. One reason is that ICDPs are very new. Few projects evaluated by the World Bank study had been in place for more than five years; yet it is widely accepted that ICDPs require long term effort to be successful. Other reasons given included: lack of effective government institutions; laws, policies, and economic changes that cannot be influenced by local ICDPs; and the difficulty of eliciting true participation in decision-making.

Despite the dearth of ICDP successes, the World Bank study nonetheless concluded that there are no viable alternatives to ICDPs if we are to adequately conserve tropical biological diversity. Even though traditional enforcement may continue to play a critical role, it is inconceivable that protected areas will be maintained indefinitely by what, in some cases, will amount to military force. "Innovative ICDPs which constructively address local people-park relationships are therefore an essential component of sustainable development" (Wells et al. 1990:4).

In Chapter 14, Molly Kux provides an appropriate conclusion to the discussion of links between conservation and development. She addresses the question of how development agencies can meet conservationists' concerns as well as respond to the economic aspirations of developing nations. She examines specific donor-funded projects, including the innovative Wildlands and Human Needs program developed in cooperation between World Wildlife Fund-U.S. and the U.S. Agency for International Development (A.I.D.). Kux, who has long been involved in addressing biodiversity conservation and rural development issues, clarifies the approaches currently being used by development agencies, and identifies the key elements that contribute to their success.

The innovative projects of bilateral and multilateral development agencies around the world, however, remain a tiny minority among all ongoing projects. A.I.D. personnel, for example, continue to question the need for biodiversity conservation. They generally judge proposed projects based on local resource management systems or objectives as unsupportable. For example, A.I.D. recently rejected an innovative caterpillar management and reforestation project proposed for a region in Africa where caterpillars provide a major source of protein, because it was considered "backward" (K. Newman, pers. comm.). Yet, most other bilateral donor agencies and the multilateral development banks lag far behind A.I.D. in the environmental sector. Few have A.I.D.'s capability or commitment to evaluate environmental assessments and otherwise assess the ecological sustainability of proposed projects.

The ecodevelopment ball is beginning to roll in a few circles, and ecodevelopment has become the "star" of many international conferences. But, where the "rubber meets the playing field" in the "development game," it is still "business as usual" in 1991.

REFERENCES

Alcorn, J.B. 1991 (in press). Ethnobotanical knowledge systems: A resource for meeting goals in rural development. *In* D.M. Warren, D. Brokensha, and L.J. Slikkerveer (eds.), *Indigenous Knowledge Systems: Cultural Dimensions and Development*. Kegan Paul, London.

Allegretti, M.H. 1990. Extractive reserves: An alternative for reconciling development and environmental conservation in Amazonia. Pages 252-264 *in* A. Anderson (ed.), *Alternatives to Deforestation*. Columbia University Press, New York.

Altieri, M.A. and S.B. Hecht, Eds. 1990. *Agroecology and Small Farm Development*. CRC Press, Boca Raton, Florida.

Anderson, A.B., Ed. 1990. *Alternatives to Deforestation*. Columbia University Press, New York.

Browder, J.O., Ed. 1989. *Fragile Lands of Latin America: Strategies for Sustainable Development*. Westview Press, Boulder, Colorado.

Clay, J.W. 1988. *Indigenous Peoples and Tropical Forests: Models of Land Use and Management in Latin America*. Cultural Survival, Inc., Cambridge, Massachusetts.

Dahlberg, K.A. 1987. Redefining development priorities: Genetic diversity and agroecodevelopment. *Bulletin of Science, Technology, and Society* 7:367-382.

Daly, H. 1990. Towards an environmental economics. Paper presented at the Conference on "The Ecological Economics of Sustainability Conference," World Bank, Washington D.C.

Dasmann, R.F. 1984. An introduction to world conservation. Pages 16-24 *in* F.R. Thibodeau and H.H. Field (eds.), *Sustaining Tomorrow: A Strategy for World Conservation and Development*. University Press of New England, Hanover, Massachusetts.

Fearnside, P.M. 1989. Extractive reserves in Brazilian Amazonia. *BioScience* 39:387-393.

Forte, G.A. and R. Maguire. 1988. Qualifying success. *Grassroots Development* 12(3):11-19.

Gliessman, S.R., Ed. 1990. *Agroecology: Researching the Ecological Basis for Sustainable Agriculture*. Springer-Verlag, New York.

Jiang, X. 1990. Some eco-economic principles and their applications in agricultural production–introducing a Chinese ecological farm. Department of Agricultural Economics, Kansas State University, Manhattan.

Lynch, O. and J.B. Alcorn. 1991. *Empowering Local Forest Managers*. World Resources Institute, Washington, D.C.

Lynch, O. and K. Talbott. 1988. Legal responses to the Philippine deforestation crisis. *New York University Journal of International Law and Policies* 20(3):679-713.

MacDonald, J.F. 1989. *Biotechnology and Sustainable Agriculture: Policy Alternatives*. National Agricultural Biotechnology Council, Boyce Thompson Institute, Ithaca, New York.

Reganold, J.P., R.I. Papendick and J.F. Parr. 1990. Sustainable agriculture. *Scientific American* 262(6):112-120.

Richards, P. 1985. *Indigenous Agricultural Revolution: Ecology and Food Production in West Africa*. Westview Press, Boulder, Colorado.

Salmen, L.E. 1988. Determining success in development. *Grassroots Development* 12(3):41.

Santana, C.E., M.R. Guzman and P.E.Jardel. 1989. The Sierra de Manantlan Biosphere Reserve: The difficult task of becoming a catalyst for regional sustained development. Pages 212-222 *in* W.P. Gregg, Jr., S.L. Krugman and J.D. Wood, Jr., (eds.), *Proceedings of the the Symposium on Biosphere Reserves*. U.S. Department of the Interior, National Park Service, Atlanta, Georgia.

Warren, D.M., L.J. Slikkerveer and S.O. Titilola, Eds. 1989. *Indigenous Knowledge Systems: Implications for Agriculture and International Development*. Studies in Technology and Social Change, Publ. No. 11. Technology and Social Change Program, Iowa State University, Ames.

Wells, M., K. Brandon and L. Hannah. 1990. *People and Parks: Linking Protected Area Management with Local Communities*. World Wildlife Fund, Washington D.C.

Western, D. 1984. Conservation-based rural development. Pages 94-108 *in* F.R. Thibodeau and H.H. Field (eds.), *Sustaining Tomorrow: A Strategy for World Conservation and Development*. University Press of New England, Hanover, Massachusetts.

Wilken, G.C. 1987. *Good Farmers: Traditional Agricultural Resource Management in Mexico and Central America*. University of California, Berkeley.

10

Ecological Basis of Traditional Management of Wetlands in Tropical México: Learning from Agroecosystem Models

Stephen R. Gliessman

INTRODUCTION

The need for the development of sustainable agriculture has become a topic of considerable discussion and concern among a broad spectrum of people involved in the multiple components of the production, processing and distribution of food, feed, and fiber around the globe (Knorr 1983, Douglass 1984, Jackson et al. 1984, Edens et al. 1985, Dover and Talbot 1987, Agency for International Development 1987, Allen and Van Dusen 1988). This is especially true for agriculture and natural resource management in the humid tropics. Faced with the need to meet the demands of both growing populations and increasing foreign debt, most tropical countries look to agricultural development to feed their people and solve serious cash-flow problems. Based on the experience of development projects carried out during recent decades, though, most agricultural projects in tropical regions have not delivered the promised results (Barkin 1978, Lappe and Collins 1982). Too often it appears that agricultural development in the tropics contributes more to the loss of biological and cultural diversity than it contributes to the long-term well-being of the peoples involved and the environments affected (Altieri et al. 1987). For the establishment of truly sustainable agroecosystems to occur, grassroots approaches must be employed that build upon local knowledge (Klee 1980, Gliessman et al. 1981, Altieri and Anderson 1986, Wilken 1988) and the agroecological basis of production that has contributed to the stability of these systems over the long term (Gliessman 1984, Altieri 1987).

THE WETLANDS OF SOUTHEASTERN MEXICO

One of the major limiting factors to agricultural production in the tropical lowlands of southeastern México is the excess of water, either as rainfall or runoff (West et al. 1969). Poorly drained soils, low-lying topography, and excess water are limitations that can be very difficult for food production systems to overcome. Local inhabitants currently manage a remarkable array of agroecosystems which have been adapted to this ecological situation, and considerable evidence has accumulated over the past several years that these regional, traditional agroecosystems have their roots in prehispanic times (Harrison and Turner 1978, Gliessman et al. 1985, Gómez-Pompa and Jimenez-Osornio 1987). The challenge, then, is how to learn from these systems in order to design and manage present day systems that maintain an independence from large quantities of imported purchased inputs, and that can be adapted to varying environmental and cultural conditions. This is a process that takes time and resources, and cannot be expected to be successful in all aspects in short periods of time. Agricultural development is as much an ecological and evolutionary process as it is an economic one. Experimental model agroecosystems can be useful tools in this complex process. Which components of the agroecosystem we choose to include in our models, as well as our mode of analysis, will determine in the long run the level to which beneficial change will come about. For the purpose of this discussion, I will focus my review on agroecosystem management which has been used in the southeastern lowlands of tropical México. There have been successes and failures, but I believe that the basis for a more integrated approach to agricultural development is within reach.

PREHISPANIC WETLANDS MANAGEMENT

The indication of how low areas were initially managed in tropical México comes from recent evidence for prehispanic use of wetlands. Recent studies have demonstrated that prehistoric Maya of the Yucatan peninsula region developed several types of technologies that enabled them to cultivate wetlands (Turner and Harrison 1981, Siemens 1982, Darch 1983, Gliessman et al. 1985, Gómez-Pompa and Jimenez-Osornio 1987). Observations of archeological ground patterns suggest that large areas of wetlands were devoted to cultivation which relied on systems of canals or drains, raised platforms, and other structures that permitted water manipulation.

In a study of a large depression (*bajo*) in Morocoy, southern Quintana Roo, México, a wetland system that encompassed up to 20,000 ha was examined (Gliessman et al. 1985). Topographic surveys and soil analysis demonstrated the presence of an intricate system of depressions and raised surfaces that formed a complex and interconnected network of canals and platforms that were constructed as early as the Late Preclassic Maya period (800 BC-150 AD) and utilized through the Late Classic period (600-850 AD) when they were abandoned for unknown reasons. Evidence suggests that this system was used continuously for over 1000 years.

Dark gray soil (layer 3), found only on the platforms, is the remnant of the ancient planting bed excavated from the lower depressions (canals) and mounded up to form a planting surface, available for planting even at the peak of the wet season (Table 10.1). This soil layer probably formed the original bed of the *bajo*, which was excavated down to the parent material and mounded up to the 60-70 cm height the beds probably had during the time of their use. Even given the accumulation of surface sediments during the centuries since abandonment, a 30-45 cm difference still exists between the platform and canal surfaces. The similarities between the results of analysis of the surface soil layer and the dark gray layer on the platform supports the argument that the original *bajo* surface soil was used for mound construction. Acidic pH, considerable residual organic matter, higher levels of phosphate, and lower levels of calcium are all differences that would have been established by the mounding up of the original topsoil, and then maintained by a combination of periodic canal cleaning and agricultural practices which promoted organic matter inputs. Higher nitrogen levels in the surface soils of the canals may be more a result of fertilizer applications to rice planted during recent years rather than a reflection of past activities.

Very limited evidence exists on the types of cultivation practices associated with these relict farming systems. Work in Belize indicates that maize and cotton may have been important species (Turner and Harrison 1981). It is also unknown with what frequency the canals may have been cleaned for mounding up the platforms. The intensity of cropping of the platforms, as well as yields, would have been influenced greatly by this practice, much as we now observe in raised field systems (Wilken 1988). Otherwise, only by studying present-day traditional use of wetlands in southeastern México (Gliessman et al. 1981, Gómez-Pompa 1987), perhaps derived in part from practices used during Maya times, inferences as to past use may be drawn.

TRADITIONAL USE OF WETLANDS

An interesting and productive use of wetland areas for the production of corn and other crops has been observed in Tabasco (Orozco

TABLE 10.1
Average Depth, Color, and Analysis of Soil Horizons on Prehispanic Raised Fields*

Horizon	Color†	pH[a]	O.M.[b]	PO_4[c]	N[d]	Ca[e]	Mg[e]	K[e]
Surface:								
0-25 cm	Dk Br	5.8	3.4	0.70	.153	51.2	19.2	.41
Platform:								
25-45 cm	Gr-Br	5.8	3.6	2.79	.091	53.9	18.4	.39
45-75 cm	Dk Gr	5.8	2.5	1.87	.041	54.6	19.0	.28
75-100 cm	Lt Gr	7.4	1.6	2.17	.021	64.2	20.5	.19
Canal:								
25-45 cm	Gr-Br	6.5	3.2	1.08	.140	58.4	18.0	.56
45-105 cm	Lt Gr	7.1	1.3	0.96	.034	59.0	19.6	.28
Bedrock:								
>105 cm	Lime	7.3	Tr	0.78	.025	78.3	23.2	.23

Source: Modified from Gliessman et al. (1985).
*Fields are sited near Nicolas Bravo, Quintana Roo, México.
† Dk=dark; Lt=light; Br=brown; Gr=gray; Lime=limestone.
[a] pH below value neutral (7.0) indicates more acidic soil conditions.
[b] Percent organic matter (O.M.) in the soil.
[c] Easily exchangeable phosphate in ppm (parts per million).
[d] Percent total nitrogen determined by kjeldahl digestion.
[e] Milliequivalents (me) per 100 grams of soil.

and Gliessman in press). Since this practice is most often observed in communities comprised primarily of remnants of indigenous populations, it might indicate that the practice is not one that was recently introduced, but rather may have been handed down from generation to generation. In this system, corn is planted on higher ground around flood-prone areas during the wet season months of June to December, but as water levels drop during the dry season months of February and March, farmers follow the receding water line with another corn planting, known locally as the March planting (Marceño). Similar systems have been observed as far north in México as Veracruz (Siemens 1990).

During much of the year the low areas are inundated from a depth of a few centimeters to as much as a meter. Marsh vegetation densely covers the area during the wet season and is felled quickly with machetes as the water level recedes. A very dense mat of organic matter is produced by this process (10-20 cm), and seed is planted into a hole made with a pointed stick driven into the mat. About a week following planting, fire is used to burn part of this organic mat, as well as kill back any weed seedlings or sprouts of the marsh plants. The timing of burning is very important in order to ensure that the fire burns only the dry leaves on top of the mat and does not reach the soil. The corn seed, planted 10-15 cm below the surface of the soil, is not harmed by the fire. Local short-cycle varieties of corn (2-3 months from planting to harvest) are most frequently used. The practice of using seed from the previous harvest for the subsequent planting favors the use of local varieties, rather than purchasing what today is usually hybrid or improved seed produced at distant locations. The use of a 2.5 month variety known as *mejen*, a Maya word meaning "precocious" or "early maturing", again shows the link to the past that many of the present day systems may have.

The corn grows very quickly in this system, and when fire is not used excessively and flooding is allowed to occur every year, weeding is usually not necessary. After approximately 2.5 months, the mature corn stalks are "doubled-over" just below the corn ear, facilitating final drying of the grain for another 2-4 weeks before harvest. Yields between 4-5 tons/ha of dry grain are common, with some reaching 10 tons/ha. This is many times the average yield of 1-1.5 tons/ha for mechanized production on lands that have been cleared and drained in the same region. These greater yields are obtained at a fraction of the input costs and labor invested in mechanized production systems (Amador and Gliessman 1990).

Following the harvest, all crop and non-crop residues end up on the soil surface. This contributes to a key element in the productivity of the system–the soil. Soil profiles demonstrate the presence of a thick organic soil to a depth of 30-40 cm below the surface. During inundation, organic matter produced by the marsh plants or left by the previous cropping cycle is incorporated into the soil and conserved below water. Nutrient minerals that enter the system with surface drainage are captured by the highly productive aquatic sector of the ecosystem, leading to the formation of a soil that has organic matter levels over 30%, total nitrogen (N) as high as 3%, and high levels of other important plant nutrients (S.R. Gliessman, unpubl. data). The key element, then, in the management of this system is to take advantage of the benefits gained from a period of inundation. When the system is drained artificially to extend the cropping season, the organic soil layer can be reduced to 5 cm in less than two years (Mejia and Gliessman, unpubl. data), and yields drop dramatically.

Based on the knowledge of the *Marceño* planting, it can be hypothesized for the prehispanic raised field systems that a similar dry season planting may have been practiced in the lower canal areas, since it is doubtful for this particular region that there was standing water in the depressions during the entire dry season. In an exceptionally wet year, two crops may have been possible on the platforms; in a dry year, the canal would have presented yet another planting opportunity. There must have been considerable information in the local culture about how to work with, as well as manipulate, the seasonal patterns of rainfall and water flow, just as there is considerable knowledge in present day traditional agriculture.

EXPERIMENTAL WETLAND MANAGEMENT

When experimental wetland management systems were designed and implemented in Tabasco in the mid-1970s, the information presented above was not yet available. Therefore, systems were designed that were based on the *chinampa* systems of the central valley of upland México (Gómez-Pompa 1978). It seemed reasonable to assume that an agroecosystem that had evolved under conditions of high water tables and poorly drained soils could be adapted to the lowland tropics where the same two factors were limiting to agricultural production. The ecological basis of *chinampas* was fairly well understood (Venegas 1978), and it was hypothesized that, with modifications for lowland tropical environments, an appropriate model could be found. In many of those upland regions, chinampa or raised field agriculture is continuing to function and expand (Crews 1985, Gonzalez 1986); it is not "literally drying up," as often reported (Chapin 1988; see also Ch. 11 this volume). A system that was continuing to evolve along with environmental and cultural change (Venegas 1978) offered useful information for adaptation to other regions of the same country.

One of the first attempts at this transfer began in 1975 near the town of Nacajuca, Tabasco, at a technical school for students from nearby communities that were made up primarily of descendants of indigenous Chontal peoples. In an attempt to help local people develop production systems that were adapted to the flooded or flood-prone conditions of the region, as well as take advantage of local resources, the school constructed small raised platforms (20 m x 20 m) by mounding up soil by hand and layering it over swamp vegetation, similar to the way raised fields are made today not only in Xochimilco, but in other parts of the central plateau of México (Venegas 1978, Wilken 1988). Canals that were formed around the platforms soon were populated by native fish and aquatic vegetation. Initial experimentation was carried out with local subsistence crops as well as crops with potential market value. Local people, except the director of the school who was from another part of México, implemented the project. Four platforms were soon constructed, and plans were laid out to construct at least another twelve to cover an area of about 1 ha. The development process began slowly with opportunity for local trial and error.

It was very soon thereafter that México's Instituto Nacional de Investigaciones sobre los Recursos Bióticos (INIREB) began experimental studies on similar tropical *chinampas* in locations in Tabasco and Veracruz (Gómez-Pompa and Venegas 1976, Jimenez-Osornio and Gómez-Pompa 1987). These projects depended upon financial support from outside the region, but they had the distinct advantage of including inhabitants from a chinampa region in upland México who advised people hired locally in the laborious task of building the platform and canal systems. Experience was slowly gained on how to successfully modify the upland system to lowland conditions.

In early 1977, the Ecology and Plant Pathology Departments at the Colegio Superior de Agricultura Tropical (CSAT) in Cardenas, Tabasco, initiated experiments with raised field agriculture as part of a 12 ha project at the college's field station. These experiments involved the construction of production units with annuals, perennials, and forest shelterbelts in concentric belts around a central tank filled with water for the production of fish, ducks, and aquatic vegetation (Gliessman et al. 1978, 1981). A narrow belt of raised beds for annual crops was constructed close to the tank and managed in a way similar to *chinampas* by incorporating aquatic vegetation for soil improvement and irrigating through capillarity. The project served for many years as an important location for experimentation on the management of local crops using traditional knowledge and practices, something that is not very common on most agricultural research stations in the tropics, or elsewhere for that matter. A modified version of the modular system was installed by CSAT in collaboration with local rural farmers (*campesinos*) in a community near Teapa, Tabasco. The Teapa version incorporated several more modifications of wetland management strategies (Gliessman et al. 1981).

In late 1977, INIREB assisted CSAT with the construction of another system of experimental raised fields similar to *chinampas* for research and demonstration purposes. Located on the grounds of the college, close to classrooms and laboratories, this system was intended to be a proving ground for the processes that would need to be carried out in order to adapt *chinampas* to the lowland tropics and train people in their use. These were the first experimental raised-bed systems constructed with mechanized equipment. Pressed sugar cane fiber (bagasse) from a nearby sugar refinery was layered with soil backhoed from parallel canals on each side of a platform that was 8m wide and 500 m long along what had previously been an agricultural drainage ditch. The system was divided into smaller platforms 20 m long by hand-digging smaller cross canals across the platform in order to improve water circulation throughout the system. Crops characteristic to the region were combined with introduced species, but the cropping patterns were primarily based on traditional multiple cropping and rotations. These experiments produced information on design considerations for construction (Vandermeer and Gliessman 1980) and the use of green manures for soil improvement (Boucher et al. 1983), as well as other information reported in thesis projects and reports at the college, but much of this information is as yet unpublished. One of the primary findings was that it took two years for adequate levels of organic matter and available nutrients to form on the platforms, and at least another three years for good agricultural soil to develop (Gliessman, unpubl. data). It was determined that it was possible to produce up to 15 kg of dry matter per m^2 of canal surface per year for application to the platform surface. Two full-time salaried workers were able to maintain a third of a hectare of platform surface, and a similar area of canals, in year-round production.

In late 1977, INIREB and CSAT approached the director of the school in Nacajuca about collaboration and experimentation on the platforms that the students had constructed. Shortly thereafter, with more resources, more platforms of different designs and dimensions were constructed. Many of these were constructed by using machinery. Attempts to selectively manage fish in the canals were begun, experiments with different crops and rotations were initiated, and different ways of building soil productivity were started. Given the participation of local students, it looked like the opportunity to combine traditional knowledge with experimentation was very great. The students also formed a potential conduit for the transfer of the most appropriate raised field designs out to local communities.

It was at this time that the Chontal community began to put pressure on the local government, especially through the Instituto Nacional Indigenista (INI), for greater economic assistance to their region. Other parts of Tabasco were prospering from the regional the oil-boom. Why not this region as well? Government officials eyed the fledgling experiments in progress in different parts of the state, especially those at the school located in the center of the Chontal region. They made the decision to initiate a large-scale program designed to build large areas of raised fields (*camellones chontales*) near several Chontal communities. Although cautioned by both INIREB and CSAT researchers that more information was needed before a large scale raised field project should be initiated, the government used a large marine dredge to construct a system that eventually covered 60 ha near the town of Tucta, Tabasco. Additional systems of similar size and construction were installed at two other communities in the nearby region within a few more years.

The problems that the management of the *camellones chontales* encountered have been reviewed elsewhere (Gómez-Pompa and Jimenez-Osornio 1987, Chapin 1988). Soils that had been dredged from several meters below the surface presented a major barrier to successful crop production (Table 10.2). They were extremely heterogeneous and low in organic matter and available nutrients. Crops that were not normally grown by local farmers formed the main focus of the first several years of planting. Economic projections that planners had made for the system were based on the production of crops poorly suited for the lowland tropics on one hand, and designed to meet the demands of distant markets on the other. By the time better agricultural soils finally began to form, and farmers began to plant crops that were more adapted to the region, almost all economic support and technical assistance had dried up.

While the *camellones chontales* were in their early stages, yet another experiment was begun on raised fields at CSAT (Amador 1980). It was designed more for the large areas of the lowland tropics that are subject to periodic annual flooding, rather than the more restricted areas that are permanently inundated. Although the CSAT researchers were aware of chinampa technology from the uplands, they relied more on a combination of research on the prehispanic raised fields of Quintana Roo and observations of the *Marceño* planting systems which took advantage of the periodic flooding characteristic of much of the region. Experience dictated that it would take years before good yields could be produced on platforms formed from excessively deep excavation of soil. At the same time, research was needed to find a way to ensure that a planting surface would be available regardless of the intensity of the rainfall. Thus, a system of platforms (called *bancales tipo CSAT*) was constructed using a road-grader as the principal machine. The blade of the machine was used to lift approximately 30 cm of soil from what would become the canals of the system, and it was placed to the side to form what would become the elevated platform. Upon completion of construction, the difference in height between the highest and lowest part of the system was 60 cm. The platforms were 5 m wide, and the canals 4 m, with about 0.5 m on each side of the platform forming a sloping border into the canal. A day's work by the road-grader could construct a hectare of *bancales*, leaving a system that could then be maintained either by hand or with normal agricultural machinery.

TABLE 10.2
Soil Analysis of Raised Beds of *Camellones Chontales* of Tucta, Nacajuca, México

Soil Analysis	Soil Depth (cm):		
	0-15	15-30	30-45
% Organic Matter	1.30	1.13	1.56
pH	7.60	7.80	7.80
Total Nitrogen (%)	0.04	0.04	0.03
Available Phosphorus (ppm)	35.12	43.90	21.84
Ca^{++} (me/100 g)	21.57	23.28	23.02
Mg^{++} (me/100 g)	10.10	8.15	10.10

Source: Informe de Actividades, Departamento de Ecología, Colegio Superior de Agricultura Tropical (1979, unpubl. data).
*Sample was taken 18 December 1978—one year after construction of the platforms.

In initial experiments, 1 ha of *bancales* was constructed at CSAT, followed the next year by the construction of another 4 ha.

Even though soil was only removed down to a depth of 30 cm, this disturbance affected the quality of the soil forming the surface of the canal (Table 10.3). Lower levels of organic matter, total N, and interchangeable potassium (K) and calcium (Ca) were very noticeable the first year. But if the model was correct, with time nutrients deposited and captured in the canals would quickly reduce the differences in these soil characteristics. Water was maintained at a depth of 15 cm in the canals during the entire wet season, with only the excess water from occasional heavy rains allowed to drain from pipes installed at the ends of the canals that were connected to lateral drainage ditches. This simulated, then, the nutrient trapping ability of the marshes of the *Marceño* system during the period of inundation, but rather than be in fallow, the canals were planted to rice. The platforms were planted to corn during the wet season using varieties local farmers would employ at that time of the year for planting what they called high ground (*tierra alta*).

Yields for the first season, considering the newness of the system, were quite impressive. The average yield of 1.18 tons/ha of dry grain for corn on platforms to which locally recommended chemical fertilizer levels were applied was slightly higher than the regional level for mechanized corn planting (1.13 tons/ha). With no fertilization, yields

TABLE 10.3
Soil Analysis of Platforms and Canals in the Raised Field, *Bancales Tipo CSAT*, Tabasco, México*

Horizon	pH	O.M.[a]	N[b]	K[c]	Ca[c]	Mg[c]
Platform:						
0-15 cm	6.5	3.5	0.176	.51	23.81	11.34
15-30 cm	6.6	2.3	0.121	.39	22.61	11.58
Canal:						
0-15 cm	6.8	1.36	0.071	.29	18.24	14.35
15-30 cm	6.6	1.21	0.067	.29	17.32	15.17

Source: Modified from Amador (1980).
*Soil sample was taken in May, 1980; this site is also known as the "tropical *chinampas*" at el Colegio Superior de Agricultura Tropical at Cardenas in Tabasco.
[a]Percent organic matter (O.M.) in soil.
[b]Percent total nitrogen determined by kjeldahl digestion.
[c]Milliequivalents (me) per 100 grams of soil.

were lower (0.64 tons/ha) indicating that a certain amount of soil improvement would be necessary before achieving the average 1.5-2.5 tons/ha typical of local traditional corn plantings. Green manure cropping during the dry season on the platforms, coupled with periodic "mucking" from the canal bottoms, was proposed as a means of achieving soil improvement over time. At the same time, another corn crop could be planted in the canal during the dry season using the *marceño* system once the soil had been improved.

TEN YEARS LATER

There have been many changes in Tabasco since the experimentation with wetland management and raised fields began over ten years ago. According to some sources (Chapin 1988), most of the experimental systems are no longer in existence, and nothing of value has been gained from the agroecosystem models that have been described above. But more remains than can be seen during very short visits to a region. To gain a realistic perspective and background on the development process in the evolution of any agricultural or natural resource management system, one needs to look further into what has been learned from these experiments.

The experimental tropical *chinampas* of INIREB no longer exist in the communities where they were built. These systems were severely limited by the fact that economic support for their management was reduced when the large-scale *camellones chontales* projects began. They were not allowed to function long enough for either the ecological or cultural processes of adaptation to proceed far enough for them to survive on their own. Modifications in their construction and management could not be made until the experience of several seasons were gained, but this really did not happen. At the same time, the initial interest and involvement of local people, at first maintained by wages and other inputs, had not developed enough to survive on its own after such support was removed. Nevertheless, valuable information and experience had been gained.

The raised fields at the school near Nacajuca suffered much the same fate. As more support was invested in the large-scale systems near Tucta and other communities, much less support and interest was forthcoming for the school. The systems fell into disuse, although the network of canals and platforms still exists. The school continues to teach and train students from the local communities, but a large share of the direct experience with managing the raised field models comes from work on the camellones. A lot can be gained through a program that integrates local training and experimentation into development projects.

The experimental systems at CSAT enjoyed a much longer and productive period of management and testing. Up until July of 1985, when the College was abruptly closed because of budgetary and political problems, the modular production unit, the experimental tropical chinampa, and the *bancales* all were utilized for a broad array of research and training projects. Unfortunately, most of the information produced during their management had not been formally published, and much was lost during the two years that the school was closed. The systems were completely abandoned following closure, and no further measurements were taken. The installation was reopened in 1987 as a regional center for the Colegio de Postgraduados in Chapingo. Known as Centro de Enseñanza, Investigación y Capacitación para el Desarrollo Agropecuario, Forestal y Agricola del Sureste (CEICADES), the center's activities focus on graduate student research and training in tropical agriculture, aquaculture and natural resource management. The process of placing these experimental systems into production once again is underway. The important thing is that the basic structures of the systems still exist.

The models continue to be of value at CEICADES. During a graduate tropical agroecology course held at the center in July, 1988, students analyzed several characteristics of the tropical *chinampa* and *bancal* systems. They found that soils on the platforms showed many characteristics of good agricultural soils in the region. Good organic matter layers had formed, soil aggregate formation was well advanced, and soil macrofauna (e.g. earthworms) were very abundant (Gliessman unpubl. data). Especially important were the findings of muck formation and organic matter deposition on the bottoms of the canals, as noted in Table 10.4. Up to 8 cm of muck (*agualodo*) had accumulated, and partially decomposed organic debris were present on top of that layer (approximately 0.63 kg/m^2 and 1.33 kg/m^2 dry weight in the *chinampas* and *bancales* systems, respectively). Neither system had yet had the muck layer harvested for application to the platform surfaces, providing a potentially very important internalized input that is known to be valuable in the upland chinampa systems (Wilken 1988). These systems have been able to stabilize in the eight to ten years since their initial construction, offering an important and unique opportunity to continue experimentation that will be critical to establishing guidelines for sustainable management of such systems in the future.

There is much greater confusion, though, about what has happened to the larger scale projects. A proposal to install several model *bancales tipo CSAT* projects in communities of the collectivized agricultural development area known as the Plan Chontalpa (Amador 1985) led to the construction of at least two *bancal* systems covering several hectares each. They are being used for the mechanized production of grain crops, such as corn and rice, and depend heavily on agricultural chemicals. They have achieved varying degrees of success (M.F. Amador, pers. comm.). Attempts at establishing a more ecologically based management strategy for the *bancales* in these communities are currently very limited.

The modular production unit constructed in the community in southern Tabasco continued to function very well, but with heavy outside subsidy, for over five years until 1983. Then a decision was made internally to dedicate almost all of the collectivized land to raising cattle and bananas, with increased reliance on wage earning outside the community. The diverse structure of the modular unit was then simplified greatly, although the trees planted 12 years ago continue to provide important sources of fruit, firewood, and construction material. The canal network of the raised field systems is still used for limited aquaculture. There is interest in the community to diversify the unit once again. Could this be another step in the process of developing this system?

After the initial difficulties and setbacks of the first years of the management of the *camellones chontales*, the Chontales of Tucta began to diversify the cropping systems on the *camellones*. As external

TABLE 10.4
Organic Matter Layers in Tropical *Chinampas* and Raised Fields, *Bancales Tipo CSAT*, Tabasco, México

Location	Muck Depth (cm)	O.M.†
Tropical *Chinampas*:		
Canal site #1	5.0	.426
Canal site #2	8.0	.716
Canal site #3	6.0	.740
Bancales Tipo CSAT:		
Canal site #1	5.0	1.21
Canal site #2	5.0	1.24
Canal site #3	5.0	1.54

*Soil sample was taken 13 July 1988, with sample sites chosen at random along accessible canals.

†Decomposed organic matter (O.M.) is expressed as kg/m^2 dry weight after 48 hours at 70°C.

support and direction was reduced, collective management of the entire system was shifted to individual family management of separate platform sections, thus giving local farmers much more say in what was planted. Today, it is not uncommon to find up to 35 different useful species planted along a 100 m section of platform, ranging from vegetables such as onion and chilies, to basic crops like beans, corn, and bananas, to tree crops for fruit or firewood such as citrus and willow, to cash crops such as papaya and cacao. Soil studies currently in progress (A. Mejia, pers. comm.) are showing that many of the problems that had been initially encountered when the platforms were dredged up have now been overcome. The soils are much more homogenous, drainage has improved, good structure has formed, and the nutrient status makes the system much less dependent on chemical fertilizers. As the limitations to production†which range from poor soils to inefficient labor organization and unclear tenure†are reduced, the process of adaptation and evolution of the system can continue. It is important that research be carried out as the process of adaptation and change occurs†not only to aid in developing viable and productive local systems, but also to help researchers better understand how the development process itself takes place.

FUTURE DIRECTIONS

There are two perspectives from which to examine the results of the above experimental wetland agroecosystem models. One is from the point of view of assessing if the original goals of the models were achieved from either economic, ecological or cultural perspectives. If those goals were not achieved for the greater part, we can take the position that the projects were failures, and that in some way we have been seduced by the models (Chapin 1988). This conclusion implies that the projects should not have been initiated or funded in the first place. On the other hand, we can assess what we learned from testing the models, and then use the knowledge gained to work with local people (if the projects are located in their communities), or continue to adjust and experiment (if the projects are located at experiment stations). The incomplete success of the models is not due to poor design or incorrect application. Rather, as so often is the case, a useful model was subjected to the myriad pressures and problems that most development projects in the tropics have had to bear. Short term financial returns were expected, while the slowness of the necessary ecological processes and the necessity of developing culturally appropriate management practices were practically ignored. Key components of sustainability were missing or understressed. Most importantly, the process has only gone through a small part of the time necessary for a new agroecosystem to evolve and adapt.

In my opinion, these models in the lowlands of southeastern tropical México offer a unique opportunity to continue the process of developing more sustainable agroecosystems that can contribute greatly to maintaining both the cultural and biological diversity of the region. Detailed baseline data from more than ten years ago exist for soils, crops, native plants and animals, community organization, and cultural diversity. All of these components of an agroecosystem have continued to coevolve and develop, with much of the more recent development taking place without outside support or intervention. At the same time, Tabasco had gone from the heady times of oil development to the difficult times of a depressed market for oil worldwide. The need to develop food production systems that are more self-sufficient, less dependent on costly imported inputs, and adapted to the local environment, has never been greater.

Wetland ecosystems are known to be some of the most productive in the world from a purely ecological perspective (Whittaker 1975). Humans in many parts of the world have learned how to take advantage of this productivity with the development of a great diversity of agroecosystem designs and management strategies. Southeastern México is an area known for its richness of traditional agriculture, having gone through this adaptation process for many centuries. But as the complex issue of sustainability becomes more important (Brown et al. 1987), new systems will need to have a greater focus on maintaining productivity on the long-term, rather than an exclusive focus on the maximization of yields (Gliessman 1985). The raised bed agroecosystem model in its many forms is still a potentially very viable model from which to work (Gómez-Pompa 1978, Redclift 1987). It is a key time to expand support for these systems.

ACKNOWLEDGMENTS

The author gratefully acknowledges the assistance, collaboration, and tolerance of a large number of colleagues, students, and local farmers while on the faculty of the Colegio Superior de Agricultura Tropical. Special thanks and respect are extended to Alba Gonzalez Jácome, Roberto García Espinosa, Moises Amador Alarcón, John Vandermeer, Armando Mejia Nuñez, Ramiro Sanchez, and Arturo Gómez-Pompa. Support for the preparation of this report was provided in part by the Kellogg National Fellowship Program, and the Alfred Heller Chair in Agroecology.

REFERENCES

Agency for International Development. 1987. *The Environment: Managing Natural Resources for Sustainable Development*. Special Report, Fall 1987. Agency for International Development, Washington, D.C.

Allen, P. and D. Van Dusen. eds. 1988. *Global Perspectives on Agroecology and Sustainable Agriculture*. University of California, Santa Cruz.

Altieri, M.A. 1987. *Agroecology: The Scientific Basis of Alternative Agriculture*. Westview Press, Boulder, Colorado.

Altieri, M.A. and M.K. Anderson. 1986. An ecological basis for the development of alternative agricultural systems for small farmers in the Third World. *American Journal of Alternative Agriculture* 1(1):30-38.

Altieri, M.A., M.K. Anderson and L.C. Merrick. 1987. Peasant agriculture and the conservation of crop and wild plant resources. *Conservation Biology* 1:49-58.

Amador, M.F. 1980. El uso de bancales como alternativa en los suelos mal drenados. (Unpubl. Manucript). CSAT, Cardenas, Tabasco, México.

Amador, M.F. 1985. Los Bancales. Una alternativa para los suelos inundables de Tabasco. SECUR. Villahermosa, Tabasco, México.

Amador, M.F. and S.R. Gliessman. 1990. An ecological approach to reducing external inputs through the use of intercropping. Pages 146-159 *in* S.R. Gliessman (ed.), *Agroecology: Researching the Ecological Basis for Sustainable Agriculture*. Springer-Verlag, New York.

Barkin, D. 1978. *Desarrollo Regional y Reorganización Campesina: La Chontalpa come Reflejo del Problema Agropecuario Mexicano.* CECODES. Editorial Nueva Imagen, México.

Boucher, D.H., J.M. Espinosa, S.B. Romero and S.R. Gliessman. 1983. Out-of-season planting of grain legumes as green manure for a tropical raised-field agro-ecosystem. *Biological Agriculture and Horticulture* 1:127-133.

Brown, B.J., M.E Hanson, D.M. Liverman and R.W. Merideth, Jr. 1987. Global sustainability: Toward definition. *Environmental Management* 11:713-719.

Chapin, M. 1988. The seduction of models: Chinampa agriculture in Mexico. *Grassroots Development* 12:8-17.

Crews, T. 1985. Raised field agriculture, Tlaxcala, Mexico; with a focus on nutrient cycling. B.A. Thesis, Environmental Studies, University of California, Santa Cruz.

Darch, J.P. 1983. *Drained Field Agriculture in Central and South America*. BAR International Series 189, London.

Douglass, G.K. 1984. *Agricultural Sustainability in a Changing World Order*. Westview Press, Boulder, Colorado.

Dover, M.J. and L. M. Talbot. 1987. *To Feed The Earth: AgroEcology for Sustainable Development*. World Resources Institute, Washington, D.C.

Edens, T.C., C. Fridgen and S.L. Battlefield, eds. 1985. *Sustainable Agriculture & Integrated Farming Systems*. Michigan State University Press, East Lansing.

Gliessman, S.R. 1984. Resource management in traditional tropical agroecosystems in Southeast Mexico. Pages 191-201 *in* G. Douglass (ed.), *Agricultural Sustainability in a Changing World Order*. Westview Press, Boulder, Colorado.

_______. 1985. An agroecological approach for researching sustainable agroecosystems. Pages 184-199 *in* H. Vogtmann, E. Boehncke and I. Fricke (eds.), *The Importance of Biological Agriculture in a World of Diminishing Resources*. Verlagsgruppe, Germany.

Gliessman, S.R., R. Garcia E. and M. Amador A. 1978. *Modulo de Producción Diversificada*. Colegio Superior de Agricultura Tropical, Cardenas, Tabasco, México.

_______. 1981. The ecological basis for the application of traditional agricultural technology in the management of tropical agroecosystems. *Agro-Ecosystems* 7:173-185.

Gliessman, S.R., B.L. Turner, F.J. Rosado-May and M.F. Amador. 1985. Ancient raised field agriculture in the Maya lowlands of southeastern Mexico. Pages 97-111 *in* I.S. Farrington (ed.), *Prehistoric Intensive Agriculture in the Tropics. BAR International Series 232*, London.

Gómez-Pompa, A. 1978. An old answer for the future. *Mazingira* 5:50-55.

_________. 1987. On Maya silviculture. *Mexican Studies* 3:1-17.

Gómez-Pompa, A. and A. Venegas. 1976. La chinampa tropical. *INIREB Informa* 5:1-4.

Gómez-Pompa, A. and J.J. Jimenez-Osornio. 1987. Some reflections on intensive traditional agriculture. Paper presented at the 1987 Meeting of the Society of Economic Anthropology, University of California, Riverside. (Available from A. Gómez-Pompa)

Gonzalez J. A. 1986. Agroecología del suroeste de Tlaxcala. Pages 201-220 *in* Anónimo (ed.), *Historia y Sociedad en Tlaxcala*, Memorias del primero simposio internacional de investigaciones sociohistoricas sobre Tlaxcala, Octubre 1985. Gobierno del Estado de Tlaxcala, México.

Harrison, P.D. and B.L. Turner. 1978. *Prehispanic Maya Agriculture*. University of New Mexico Press, Albuquerque.

Jackson, W., W. Berry and B. Colman, eds. 1984. *Meeting the Expectations of the Land: Essays in Sustainable Agriculture and Stewardship*. Northpoint Press, Berkeley.

Jimenez-Osornio, J.J. and A. Gómez-Pompa. 1987. *Las Chinampas Mexicanas*. Paper presented to El Simposio Sobre Ecología y Recursos Naturales: Destrucción, recuperación y desarrollo en Iberoamérica. Sevilla, 5-9 de Octubre de 1987.

Klee, G. 1980. *World Systems of Traditional Resource Management*, Halstead, New York.

Knorr, D. 1983. *Sustainable Food Systems*. AVI Publishing Company, Westport, Connecticut.

Lappe, F.M. and J. Collins. 1982. *World Hunger–10 Myths*. Institute for Food and Development Policy, San Francisco, California.

Orozco-Segovia, A. and S.R. Gliessman. In press. The Marceño in flood-prone regions of Tabasco, Mexico. *In* A. Gonzalez J. and S.R. Gliessman (eds.) *Pasado y Presente en los Sistemas Agricolas en México*. Instituto Nacional de Antropologia e Historia, México.

Redclift, M. 1987. "Raised Bed" agriculture in pre-Columbian central and South America: A traditional solution to the problem of sustainable farming system? *Biological Agriculture and Horticulture* 5:51-59.

Siemens, A.H. 1982. Prehispanic agricultural use of wetlands in Northern Belize. Pages 205-225 *in* K.V. Flannery (ed.), *Maya*

Subsistence: Studies in Memory of Dennis E. Puleston. Academic Press, London.

________. 1990. Reducing the Risk: Some implications regarding prehispanic wetland agricultural intensification from contemporary use of a wetland/terra firma boundary zone in central Veracruz, Mexico. Pages 233-250 *in* S.R. Gliessman (ed.), *Agroecology: Researching the Ecological Basis for Sustainable Agriculture.* Springer-Verlag, New York.

Turner, B.L. and P.D. Harrison. 1981. Prehistoric raised-field agriculture in the Maya lowlands. *Science* 213:399-405.

Vandermeer, J.H. and S.R. Gliessman. 1980. Agricultura sobre campos elevados en el trópico: Algunas consideraciones elementales de diseño. *Agricultura Tropical (CSAT)* 2:33-40.

Venegas, R. 1978. Las *chinampas* de Mixquic. Thesis Profesional. Facultad de Ciencias, Universidad Nacional Autónoma de México, México.

West, R.C., N.P. Psuty and B.G. Thom. 1969. *The Tabascan Lowlands of Southeastern Mexico.* Louisiana State University Press, Baton Rouge.

Whittaker, R.H. 1975. *Communities and Ecosystems,* 2nd Edition. McMillan, New York.

Wilken, G.C. 1988. *Good Farmers: Traditional Agricultural Resource Management in Mexico and Central America.* University of California Press, Berkeley.

11

The Practical Value of Ecodevelopment Research

Mac Chapin

INTRODUCTION

As the condition of the natural world throughout Latin America deteriorates at an ever-increasing rate, with national economies sliding into bankruptcy across the board, people have begun to react. Virtually every country now boasts at least an incipient conservationist movement, and environmental issues are rapidly becoming a legitimate theme in public debate. Thus far, however, discussion has far outstripped concrete action. This is perhaps to be expected, since the problems we face are complex, and there is still widespread disagreement regarding the mix and relative weight of the multiple factors propelling us over the red line into the danger zone. Without consensus as to the causes and how they fit together, the task of arriving at coherent and workable recommendations is clearly difficult, especially in a political environment which is seldom amenable to environmental concerns.

The once-heralded Green Revolution, with its monocrop systems, intensive use of machinery and chemicals, and the concentration of wealth it brought in its wake, is now seen in retrospect not only as having fallen far short of expectations, but also as a major contributor to environmental devastation (Pearse 1980). At the same time, it is apparent that real alternatives to present patterns of exploitation have not been forthcoming. The truth of the matter is that most attempts to find ecologically sound production systems have been arrested at the conceptual, or at best the experimental, stage. Applied work with peasant farmers in Latin America has been characterized by lack of coordination among funders and technical assistance agencies, poor understanding of the needs and concerns of rural people, and massive confusion over short- and long-term objectives. While good intentions often underlie the tiny handful of field projects one finds in Latin America, tangible results are negligible.

This lack of success in field research is in sharp contrast to the massive flow of articles, position papers, and even books that have appeared in the short span of a few years within the area currently referred to as "sustainable development" or, alternatively, as "ecodevelopment"[1] in Latin America. Conferences on one or another theme surrounding the much talked about alliance of conservation and development are being held at a dizzying rate (especially in northern industrialized countries)—to the point where the topic has developed into what can only be termed a full-blown fad. Unfortunately, a large part of the literature piling up at high speed is theoretical, unrelentingly general, and only vaguely informed by critical information from the field. Aside from the usual raft of promotional pieces and cries for action, much recent work is cerebral and academic, consisting of conceptual models of ideal systems and programmatic statements of how things might or should be done. What is lacking is careful analysis of ongoing projects to find out what works and what does not work, and the reasons for success or failure. Only with this information in hand can useful strategies be formulated that will combine conservation and development.

In this essay, I confront the matter of traditional knowledge and its value in the search for ecologically sound production systems. However, from the start I want to emphasize that my definition of "traditional knowledge" is not limited to ethnobotany or particular farming techniques. Rather, it includes these within a wider context that also encompasses the social, economic, physical, technical, and political environment within which peasants and indigenous peoples live. My reason for insisting on this definition is that specific technologies do not exist as isolates, suspended in space, as it were, in the absence of a living context. By the same token, particular technologies are neither inherently good nor bad; they tend to thrive in certain environments, but wither in others. Only if we consider them in their natural environment, as an integral part of a coherent social, economic, and political system, will we come to understand what we are dealing with.

Also crucial to my argument is the assumption that all applied research in ecodevelopment—or, more specifically, "sustainable agriculture"—should by definition have utility for small farmers. This may seem too obvious to merit mention, but the truth is that a large portion of research that purports to be in some sense "applied" either has no particular application in the field or is only of peripheral or insubstantial interest to peasant farmers. Researchers often harbor the persistent notion that they "know" whereas rural people do "not know."[2] Few of these outsiders carry with them the guiding influence of even the most rudimentary understanding of what it means to be poor and rural in the Third World. And, of course, when they arrive in the field with an alien frame of mind and their own objectives and agendas, they inevitably emerge with findings that hold no practical value to anybody except, perhaps, fellow academics. Thus far, ecodevelopment work in Latin America has been carried out overwhelmingly by biologists. Given most biologists' concern for the deteriorating state of the planet's natural endowment, their prominent role in the search for alternative agricultural strategies is to be expected.[3] Beyond this, many biologists are academics who have "entered the fray" because they are alarmed about the damage being visited upon the natural world they study.

ECODEVELOPMENT IN MEXICO

In June and May of 1987, I spent seven weeks in México visiting close to 20 small-scale agricultural projects which purported to fall within the boundaries of ecodevelopment. During the course of my travels, I had ample opportunity to observe a variety of strategies in operation, and to discuss at some length the successes, failures, frustrations, and other concerns of people who worked at the community level. Brief accounts of two of these projects (at the two study sites indicated in Figure 11.1) are herein presented as a basis for discussion of a number of points that arose during this study which I consider crucial to successful application of traditional knowledge among small farmers in Latin America. These two case studies are meant to be illustrative of issues raised in the concluding section.

The Chontal Raised Beds of Tabasco

In the mid-1970s, México's Instituto Nacional de Investigaciones sobre los Recursos Bióticos (INIREB) launched a program that showcased the construction of *chinampas*, or "raised beds" (see Chapin 1988), in the swampy regions of Veracruz and Tabasco, along the east coast of México. *Chinampa* is a Mexican term derived from the Nahuatl word *chinamitl*, which literally means "enclosure of cane."[4] *Chinampas* are rectangular agricultural plots built in shallow lakes or swamplands of layers of organic material and mud. Given optimum conditions, *chinampa* agriculture stands as one of the most intensive and productive farming systems ever developed by man. At the time of the Spanish conquest, a vast *chinampa* network covered the southern reaches of the marshy Valley of México and constituted the productive backbone of the heavily populated region. Even today,

FIGURE 11.1 Map of México showing the two major research sites at Oaxaca City in the state of Oaxaca and Villahermosa in the state of Tabasco.

chinampa agriculture, using basically the same techniques as those of preConquest times, persists in a few, small–and shrinking–areas of the Valley of México. The best known of these is Xochimilco, the site of the famous "floating gardens" which, of course, do not float.

Chinampa beds were, and are still today, built of alternating layers of aquatic weeds, bottom muck, and earth packed inside rectangular cane frames firmly rooted to the lake/swamp floor. These "artificial islands" vary in size, ranging from 3 m to 8 m in width and 30 m to 100 m in length. *Ahuejote* trees (a type of willow) are planted on the edges of the *chinampas* to provide shade, and their roots form living fences that anchor the beds more securely to the lake bottom. The narrowness of the beds assures that the water in the surrounding canals filters evenly through the plots at root level. Soil fertility is maintained through regular applications of swamp muck, aquatic plants, and manure. Canals (1 m to 3 m wide) separate the *chinampas*, forming a network of isolated islands reachable only by water.

Considerable literature exists on the history and agronomic value of *chinampa* agriculture (e.g., see Coe 1964, Armillas 1971, Gómez-Pompa and Venegas 1976, Gómez-Pompa 1978, Venegas Cardoso 1978, Rojas Rabiela 1983, Gómez-Pompa and Giddings 1986, Chen 1987). In a very real sense, the *chinampa* form of agriculture has represented a self-contained and self-sustaining production system for centuries. Until the last several decades, it demanded no significant capital inputs, yet maintained an extraordinarily high level of productivity year after year. A wide variety of crops, ranging from staples such as corn and beans to vegetables and flowers for the market, were mixed with an array of fruit trees and bushes. In addition, abundant aquatic life, such as fish, salamanders, frogs, turtles, and fowl, added protein to the agricultural diet. The system is, however, labor intensive. As in prehistoric times, the primary agricultural tool remains the digging stick. The farming plots, which ideally have considerable selection of crops in perpetual rotation throughout the year, need constant tending, and interconnecting waterways must be kept clear of mud and weeds.

The *chinampa* system as it existed in pre-Hispanic times went into rapid decline during the Conquest. Disease caused decline of the population of the Valley of México from an estimated 1-1.5 million to around 70,000 by the end of the 17th century, and the beds were abandoned and left in disarray. The Spaniards drained large portions of the Valley's lake system, and by the first decades of the 20th century, Lake Xochimilco had shrunk to one-quarter of its former size. In recent years, the few remaining *chinampas* have been subjected to heavy assaults from all sides as México City has extended out and swallowed Xochimilco. The fresh water springs that once fed the lake have been diverted for urban use and more than two-thirds of the *chinampas* in existence in the 1930's have dried up and been covered with houses and streets. The *chinampas* still hanging on are provided with low-grade recycled sewage. Many of the canals have become stagnant and contaminated with garbage, domestic waste runoff, and, increasingly, insecticides and chemical fertilizers used by farmers, called *chinamperos*, to raise new and "improved" plant varieties. The fauna once so abundant in the area has vanished, the willows are dying off, and if the present trend continues the *chinamperos* will soon follow suit into extinction (Mendoza 1980, Siller 1980, García Davish 1984, Outerbridge 1987).

But while the once-flourishing *chinampa* system from the Valley of México is literally disappearing, the model remains intact. Moreover, to Mexican researchers in the 1970s, it seemed ideal as an ecologically sound alternative for small farmers in México. It promotes biological diversity, thrives without chemical inputs, and maintains high, year-round food production. It is not an abstract conceptual theory that requires exhaustive testing, but a living, functional system that has proved over the centuries that it works. And beyond this, when its applied potential came under consideration, it had a romantic patina about it: it was announced in the mid-1970s, as "an old answer to the future"–a phrase consciously used to suggest revitalization of a remnant of the glorious Mexican past (Gómez-Pompa 1978).

INIREB carried out its first *chinampa* transfer experiments in the tropical lowland region of Tabasco in 1976, at a site called San Pedro Balancán. In 1978, after the San Pedro project was declared a success,[5] the Instituto Nacional Indigenista (INI)–México's official arm of assistance to its Indian peoples–joined with the state government of Tabasco to begin construction of a series of large "raised agricultural beds," called *camellones*, in the swamps bordering the Chontal Indian villages of Tucta, La Cruz, and Olcuatitán, a short distance from the capital city of Tabasco, Villahermosa, (Alemán et al. 1987). The project received financial backing from The World Bank and the Mexican government. INI's stated objectives were to:

- Provide permanent employment for the Chontales, who were virtually landless and earned their living from wage labor.
- Bring about self-sufficient food production in the area.
- Ensure a constant flow of vegetable production for the internal markets of Villahermosa.
- Strengthen the indigenous cultural identity of the Chontales.
- Develop an alternative agricultural system appropriate for swamplands.

At the same time, however, several unstated agendas and suppositions lurked beneath these objectives. There was an attempt to divert the Chontales away from wage labor, which was deemed degrading, through promotion of intensive cultivation of both cash and subsistence crops. There was a plan to create a production center for vegetables for nearby Villahermosa–on the model of Xochimilco's relationship to the markets of México City–and indeed this aspect was important to the project's backers as a justification for the high cost of implementation. The project was framed as a program to strengthen the ethnic identity of the Chontales by returning them to what were imagined to be their pre-Hispanic roots. And it was viewed as a pioneering scientific venture, an experiment in the reclamation of agricultural land, through the resurrection of *chinampa* technology, in marshy areas of lowland tropical México. What is significant about both the stated and unstated objectives is that they did not emerge spontaneously from the Chontal community.

One of the principal sites for the new project was Tucta, a Chontal village of approximately 300 families, located a short distance from Villahermosa. The hamlet sits on the edge of a huge open-water swamp that is boxed off on the landward side by cattle ranches owned by non-Indians. Before the *chinampas* were built, only 10% of the families had farming plots, ranging from one-fifth to one-half ha (0.2-0.5 ha) in size.

The introduction of *chinampa* technology began on a grand scale. To speed up the traditional mode of construction involving hand labor and the meticulous layering of organic matter and mud, INI brought in huge aquatic dredges. They clawed up mud from the swamp bottom and heaped it in piles measuring roughly 30 m across and 100-300 m in length. The 65 raised strips of land that resulted were enormous, but the natural order of the soils within them was inverted. Organic material was not properly integrated into the beds, and the richest soil was placed on the bottom, leaving the surface of the beds covered with sterile clay which hardened like cement following exposure to air. Although the beds had the appearance of *chinampas*, they lacked the fertility and porosity required to function like *chinampas*. At the same time, the irregular gouging action of the dredge claws left the surrounding canals with a highly irregular bottom ranging from 1-5 m in depth. This made it impossible for the Chontales to use dragnets to fish in the canal areas around the beds.

Further compounding these difficulties, INI organized the Chontales into collective work teams. According to INIREB technicians brought into the project at a later phase, this particular form of organization was entirely foreign to the people of Tucta. Project managers assumed that pre-Hispanic Indians had worked communally, and therefore communal labor was seen as the most appropriate form of labor for contemporary Indians implementing a pre-Hispanic technology.[6] Beyond this, they felt that collective action would foster social solidarity. During the first season, the farmers were paid a daily wage for their work from a fund controlled by INI. Harvest profits were then to be used to repay this money, with any surplus being divided among the workers.

Technical assistance and capital subsidies were heavy during the first year. Collaborating institutions introduced massive amounts of organic matter and fertilizer into the dense clay beds to improve the quality of the soils. Technicians worked closely with the Chontales to teach them how to grow a series of vegetables that were largely exotic to the region and totally exotic to the Chontales. In fact, everyone involved found himself confronted with new and difficult challenges. By all accounts, *chinamperos* brought from the Valley of México as advisors were unfamiliar with the volume and variety of insects they encountered in sweltering Tabasco, and large quantities of chemicals had to be used to control them. The growing seasons for specific crops proved to be different from those in the Valley of México, and adjustments had to be made. Coupled to all of these problems, the farmers were reportedly chafing under their collective yoke. However, the payment of wages and major subsidies of all sorts served to mitigate the snags and confusions, and by all accounts the first harvest was reasonably good.

The real problems started when the vegetables began to ripen. Despite the technical expertise poured into the project, no one had researched the market situation, and no arrangements had been made to transport and sell the highly perishable produce. While some of the crop rotted or was lost in the resulting disarray, wholesale buyers from Villahermosa swooped in and bought up what they wanted at cut-rate prices. The first harvest ended up in economic shambles. Disputes were soon erupting among program beneficiaries over cash and labor arrangements, and the Chontales rapidly lost interest in growing garden vegetables for the urban masses of Villahermosa.

Several efforts were made to retool the project. One of the chief complaints of the Chontales, ironically, had to do with one of the few ways in which the raised beds truly resembled *chinampas*. Isolated from each other by water, the beds were difficult to reach from the village. This defect was corrected by adding a strip of land at one end that linked the beds together in the form of a comb. The plots became readily accessible by foot, but unfortunately, the new structure cut off the natural flow of water through the canals, reducing the accumulation of silt and weeds needed to fertilize the gardens.

The project continued to slide into disorder until INIREB, which began to provide technical assistance in the early 1980s, started listening to the Chontales. Communal labor was abandoned and the 65 raised beds were taken over by individual families, a traditional arrangement they preferred. Intensive vegetable gardening was discontinued. According to one INIREB technician now working with the Chontales, even today "when someone mentions the word 'vegetables,' their hair stands on end." Presently, most families grow subsistence crops such as corn, beans, and bananas, together with some fruit. These crops are not labor or capital intensive, and can be tended while the Chontales pursue wage labor in Villahermosa–a secure source of family income. Some of the more enterprising farmers are gradually moving, with caution and on a very small scale, back into cash crops (see Gómez-Pompa and Jiménez Osornio, in press).

AGRICULTURAL IMPROVEMENTS IN THE ALTA MIXTECA OF OAXACA

In the late 1970s, the Centro de Estudios de Tecnologías Apropriadas para México (CETAMEX) and the staff from a World Neighbors program in Guatemala began a series of exchanges to discuss agricultural work among small farmers. At the time, World Neighbors had been running a successful program among the Ketchikel Indians in the province of Chimaltenango for several years. From this and other experiences stretched out over two decades, they had developed a practical methodology for what is termed "agricultural improvement" with small farmers (Bunch 1982). In the early 1980s, World Neighbors was forced to leave its base in Chimaltenango in the face of the extreme violence then running through the Guatemalan highlands.[7] They sought refuge in México, along with four seasoned Guatemalan technicians. Initially, they came to rest in Oaxaca where they entered into a collaborative agreement with CETAMEX. During the next several months, they journeyed about the country looking for potential work sites, and finally settled on the Oaxacan municipality of Magdalena Yodocono in the Alta Mixteca.

The landscape surrounding Yodocono has a stark, devastated aspect. Over the centuries, the rolling hills have been stripped virtually clean of natural vegetation; heavy erosion has laid the underlying bedrock bare, and opened up deep gullies of sterile clay. The formerly thick forests of pine and oak have been replaced by xerophytic plants such as prickly pear and organ cactus. An estimated 50% of the once arable land is now unused; and even with infusions of chemical fertilizers, the agricultural soils, often no more than 5-10 cm thick, are relatively unproductive. In fact, yields with traditional farming methods are so paltry that the bulk of Yodocono's basic grains must be bought from CONASUPO, the Mexican government's grain supply agency. Most families in the area depend on wage labor from seasonal migratory work. They work on nearby agroindustrial farms, seek small jobs in México City, or, in increasing numbers, journey to the United States for more permanent employment. Extreme poverty is the principal underlying theme of the entire Alta Mixteca area.

Another pervasive theme of the region is mistrust, especially of outsiders. The World Neighbors/CETAMEX program was first installed in 1982 with four outsiders—Guatemalan Indians—a director and three extensionists. Acceptance was gradual and conditioned strongly by the fact that it was soon evident that their mission was purely technical.[8] They managed to secure a small plot of land from the director of the local secondary school, and set up a series of basic demonstrations involving contour ditches to retain water, living barriers to stem soil erosion, and use of organic matter from compost piles and foliage gathered about the village. Everything they did was simple, was an improvement on local practices rather than a novel innovation, and was within the economic reach of local farmers. As these demonstrations unfolded, a few farmers moved closer to inspect them, became impressed by what they saw, and were incorporated into the program.

It was during this phase that program staff carried out an in-depth study of the "human and physical resources" in the communities of the municipality. This study followed a methodology which has evolved over the years within World Neighbors, and has been clearly articulated in a book called *Two Ears of Corn* (Bunch 1982). By means of informal conversation, open-ended interviews, formal surveys, and group meetings, the team took a systematic look at traditional farming practices, tenure systems, labor arrangements, natural resources and constraints on these resources, the local social, political, and economic context, consumption patterns, and local market prices.

From the start, program staff took steps to incorporate local farmers into this diagnostic process. They began with farmers interested in learning some of the techniques being offered. Those who showed particular capabilities and "enthusiasm"—a word used frequently by World Neighbors personnel—were then encouraged to work more formally as "promoters" with the program. They were given more intensive training and paid the equivalent of a local minimum wage (for two work days per week) to work with other farmers in the community. Their task was to spread "agricultural improvements" among local farmers. With time, the amount of work of these promoters in the program was increased, occasionally reaching 20 days per month—or a more or less full-time occupation. In this way, the World Neighbors/CETAMEX program spread to 18 communities in the Alta Mixteca region around Yodocono from 1982 and 1986. Two of the Guatemalan extensionists have left, while the remaining two Guatemalans have taken largely administrative jobs, turning virtually all of the promotional field work over to the 11 local promoters who have been incorporated into the program. Four other local farmers trained as promoters have been sent off to work in other regions of the country.

At the community level, the program religiously follows a comprehensive set of criteria, as expressed in the following questions (Bunch 1983:4-5):

- Does the technology respond to a felt need of the people?
- Is a significant increase of economic income achieved through this technology?
- Does the technology bring recognizable success to peasants in a relatively short period of time?
- Does the technology fit well into the existing agricultural system of farmers of the region?
- Does this technology deal with the limiting factors within the agricultural system of the area?
- Does this technology make use of the resources that the poorest people already have?
- Will this technology succeed in lowering risks for the farmers, rather than increasing them?
- Is the technology culturally acceptable for farmers and their families in the zone of impact?
- Is it simple enough to be easily learned?
- Can it be put into practice by the use of very little capital?
- Are there reliable and available marketing possibilities that will pay farmers suitably, even without access to their own vehicles?
- Will there be a sufficient market so that the prices do not drop too much at harvest time or when there is greater than average production in the area?
- Will this technology be safe for the ecology of the area?
- Is it easy to teach, and does it require little supervision, so that the program will be as effective as possible?
- Does it awaken enthusiasm among the farmers?
- Will the technology be adaptable to a sufficiently wide region?

Most of these questions are fairly straight-forward and in line with considerations that should be taken by all of those working in development at the community level. At least in theory, for example, few will disagree that proposed technologies should satisfy local needs, or that risky technologies should be avoided. However, the rationales for several points are less self-evident, and even go against the grain of conventional practice; they therefore deserve some explanation.

One of the hallmarks of the program is its strict opposition to any form of paternalism, which is so pervasive throughout the Alta Mixteca. Some 75 state and federal programs are active in the region; virtually all of them reward participating peasants with payment of cash or material goods. The World Neighbors/CETAMEX strategy, by contrast, is never to give the farmers ". . . funds, equipment, nor gratitude for work done on their own land" (Bunch 1983:5). On this point, the program differs markedly from the *chinampa* effort among the Chontales, which was replete with subsidies of all sorts, as well as from most of the ecodevelopment projects I have seen in México or, for that matter, in several other Latin American countries.

While few development specialists openly condone paternalism, many architects of ecodevelopment projects see subsidies as unfortunate yet necessary during the initial stages. This is because the most important objectives are generally long term, and tangible benefits will only begin rolling in after five or six years have passed. If the objective is soil conservation and the strategy for getting there is a combination of reforestation, terracing, and soil reconstitution through the introduction of organic matter, peasant farmers will not reap any tangible reward for their efforts until the measures have taken hold. This will necessarily be some distance down the road. With no immediate returns, farmers will have little or no interest in carrying out conservation measures. Thus, the argument goes, farmers must be helped over the "dry spell" with some sort of payment, either in wages or in kind, to keep their interest from flagging. Once the trees have grown, the soil has regained its fertility, and the water table has been restored, then production will begin to increase, and farmers will come to appreciate the reasoning of project managers. Demonstration will enlighten the farmers, and subsidies will no longer be necessary. The project will take hold and continue to run without the fuel of payment.

The CETAMEX/World Neighbors program has devised a strategy for getting around the incentive problem without relying on subsidies. It combines long term measures of soil and water conservation (contour ditches to retain water, living barriers to retain soil, green manuring, etc.) with "other techniques which increase harvests and income, so that the teachings of the programs maintain the interest of the farmers until the time when they see with their own eyes the long term results of soil conservation" (Bunch 1983:6). By providing income generating techniques from the start, World Neighbors is able to offer farmers small sums of credit to cover the cost of upgrading their operations. In Yodocono, project staff have worked to improve bean production by means of simple processes such as planting in furrows instead of scattering the seeds (primarily to control weeds), using carefully gauged quantities of chemical and organic fertilizers and insecticides, and reducing the number of plants in farm plots. Similar improvements are made with potatoes, another traditional crop. During the beginning phase of the program, project staff did not work to improve production of corn–which is, of course, one of the most important crops of the region. This was because, after initial limited experiments, they could find no way of raising production that was economically feasible for the local farmers. In this way, they hold to the tenet that introduced techniques that do not produce significant and rapid results should be dropped, so that room can be made for strategies carrying a payoff. This approach allows for extreme flexibility, allowing project staff to adapt its methods to local conditions and needs.

Another important characteristic of the World Neighbors/ CETAMEX program is that it not only talks about participation, it actually operates with participation. This point must be underlined and stressed. As all of those working with small farmers know, the rhetoric of "participation" is enjoying wild popularity throughout the field, yet examples of it genuinely occurring are unfortunately rare.[9] While initial work in Yodocono was carried out by Guatemalans, extension activities in communities were rapidly turned over to local farmers, who became staff promoters with the program. At present, one of the Guatemalans still manages the program, but all of the promoters are local Mixtec Indians. From time to time, technicians are brought in from the outside to give seminars and demonstrate new agricultural practices. But it must be noted that information gathered in these sessions is passed through the filter of local conditions and practices, and in this way its worth and appropriateness are weighed for possible use. Technical assistance assumes no more than a supporting role within the existing context of Yodocono.

The advantages of this reliance on local promoters should be obvious. There is no social or economic gulf between promoters and the people they work with, and therefore none of the "superior technician–stupid peasant" syndrome which plagues much development work. There are also no linguistic problems since the promoters speak Mixteco, the predominant language of the region. The promoters have an innate understanding of local etiquette and therefore know what is socially and politically acceptable, how to approach people, and when and where not to promote activities. Finally, the promoters are farmers themselves and, consequently, not only understand local farming systems thoroughly, but also know the work rhythm of the community. With the CETAMEX/World Neighbors program, participation is an all-pervasive reality.

CONCLUSIONS

The two programs presented here began at opposite ends of the development continuum, and therefore demonstrated very different histories. The *chinampa* project among the Chontalcs was designed and implemented by outside technicians without any significant local participation at the decision-making level. The *chinampa* project managers were preoccupied with the narrow technical task of implanting a conceptually pleasing agroecological model, which had the added attraction of carrying a venerable pre-Hispanic pedigree. However, stated and unstated objectives of project managers had little bearing on the interests and needs of the farmers, and there was no concern for the wider social, economic, and political context within which the farmers lived. As a result, there was no notion of how that model might accommodate itself within the local context, and the project rapidly fell to pieces when the "beneficiaries" failed to cooperate. It was only after the technicians began listening to the Chontales that the project fell in line with farmers' interests and, at the same time, achieved some harmony with local social and economic structures as well as Tabasco's physical environment.

By contrast, the CETAMEX/World Neighbors' program in Yodocono began with a thorough diagnostic study of the region's human and physical resources which took into account social, economic, political, technical, and environmental factors. On the basis of this study, project staff worked to find ways to upgrade existing production systems with incremental, easily understood technologies which were affordable, appropriate, and effective. The technologies introduced in this manner were aimed at increasing crop yields in the short run (to secure farmer interest in the program) and rehabilitating the natural environment in the long run (as part of the program's broader objectives). No attempt was made to impose a full blown agricultural model as a "technological fix" for poor farmers. Rather, the CETAMEX/World Neighbors methodology operated within the context of the community, picking and choosing from its store of available technologies to find those which made sense to local farmers. Social and technical factors were inextricably bound together in a single methodological weave. And any judgment regarding the worth or non-worth of a particular technical, economic, or social introduction was made by the farmers rather than outside technicians. This was assured because the CETAMEX/World Neighbors extension agents were themselves local farmers.

It is relatively easy to come forth with descriptions of traditional technologies from different parts of the world which are both productive and carry some positive ecological effect. A far more difficult task is to find applications for this sort of knowledge among the millions of small farmers of this world. Research of a practical nature must be oriented, from the very start, by the concerns of the farmers themselves. Research questions must be framed within the farmers' world, not the laboratory world of the researcher, so that the information gathered is relevant to the farmers. Random lists of ethnobotanical beliefs and primitive technological oddities collected along the lines of the scientist's agenda and ripped from the context in which they function, are of little practical value to rural people anywhere.

At the present time, many academics–primarily biologists with a sprinkling of anthropologists–have jumped onto the battlefield in an attempt to help preserve biological and cultural diversity and to put a stop to environmental devastation throughout Latin America. Unfortunately, most of them come poorly prepared for the tasks at hand. While they often have valuable technical expertise, together with an all-important ecological perspective, their formal training has seldom equipped them to deal with the reality of life in rural communities, where politics, economics, and social interaction are inextricably melded with agricultural technologies.[10] The discipline of biology–and to a lesser, yet still significant, extent anthropology–does not expose students to applied research and practical work with peasant farmers. Until this situation changes, a large portion of the ongoing ecodevelopment research, as well as the attempts at implementation that flow from it, will remain unfocused, irrelevant, and ineffective.

NOTES

1. "Ecodevelopment" (*ecodesarrollo*) is the term preferred in Latin America (e.g., see Barkin and Suarez 1985, which was published by the Centro de Ecodesarrollo, in México, and Marino de Botero and Tokatlián 1985). The term "sustainable development" is a product of the industrial north. According to Gow (1988:5), it was born "through the work of the International Institute for Environment and Development (IIED) and the discussions of the Brundtland Commission." In this paper, the term "ecodevelopment" will be used in lieu of "sustainable development", which seems redundant since all development, to be judged successful, should be in some sense sustainable.

2. Chambers (1983) provides an extensive review of the biases "outsiders" carry into the countryside with them. Most of these outsiders, academics and a variety of other professionals, appear to place their own concerns before those of the people they are studying or directly attempting to assist.

3. In his introduction to *Saving the Tropical Forest*, Michael Robinson (Gradwohl and Greenberg 1988:12) states: "So why are we scientists involving ourselves in a rainforest/biodiversity conservation campaign? . . . We . . . bring two things to bear on the problems. First, I think that we have a uniquely acute awareness of the prob-

able consequences of our present actions. I am sure that our sense of urgency about the nature of the crisis is totally unrivaled. Secondly, we should have some insights into feasible alternatives to present systems of resource exploitation, since the problems are essentially concerned with increasing the yield of natural systems of foods, fuel and raw materials. These two reasons seem to me ample justification for entering the fray, for throwing down the gauntlet to the developers."

4. Nahuatl was the predominant language in the Valley of México at the time of the Spanish Conquest. It is still spoken by Indian people throughout central México.

5. The San Pedro project was prematurely lauded as a "complete success because the area abandoned by us in 1976 was left to the care of the local people who have preserved and enlarged it since then with their own resources" (Gómez-Pompa 1978:5). However, in the early 1980s, peasant farmers likewise abandoned the area. These *chinampas* are presently unused and overgrown with weeds.

6. This was a romantic misconception on the part of project designers. Ancient Mexica (Azteca) law stipulated that ownership of land was collective, yet usufruct was individual. According to Soustelle (1961:79-80), "every adult married man had the right, and the indefeasible right, to be given a plot and to work it. He was inscribed in the rolls at the moment of his marriage, and if he had not already inherited the right of working his father's land, the *calpulli* was obliged to provide him with a piece. And at his death he left his children not the piece of land, but the right of farming it."

7. From the late 1970s until 1985, between 50,000 and 75,000 people were killed or disappeared in Guatemala; as many as 1,000,000 people became refugees internally or abroad. The focus of this violence was the Indian highlands region, and the Department of Chimaltenango was hit particularly hard, forcing World Neighbors to pull its operations out.

8. General distrust of outsiders had fueled rumors that the project staff had escaped Guatemala because it was active with the guerrillas in that country.

9. To cite a single example in the present context, INIREB's applied work with "integrated farms", which included the *chinampa* technology transfer effort, stressed community participation in programmatic documents (Gómez-Pompa 1978, Morales 1979, 1984:9-19). In actual fact, however, local participation was absent from the program. The sociologist who was the primary technician in charge, Morales (1984:84) commented retrospectively on the failed program: "Of course there was talk, everywhere, of promoting and respecting popular participation, but in practice decisions were made in a vertical manner."

10. In making his case for the prominent role of biologists in the field of ecodevelopment, Robinson (Gradwohl and Greenbert, 1988:12) said: "We have no particular expertise in the area of politics—probably the opposite." Indeed, this is a serious problem.

ACKNOWLEDGMENTS

Support for this study was provided by the Inter-American Foundation.

REFERENCES

Alemán Ramos, L., J. Lobato Cárcamo, P. González Franco and O. Pierard. 1987. *Los Camellones Chontales*. INIREB/Tabasco, México.

Armillas, P. 1971. Gardens in swamps. *Science* 174:653-661.

Barkin, D. and B. Suárez. 1985. *El Fin de la Autosuficiencia Alimentaria*. Centro de Ecodesarrollo, México City.

Bunch, R. 1982. *Two Ears of Corn*. World Neighbors. Oklahoma City, Oklahoma.

------ 1983. Management of Hillsides and Recovery of the Productive Capacity of Eroded Lands in Oaxaca: Monograph Concerning an Agricultural Program in La Alta. Mixteca, Oaxaca. World Neighbors, Oklahoma City.

Chambers, R. 1983. *Rural Development: Putting the Last First*. Longman Press, New York.

Chapin, M. 1988. The seduction of models: *Chinampa* agriculture in Mexico. *Grassroots Development* 12(1):8-17.

Chen, A. 1987. Unravelling another Mayan mystery. *Discover* (June): 40-49.

Coe, M. 1964. The *chinampas* of Mexico. *Scientific American* 211(l): 90-98.

García Davish, Francisco. 1984. Xochimilco, Virtual Zona de desastre. Uno Más Uno. 3 October.

Gómez-Pompa, A. 1978. An Old Answer to the Future. Pergamon Press, Ltd., Oxford. (Originally published in *Mazingira* 5)

Gómez-Pompa, A and L.E. Giddings. 1986. *INIREB's New Approach to Applied Research, Development, and Teaching*. World Resources Institute, Washington, D.C.

Gómez-Pompa, A and J.J. Jiménez Osornio. In press. Some reflections on intensive traditional agriculture. *In* C. Gladwin and K. Truman (eds.), *Food and Farm: Current Debates and Policies*. Westview Press, Boulder, Colorado.

Gómez-Pompa, A and R. Venegas. 1976. La *Chinampa* Tropical. INIREB Informa. INIREB, Xalapa.

Gow, D. 1988. *Beyond the Project: The Quest for Sustainability in the Third World*. DESFIL, Washington, D.C.

Marino de Botero, M. and J. Tokatlián, eds. 1985. Ecodesarrollo: El Pensamiento del Decenio. Instituto Nacional de los Recursos Naturales Renovables y del Medio Ambiente (INDERENA), Bogotá, Colombia.

Mendoza G., D.A. 1980. Las *chinampas* de Xochimilco. El Sol de México, 20 March.

Morales, H.L. 1979. La Granja Integrada: La Mancha. Boletí Bio-Aqua No. 1, Deceimbre. INIREB, Xalapa.

Morales, H.L. 1984. Análisis y Estrategias para el Desarrollo de un Programa Nacional de Agropiscicultura. CETAMEX, México City.

Outerbridge, T. 1987. The disappearing *chinampas* of Xochimilco. *The Ecologist* 17(2):76-83.

Pearse, A. 1980. *Seeds of Poverty, Seeds of Want: Social and Economic Implications of the Green Evolution*. Clarendon Press, Oxford, United Kingdom.

Robinson, M. 1988. Introduction–Beyond destruction, success. Pages 11-16 *in* J. Gradwohl and R. Greenberg, (eds.), Earthscan Publication Ltd., London.

Rojas Rabiela, T, ed. 1983. *La Agricultura Chinampera*. Universidad Autónoma Chapingo, México. Colección Cuadernos Universitarios.

Siller, D. 1980. De no tomarse medidas urgentes, en menos de 10 años Xochimilco se convertirá en un gran pantano. Uno Más Uno. 14 Agosto.

Soustelle, J. 1961. *The Daily Life of the Aztecs on the Eve of the Spanish Conquest*. Stanford University Press, Stanford.

Venegas Cardoso, F.R. 1978. Las *chinampas* de Mixquic. B.S. Thesis, Departamento de Biología, Universidad Nacional Autónoma de México, México City.

12

Cultural Survival and Conservation: Lessons from the Past Twenty Years

Jason Clay

INTRODUCTION

One of the primary difficulties in preparing this paper was in assessing the role of indigenous peoples as conservationists; or, put another way, in attempting to distinguish between those people who use resources and those who manage them. Anthropologists, who have done most research on the economic activities of indigenous peoples, as "the once and future resource managers." Biologists and conservationists, by contrast, are often adamant that unless the peoples articulate a conscious intent regarding specific resource utilization practices, these practices should fall under "sustainable resource use," not "sustainable resource management."

This basic distinction can be—and is—debated for hours. However, this polarization or endless debate is not terribly useful. The ways in which resources are utilized on all but a tiny fraction of European and North American farms are not sustainable, but yet they are highly managed. Some practices are accepted as conservationist without an understanding of their "scientific" basis. Indigenous people, because of the romanticism that surrounds what is often assumed to be their pristine lifestyle, are apparently being forced to adhere to a different standard than everyone else. What we are talking about are resource utilization systems that are in relatively sustainable stasis with the environment. It is safe to say that the lifestyle of any conservationist, at least one that lives in an industrialized part of the world, is less sustainable from an environmental point of view than that of an indigenous person who happens to live in the "front line" of environmental destruction in one of the more fragile areas of the earth.

Indigenous peoples are not conservationists in the same way as we all too infrequently try to be; for example, they rarely attempt to get other peoples' houses in order. The various world views and beliefs about the environment that distinguish indigenous peoples from us as well as from each other lead to specific systems of resource management. These systems are rarely random or even mostly opportunistic. The belief systems and the actions that are central to these systems represent attempts to maintain order in the world. Some of them are sustainable; some are not. Some are sustainable under certain conditions, but become destructive under others. Some individuals are more conserving than others in the same group. This is not so different than in our own societies.

The question that comes to mind, then, is: which conditions will encourage more indigenous people living in fragile environments to conserve resources? One of the most important issues underscored in this paper is the ability of indigenous peoples to organize themselves both to work with those who would support them and against those who would eliminate them. Strong organizations allow indigenous peoples to focus on other issues that are also key to their survival–the local autonomy and the information necessary to make decisions that affect the future; secure land rights and sources of income that complement the environment and lifestyles of the local inhabitants. Historically, indigenous peoples throughout the world have developed resource management systems in order to meet basic subsistence needs. These systems often combine root crops, cereal and vegetable crops, tree crops, and herding of domestic animals with hunting, fishing, and gathering. Some indigenous groups have added cash crops or market oriented activities and wage labor to this mix. These activities, in conjunction with growing populations and shrinking land bases, throw formerly sustainable resource management systems into imbalance, posing severe threats not only to the environment but to the future of the groups as well.

As the cash and material needs of these groups have increased, many have modified traditional resource management systems. Most modifications have taken place without advice or assistance from outsiders, and many have focused on new types of exploitation that preserve a resource base. Increasingly, however, organized indigenous peoples have, as groups, sought advice and assistance from a variety of sources. These efforts to adapt traditional subsistence systems of resource management (see Clay 1988) to include more income-generating activities while preserving resource bases deserve special attention. Traditional subsistence activities and introduced cash-generating activities are now combined into strategies that are constantly adjusted in response to environmental, political and economic changes. Having been forced into some of the most fragile regions in the world, the efforts–both successes and failures–of such groups offer many lessons not only about the functioning of these endangered areas, but also the limits to growth and colonization and possible economic activities in them.

Indigenous societies are becoming extinct at an even faster rate than the environments they have traditionally inhabited. They must not only confront outsiders attempting to exploit their areas, they must also face conservationists who attempt to restrict or exclude their activities. In the meantime, many indigenous societies are undergoing rapid change, and young people no longer learn the methods by which their ancestors lived in fragile regions. Little time remains to encourage or support the maintenance of the intimate knowledge upon which new strategies for ecologically and socially sound sustained yield development might be based.

A number of questions arise when attempting to determine the sustainability of indigenous management systems, particularly when such groups are confronted with population growth, increased material needs or a diminished land base. The Cultural Survival, Inc. (hereafter refered to as CS) case studies included in this paper raise a number of questions in this regard. Four general questions should be kept in mind. First, were the resource management systems employed traditionally by indigenous peoples ever sustainable? Second, can traditional resource management systems be sustainable given a wide variety of new constraints? Third, do indigenous practices allow for the generation of sufficient surplus production to meet the "essential" needs of the people in question or colonists? Fourth, can indigenous management practices related to specific belief systems be converted to acceptable conservation axioms that will be amenable to other indigenous peoples or colonists? Additional questions arise if one is trying to determine if indigenous management systems are appropriate models for colonist populations.

CULTURAL SURVIVAL, POLITICAL ORGANIZATION AND CONSERVATION

In the past two decades, indigenous peoples have begun to organize at both the village and national levels in order to take control of their lives and the destinies of future generations (see Clay 1984, Macdonald 1985, Posey and Balée 1989). These organizations cite land rights and development of sustainable systems of resource management as top priorities. Some programs involve the relatively straightforward but difficult task of obtaining legal title to the lands traditionally used. Others involve members working closely with outsiders to establish parks or reserves that overlap with indigenous areas, thus protecting the resource base of the group and the world as a whole. Sometimes, indigenous groups consult directly with outsiders in order to evaluate their options and propose appropriate alternatives for the present and the future.

Some project activities proposed by indigenous peoples involve modifying existing resource management systems into a more productive and conservationist mold. A number of indigenous initiatives which relate directly or indirectly to conservation and resource use are described below. Many similar projects also deserve description and analysis (some are described in detail in publications distributed by CS). The purpose of such work, however, should not be to develop models, but rather to stimulate our imaginations as well as those of indigenous peoples toward unique solutions to unique problems.

ATTEMPTS TO MODIFY DESTRUCTIVE RESOURCE-USE SYSTEMS

Many indigenous peoples are attempting to modify land use patterns that have damaged natural diversity and are jeopardizing their ability to maintain the resource base necessary for survival. These strategies often entail ending exploitative relationships with outsiders and adapting to smaller, more fragile environments as a result of being expelled from traditional homelands. In addition, strategies must be developed for the management of resources which had not previously been used for income. In the following cases, the close collaboration of indigenous communities with outside consultants has given the projects greater chances of success.

Huichol Communal Forests and Carpentry Workshops

In October, 1982, the Huichol Indians of México established a carpentry workshop and school called "Project Tuapuri," which grew out of their need to guarantee and maintain their land base and livelihood. They had previously been selling logging concessions to timber companies (Leon 1986); they realized that if present cutting rates continued, they would soon deforest their land. Under the supervision of a forestry professor from the University of Guadalajara, a dozen apprentice Huichol learned how to choose trees suitable for cutting, how to fell trees and how to saw and dry the boards. Through the program, the Huichol selectively utilize the forest, thus demonstrating the community's claim to surrounding forest resources. They also selectively rid their forests of damaged or diseased trees while increasing employment opportunities on their own land.

An Austrian chemistry and physics professor obtained funds for an experimental solar powered kiln to dry lumber. In 1983, the kiln provided sufficient lumber for the manufacture of 150 chairs commissioned by Guadalajara's Instituto Technologico y de Estudios Superiores de Occidente (ITESO). The Tuapuri workshop received 1,850 pesos per chair. One pine log, which yielded enough lumber for 50 chairs and generated approximately 92,500 pesos, would have sold for only 200 pesos when uncut. The chairs thus produced a 450% increase in gross income per log for the Huichol. Profits from finished wood projects enable trained carpenters to earn an income and remain at the workshop to train apprentices. The sale of chairs and wood products will help the workshop to become financially independent. The program has been expanded to three other communities.

Instruction at the workshop is geared to the production of high-quality goods for domestic sale. This creates a profitable, community-run industry which offers an alternative to migratory work and avoids disrupting the community's traditionally egalitarian economy. The workers are first to receive income generated from the sale of finished products; the surplus goes to the communal treasury. The workshop also serves as a focus for community mobilization. A 10% levy on all income from the sale of furniture goes toward communal projects (such as schools and health care) in the participating villages. When a recent drought created a shortage of corn in the area, a communal warehouse was erected near the site of the school workshop and profits from the workshop allowed the village to purchase 30 tons of corn obtained from the Mexican government at subsidized rates. Needy Huichol were lent the corn on the condition that they repay it after the next year's harvest, thus eliminating the need to borrow money at high interest rates.

CRIC–Environmental Restoration in Colombia

CRIC, the indigenous council in Colombia's Department of Cauca, is composed of 56 Indian community councils. The organization was formed in 1971 to protect Indian lands, culture and human rights. Cauca, a region that holds one of Colombia's highest concentrations of Indians, is located in the southwest part of the country. Cauca's original forest cover was destroyed long ago by mining and cattle ranching interests. Its eroded lands and impoverished population stand in contrast to the Cauca River valley–the site of the largest, most prosperous sugar plantations in Colombia. As the valley's sugar plantations have expanded, poor peasant farmers have moved further up the hillsides into upland frontier areas, encroaching on lands belonging to indigenous peoples. As a result, intense competition has arisen over land that is both scarce and of poor quality (Chernela 1987c:72). Conflicts over land have attracted both government troops and guerrillas. In the past 15 years more than 100 Indians have reportedly been killed in the area.

In 1984, CRIC ranked environmental restoration as its first priority. With a staff of more than 30 non-Indian specialists, CRIC ensures that the indigenous communities retain control of the development in their areas. CRIC began a reforestation program by establishing three tree nurseries, run by local communities, to supply seedlings to communities that agree to plant a minimum of 1,000 trees–all native species. One community has carried out nine reforestation projects in only two years. As Chernela (1987c:72) writes,

> Such projects are carried out by community members utilizing simple technologies. Consulting experts conduct an initial training phase in each community and visit the projects at intervals thereafter. Community trainees learn to recognize native species and their qualities, such as growth rates, required growing conditions and economic and environmental uses. Former trainees often teach others to collect, prepare and plant seeds. In addition, CRIC has prepared a simple manual for communities or individuals wishing to practice forest restoration.
>
> The reforestation program is a model for community resource management. Many formerly unskilled community members, trained initially by outside specialists, have become specialists themselves. They improve their skills by attending seminars in Bogota, and later lend their expertise to other communities. Outside specialists train community members, who then hand down those skills to the community. This process is slow, but it creates a sustainable project, eliminates problematic and costly dependencies and empowers a community with valuable skills.

Environmental education programs are important components of CRIC's approach to identify, and they present alternative forms of resource management. CRIC recognizes that reforestation and conservation measures that ensure sustained use of resources offer long term economic advantages over shortsighted, immediate returns. Currently, its major objective is the reforestation of watershed areas. In addition, CRIC is working with communities to plant shade and fruit trees as well as stands for firewood–a commodity now in short supply in many countries, and is also interested in luring wild animals back into the area through reforestation projects. Some research indicates that small mammals and birds can be attracted to forest patches with as little as four hectares (Clay 1988). If reforestation is undertaken throughout a given area in conjunction with other conservation schemes, some ecologists predict that local populations of some species of animals might even reach commercial levels (Janzen 1986).

Training and outreach are important parts of CRIC's resource management programs. Four years ago, Indians trained in resource management and reforestation were invited to a non-Indian community to assist in reforestation and farming; today, the community is outspoken in its support for the Indians' methods and results. In such artificially created forests, Chernela (1987b) writes, prospects for sustainability are good, providing (1) the patches are large enough, (2) species diversity is high, and (3) gains from conservation are greater than the immediate needs for forest products and their sale. The last point seems the most crucial and most difficult to predict. Replanted forests can provide fuelwood and protect watersheds, but what is the correct value to be ascribed to these functions? Will Indians (or any other residents) come to the same conclusions–that is, that the long term benefits outweigh short-term needs?

CRIC hopes to meet some of the needs of local Indians through its reforestation and management programs. Reforestation can protect fragile watersheds, provide fuelwood, protect soils from erosion and provide fruits and shade and materials for fencing. The measure of success of the program, as Chernela suggests (1987b:10), "will be not only the number of forests planted, and the efficacy of the forest in bringing about desired ends, but the increased demands from [neighboring] *cabildos* [communities] to participate in the program." In another, related environmental program, CRIC recently began an experimental farm in the indigenous community of Purace. Some 30 years before, a mine exploded in Purace, burning a large area and releasing toxic pollutants. Some lands remained hard and unproductive at the time the farm was founded in 1985.

Students from a local agricultural institute taught and assisted farmers in preparing organic compost for fertilizer. They used all manner of organic material, including crop residue, to restore the devastated soils. In 1987, eight communities attended rotating work days on the land; some people walking for as long as four hours to undertake the work. This year they harvested ten different crops on this previously unusable land, and sold surpluses above subsistence needs (even corn and wheat) in nearby towns.

Embera-Katio Land Reclamation and Agroforestry–Colombia

The Embera-Katio have undertaken a land reclamation project for deforested tropical lowlands. The Embera-Katio, some of Colombia's poorest Indians, live on slopes of the western lowlands, the Choco, which contain some of the and most diverse wettest tropical forests in the world. Nevertheless, as in many lowland areas, forest has already been cleared for pasture. In this case, the Embera had cleared the land themselves, apparently for two reasons. First, in Colombia as in many other countries with rain forests, clearing the land established their claims to it. Second, clearing the land for pasture would give them credit for purchasing cattle, which they hoped would become a lucrative cash crop. The Embera, however, could not obtain the credit because they did not hold individual title to the land. As a result, they rented their pasture to non-Indian cattle owners.

To secure their hold on the land as well as to increase income from it, they finally obtained funds to purchase cattle for fattening. This, however, has proven to be a disaster. The Embera-Katio want the pastures to be fallowed and more environmentally sound land use projects to be designed. At present, an agronomist is working with them to develop alternative land-use patterns. For the Embera-Katio, the bottom line of any land-use substitution project is that it must create at least as much income as cattle. If successful, the resource management projects developed with this group can serve as positive examples of what can be done in other similarly affected tropical lowlands.

Mapuche Extension Service and Watershed Management—Chile

This project supports the Sociedad Mapuche de Servicio, Asesoría y Investigación 'Lonko Kilapan', a Mapuche Indian-controlled resource management program. In south central Chile, technical assistance to Mapuche Indian farmers is limited. Some assistance comes from non-Indian organizations, which are often technically skilled but are regarded with suspicion by Indians due to the widespread violence against Mapuches following the 1973 coup. Other assistance is provided by Indian organizations that lack adequate training and technical expertise. Lonko Kilapan, however, is composed of well-trained Mapuche who have formed an independent Indian organization yet have maintained strong ties with non-Indian technicians who continue to provide indirect assistance.

At present, Lonko Kilapan members are working primarily with small farmers in communities within a short distance of the city of Temuco. However, as their work expands, they expect to respond to requests from more isolated high Andean communities near the Bio-Bio River headwaters. Such assistance will become essential if a planned hydroelectric dam on the upper Bio-Bio receives funding from the World Bank. In this case, the communities will require extensive assistance in forest control and other aspects of watershed management. Under the terms of the World Bank project, no aid exists for these isolated communities or their fragile resource base.

LAND TENURE AND CONSERVATION

Throughout the world, it has been noted that a people whose future is tied to certain resources will tend to protect them. Conversely, insecure land tenure produces environmental degradation. For indigenous peoples, secure land title is a first and essential step in the conservation of vast areas of fragile habitats. The following projects indicate the close relationship between land tenure and the development of sustainable resource management systems.

Siona-Secoya, Cofan, and Huaorani Land Demarcation–Ecuador

In 1980, in response to Amazonian requests for secure land tenure in a developing region, the Ecuadorian government established a commission consisting of representatives from CS, the National Agrarian Reform Institute, the Forestry and National Parks Department, a regional development agency and 22 Indians representing three groups. The commission spent three months in the Amazon studying the land needs of Ecuador's most seriously threatened Indian groups–the Siona-Secoya, Cofan, and Huaorani. The commission's re-port documented land requirements and recommended specific bor-ders for land holdings to satisfy these needs. A project report has been published and distributed. Formal demarcation is underway.

Aside from the land guarantees for indigenous peoples, the project also includes land set aside in national parks. Territory for hunting and fishing would have required large expanses of virgin forest, amounts difficult to justify considering the anticipated population pressure in the area. To resolve this problem, communal land holdings were established adjacent to existing national parks and/or faunal reserves. This relationship served three functions:

- It provided the Indians with access to extensive forest areas for subsistence hunting and fishing, thereby allowing them to maintain a traditional lifestyle while evaluating options for coping with changing circumstances. In short, it bought them time.
- It established an inhabited "buffer zone" between the parks and encroaching non-Indian colonists who, in other areas, had done irreparable damage to park land and fauna.
- It provided indigenous people with titled land suitable for market-oriented economic activities.

In addition to formal titling, the activities undertaken focused on the development of rational, sustained yield, tropical forest agroecosystems. Many officials feared that titled land would lie untouched while Indians obtained essential cash through overexploitation of the national forests and faunal reserves. Others feared that titled land would be converted to pasture, a practice now recognized as unsuitable for sustained yield production. To provide an appropriate land use scheme, newly titled Indian lands were incorporated into a large forestry development and management project in Ecuador. In this way, Indians will utilize their titled land in programs of controlled timber extraction, agriculture and agroforestry.

The Ecuadorian land projects stand as a model for international and interinstitutional collaboration, respect for indigenous land rights, protection for national parklands, and sustained yield forestry and agroforestry programs in tropical rain forests.

The Awa Ethnic and Forest Reserve–Colombia and Ecuador

The Awa-Coaiquer Indians occupy a tropical rain forest that straddles the Colombia-Ecuador border. In the early 1980s, the World Bank was considering funding a road-building and colonization project in the area. In 1980, CS supported a land demarcation feasibility study with the Awa. By 1983, regional representatives from 13 Ecuadorian government agencies established an interinstitutional commission for development in the area. One of the commission's first acts was the demarcation of the Awa-Coaiquer Reserve. According to CS's projects director, Theodore Macdonald, CS agreed in September 1984 to provide support for the following:

- Aid to the Awa-Coaiquer and Ecuadorian Agrarian Reform Agency (IERAC) for demarcating and titling traditional Awa land (approximately 80,000 ha).
- Obtain citizenship identification certificates for the adult population and thus permit their right to land ownership.
- Assist CONACNIE (National Coordinating Council of the Indian Nations of Ecuador) in establishing closer links with Awa communities.
- Provide logistical aid to CONACNIE and other support groups involved in training sessions aimed at informing the Awa of their rights within the national society and developing a panethnic organization to exercise these rights.
- Provide modest indemnification for agricultural production introduced onto Awa lands by colonists. The overall goal of the first phase of this project was simply to obtain legal title to the land Awa communities already use.

What results has the project produced? All 11 previously unorganized communities met and elected the Awa's first governing council. More than 1,100 citizenship cards were distributed. All but one small, isolated sector of Awa territory had been formally surveyed by March 1986. In the past three years, Awa Indian communities, the national Indian organization and several government agencies have peacefully collaborated to obtain land for one of the nation's smallest and most isolated societies. With only limited assistance from outside advisors, "the Awa have organized themselves into a political entity essential for the defense of their territory" (Macdonald 1986:34). Macdonald continues:

> Obtaining title and establishing a local political body to defend it is an important step toward cultural survival; guaranteeing permanent land tenure is another matter. Land that outsiders perceive as "idle" (i.e., not put to some obvious economic use) becomes a coveted parcel and an easy target once outsiders learn of its existence. In order to prevent confiscation of "idle" land, a management scheme or "development" project is essential.
>
> The Awa have already taken the first steps toward this end in order to protect their territory; a botanical research team from the University of Aarus (Denmark) undertook the first botanical survey of the area . . . . The territory is by and large ecologically undisturbed, and contains a wide variety of species whose distribution is limited to this particular area. Thus the area can be considered a "refuge zone" (i.e., one of several "islands" of tropical forest which remained as tropical forest during the driest period of the Pleistocene).
>
> Such refuge zones are considered extremely important research areas and, for that reason alone, it has been recommended internationally that they remain unmodified. Such arguments buy essential time for the local population to consider and develop a program of sustainable forest use and management, rather than simply concede to national development priorities that usually lead to extensive deforestation and cattle raising.

In 1985, Ecuador and Colombia agreed to collaborate on an integrated program that would both preserve the fragile environment in the area and aid the Awa-Coaiquer. Each country has proposed programs that would combine Indian holdings into large reserves as well as a management program that would protect their habitat. In this instance, each country's expertise complemented the other's. As a result of intense contact with indigenous communities in the area, land rights in Ecuador were first guaranteed; only later did those involved begin to grapple with the problems and land use programs. In Colombia just the opposite has happened: La Planada, a nature reserve, already existed as a major research site concerned with resource management, but it lacked the connections with indigenous communities that would allow its promoters to assess land use patterns and needs. The project has evolved into an example of international collaboration with regard to Indian rights and the controlled development of fragile ecosystems. The project demonstrates that Indian human rights, resource conservation and national economic development are compatible when combined in a program that is sensitive to the needs of the affected groups.

Achuar Land Demarcation–Peru

The Achuar, who number approximately 7,000, live in the lowland forests on the Peru-Ecuador border. Although they were previously isolated from the settlement and development taking place in the highland areas, their lands are now threatened by recent development of oil production and the pipelines that carry oil to the coastal ports.

Access to land and productive resources is critical to indigenous peoples' capacity for self-determination and the protection of their resource base for future generations. For this reason, CS seeks out and supports projects that demarcate and title lands for indigenous groups. Since 1980, CS has assisted Peruvian anthropologists' efforts to delimit and title Achuar Indian lands in Peru. In 1984, CS extended support to an Achuar organization to allow Indians to assume control of the land demarcation process.

CS previously supported efforts by Indian assistance groups for Achuar censuses and land measurement as the first steps to acquire legal recognition of their land areas. Recently, the Achuar have organized a Regional council to assume responsibility for the work. This federation, Organización Achuar Chayat (ORACH), is one of a growing number of such organizations formed by Indians as a means of promoting their self-determination. Such organizations also give Indians access to the practical skills of leadership and financial management, provide information about national bureaucracies and offer assistance in dealing with them. ORACH now receives support to develop the institutional structure and management skills necessary for assuming control of the efforts to demarcate and title their lands.

Extractive Reserves–Brazil

In October 1985, rubber tappers (*seringueiros*) from Brazil's western Amazonian states of Acre and Rondonia met to discuss new ways to protect Brazil's rain forests and the cultures they harbor. At the meeting, the attendees drafted a proposal to create "extractive reserves" based on the indigenous reserve model. These reserves would be set aside for rubber tappers and/or indigenous peoples to collectively manage. This proposal has been heralded as the first grassroots tropical forest conservation initiative to come out of Amazonia (Schwartzman 1986).

In the past few years, rubber tappers and indigenous communities have overcome considerable differences and past conflicts to come together, most recently in support of extractive reserves. Rubber tappers first arrived in the western Brazilian Amazon during the rubber boom, between 1870 and 1910. After the boom, Schwartzman (1986:41) notes that they stayed on and "supported themselves with mixed crop agriculture, gathering, hunting, fishing and selling crude latex." He also observed:

> [Rubber tappers] are the nucleus of a growing rural union movement in Acre and Amazonia. Many indigenous communities in this area also produce rubber for sale, and in both indigenous and Brazilian communities, producers' and consumers' cooperatives are emerging to eliminate exploitative middlemen.
>
> A long history of violence and mistrust exists among Indians and *seringueiros* (rubber tappers). During the rubber boom, rubber barons organized *corridas* in Acre—seringueiros were sent on expeditions to massacre indigenous communities to drive them out. Many seringueiros still regard Indians on par with malaria, snakes and jaguars—dangers of life in the forest. But the seringueiros learned how to live in the forest from the Indians, and today the free seringueiro communities are economically similar to indigenous communities.

Brazil's current constitution guarantees Indians rights to the land they occupy, and promises that all lands will be formally demarcated by 1993. Currently, less than 25% of the land thus far identified as "Indian" has even a minimal legal guarantee. Rubber tappers currently have no land rights under the constitution. During the past three decades, Indians and rubber tappers alike have seen the failure of many government colonization schemes in the Amazon. Not only have the poor and landless people brought into the region seen no benefits; they have, in addition, displaced local populations that depend upon forest resources through sustainable and productive use. Through the establishment of designated extractive reserves and group rights to the use of the areas, both Indians and rubber tappers have a stake in the long term economic productivity and sustainability of these areas. The number of people involved in these activities is not insignificant. In 1980, more than half a million people made their living from extracting resources from the forests (for example, rubber, Brazil nuts, palm nuts and ipecac). The value of the materials collected in 1989 has been estimated at more than U.S. $70 million (Schwartzman 1986:42).

Recently, the head of the Brazilian national miners' union proposed an alliance whereby Indians and indigenist groups would support mining reserves and miners would support Indian reserves. Although such unilateral support may never materialize, or may only occur in the distant future, some Indians have already forged agreements with groups of miners working in their areas. The Kayapó, for example, have made their own arrangements with independent miners who work on their reserves. As a result, the Kayapó now have more control over the miners' activities (in terms of their location and scope) and make more money from the mining activities than they did when the government's Indian organization was making the arrangements with the miners.

INDIGENOUS PEOPLES AND RESERVE MANAGEMENT

In the past, indigenous peoples were not only excluded from the management of parks and reserves but also from living in and utilizing the resources as well. In the last 20 years, however, some indigenous groups, recognizing the protective value of such designated areas, have set up their own. Park planners have recognized the utility of incorporating the concerns of indigenous peoples into overall management strategies. The cases below demonstrate the value, indeed the necessity, of including indigenous peoples in the management of areas that they traditionally claim.

The Kuna Indigenous Reserve–Panama

The Kuna Indians of Panama recently established the world's first internationally recognized forest park created by an indigenous group (Chapin and Breslin 1984, Chapin 1985). This reserve, established with the assistance of the Inter-American Foundation, the Smithsonian Tropical Research Institute (STRI), CATIE, World Wildlife Fund/U.S., and the U.S. Agency for International Development (U.S.A.I.D.), not only provides revenue to the Kuna from tourists who come to learn about the rain forest and from the sale of research rights to scientists, but protects and preserves an important part of the Kuna's heritage.

The Kuna Park has become a cause celebre among environmentalists and indigenous peoples' rights advocates. However, there are numerous reasons to be cautious about using the Kuna example as a model for other groups or areas. First, the 30,000 or so Kuna– living in more than 60 villages on small nearshore islands–never depended on forest areas to provide significant income or even space for subsistence agriculture. They depend on the ocean, trade, nearshore agriculture and, more recently, employment in U.S. armed forces bases in the Canal area, for their subsistence and cash needs. Furthermore, the Kuna already prohibited agriculture, hunting, and cutting trees in a number of spirit sites near their villages, although they did permit collecting of items to be used for food and medicine. Thus, in contrast to the experiences of many indigenous peoples, they experienced no fundamental conflict over restricting land use in rain forest areas.

Second, unlike most indigenous peoples, the Kuna already had title to the land in question. In the 1930s, through a series of armed uprisings, the Kuna forced the government to set aside this land for them. Since that time, they have maintained a strong political organization and united front in their dealings with outsiders. The Kuna have controlled, from the outset, the design, implementation and management of the park.

Third, by the 1970s the Kuna had become intensely interested in protecting their resources and their autonomy from outsiders who were encroaching on their lands. The new park–created as it was from relatively undisturbed rain forest and, better still, financed by a number of international organizations–served as an effective buffer zone between Kuna and outsiders.

Fourth, the Kuna are politically astute. Their numerous Western friends and advisors–some of whom they retain on contract–advised them about the money making potential of creating a rain forest reserve. With their help, the Kuna obtained grants to help establish the resource base that would ensure income for decades to come. Yet through all the negotiations, the Kuna remained in charge. An indication of their adroit ability to use the international concern for rain forests to their own advantage is their current desire to begin a campaign for endowment that will provide them with the resources necessary to develop a number of alternative income-generating schemes in tropical rain forests.

One area in which the Kuna have excelled is in controlling access to research sites and demanding reports from researchers. The Kuna now require payment from scientists who want to undertake collections or research in their area. Other indigenous groups, or even countries, could learn from their experience. In addition, they require that all scientists leave reports of their activities and findings behind before they leave the area. Because STRI works closely with the Kuna and most scientists work in conjunction with STRI, they must comply. The Kuna ask that subsequent analysis and reports based on the information collected are sent to them as well. They require that each scientist take a paid Kuna assistant with them throughout their stay. The Kuna can thus monitor the scientists' activities, ensure their safety and learn how different scientists collect information and observe remote areas of the reserve in order to assess conditions there.

Puerto Rastrojo Natural Resource Management–Colombia

In October 1986, CS funding permitted local Indians to become involved in the development and implementation of a conservation and management program to protect the endangered giant river turtle (*P. expansa*) in the Río Caquetá area of the Colombian Amazon (Leon 1986:78-80). The overall program functions under the auspices of the Fundación Estación de Biología Puerto Rastrojo (FBPR), an applied biological research institute that has been performing biological, ecological and anthropological research in the Colombian Amazon for a number of years. Since 1983, the World Wildlife Fund has helped FBPR to undertake research and to establish a national park along the Caquetá River. Despite existing protection laws, the giant river turtle is a highly endangered species due to the overexploitation of its eggs and meat for commercial purposes. The largest freshwater turtle in the world, its meat has long been used by both Indians and mestizos throughout the Amazon. The Caquetá River is now practically the only major river in the Colombian Amazon in which this turtle still exists.

Using Indian assistants, researchers are now collecting information on the history and current levels of turtle use in the area as well as on the cultural significance of the turtles to the Indians. Local Indians are also being trained to play an active role in the management program. The national park was created, at least in part, to protect the turtles' nesting areas. The project is designed to integrate park and resource management into the Indians' traditional forms of social and political organization and to reinforce existing Indian community resource management systems. This will increase the program's likelihood of acceptance by local Indian communities and their willingness to become involved in the long term, sustainable management of the turtles. It is also presently searching for ways in which Indians can benefit from the limited commercial exploitation of the turtle by eliminating the white intermediaries who currently monopolize such activities. Other income-generating activities also must be developed to provide profitable alternatives to the dependence on giant turtle trade.

!Kung San Development Project–Namibia

Over the past century, the !Kung San (Bushmen) of Namibia have been reduced from a self-sufficient existence, maintained through hunting and gathering in the Nyae Nyae region of the Kalahari Desert, to a life dependent on welfare. From 1900 to 1970, their population of 100,000 declined to 21,000 (1970 census, Namibia). In seventy years' time, the government or private farmers took more than 85% of their lands, forcing most !Kung San to move onto white- or black-owned farms to work as wage laborers. Since 1978, many young !Kung San men have joined the South African army to fight against SWAPO (South African People's Organization). Those that remained have been moved to a government post where they live in demoralizing poverty, dependent on government handouts. They suffer from malnutrition and alcoholism, high infant mortality rates, unemployment and internal violence—conditions unknown to them, for the most part, in their previous way of life.

The government is now attempting to create a game reserve with the remainder of the !Kung San's permanently inhabitable land, 6,000 sq. km. in eastern Bushmanland, inhabited by some 2,000 Ju/Wasi !Kung San. The government intends for a few !Kung San to remain in the park and live as hunters-and-gatherers, serving as the park's main tourist attraction. Some of these Ju/Wasi live by a mixed economy of cattle husbandry and subsistence agriculture, supplemented by hunting and gathering. !Kung San lack the land resources and the skills to live exclusively by hunting and gathering. Moreover, they do not wish to return to their traditional way of life, which they see as a dead end. They want to retain many of their old values, but they also have a number of material needs for which their traditional hunting-and-gathering way of life no longer provides. The !Kung San believe that their best chance to improve their lives is to return to traditional waterholes and take up a mixed subsistence economy, supplementing hunting and gathering with animal husbandry, agriculture, wage labor and, at least initially, assistance from the government. Many !Kung San have experiences working on farms and believe they can now use their skills to improve their lives.

To aid the !Kung San's efforts, the !Kung San Foundation was established in Namibia in 1978. CS coordinates U.S. efforts to aid the organization, which also receives private foundation and public European assistance. The foundation raises funds to assist Ju/Wasi !Kung San and those who are trying to help them in their self-development. It provides basic training in agricultural and herding techniques and direct material assistance (including livestock, equipment and medicine) to Ju/Wasi !Kung San who are trying to resettle near traditional waterholes and recently drilled shallow boreholes. As of 1987, 14 Ju/Wasi groups of about 60 people each had resettled successfully, and established cattle posts with 60 head of livestock each. The !Kung San are improving their economic condition, increasing their independence and continuing many aspects of their traditional social and cultural life. Other groups would like to follow their example.

The Ju/Wasi vehemently oppose the government's plans to establish a game reserve. If the reserve were to become a reality, all of the traditional waterholes would be expropriated. All but a handful of Ju/Wasi would be evicted with their livestock and would be forced to move back to the government post or to relocate farther to the west to expensive boreholes they could not afford to maintain. This would force them to depend on government handouts. Subsistence farming would be prohibited in the reserve, and the role envisioned for the few Ju/Wasi who would remain would be as hunter-gatherer curiosities for tourists visiting the park. If the Ju/Wasi are not allowed to continue their efforts to support themselves at traditional waterholes, their population will continue to decline in the future.

CULTURAL CHANGE AND RESOURCE MANAGEMENT

As indigenous peoples encounter other economic, political and belief systems, their own values change. Nowhere is this more evident than in the case of children educated by outsiders. Indigenous belief systems are the basis of traditional management systems; as these beliefs change, the management systems are thrown off balance. As the following case indicates, we need to learn how to reinforce the traditional values that produce sustainable management systems as well as how to translate these values into conservation axioms that others can understand.

Tukano Fishing Practices–Brazil

A number of indigenous peoples have developed intricate systems of resource management that maintain a sustainable balance between harvest rates and yields of areas. For most of these peoples the resource management systems are deeply embedded in their cultures. Chernela has written about a fascinating system of fish management employed by the Tukano Indians of the northwest Amazon. The Tukano reserve the forested river margin for fish and fishing; yet, as Chernela writes, "fishing may be restricted to as little as 38% of the total river margin available. The result is a management system which allows for, yet distinguishes, human use areas and animal refuge areas" (1987a:50). For generations the Tukano Indians of the Upper Río Negro have chosen to preserve the riparian forests rather than to deforest this area for agriculture. The Tukano believe that the long term economic benefits from fish production in the standing flooded areas of the forest are greater than those to be acquired from the agricultural use of the same area.

The Tukano, who number about 10,000, depend primarily on root crop cultivation and fishing for their subsistence needs. As Chernela (1987a:50) has stated: "Manioc is the principal carbohydrate source and primary crop, comprising about 85% of garden yields. Apart from a few small mammals and birds, fish provide almost all animal protein."

Because the Vaupes basin releases little inorganic material to enrich the water, its waters are deficient of nutrients. Not only does the Vaupes lack the basic nutrients that form the basis of the food chain upon which fish depend, but its high acid and metal content is actually deleterious to fish. Fish in such rivers derive their food from substances that fall into the river from the banks, or the food that is derived from forest areas adjacent to the river during two annual floods. At these times "the rivers swell and overflow their banks, merging the aquatic and terrestrial realms. Fish gorge themselves–spreading through the flooded forest and feeding on the abundance of forest foods only then available" (Chernela 1987a:50).

> As fishermen dependent upon these river systems, the Tukano are aware of the relationship between their environment and the life cycles of fish, particularly the role played by the adjacent forest in providing nutrient sources that maintain vital fisheries. Strong Tukano proscriptions prohibit deforestation or cultivation of the river margin which is viewed as part of the aquatic system (1987a:50).

Because the Tukano believe that the forested riverbanks belong to fish rather than humans, their forests cannot be cut. Furthermore, the riverbank soils are as poor as those further inland, so farming them brings no particular advantage. More food can be gleaned from the areas by letting the fish use them than by farming them.

Taboos are key to the Tukano system of sustainable resource management, and more fishing taboos abound during the flooding seasons than at any other time. Tukanoan taboos rest on two important concepts–boundedness and reciprocity. Tukano believe that distinct groups of humans and animals (in this case, fish) have proper places for sitting and breathing. Because the ancestors rest there and souls are recycled in alternate generations there, such locations should not be disturbed. Thus, the fish own the margins. Tukano also believe in reciprocity; in the case of fishing taboos, they believe that if fish are taken from restricted areas, the ancestors of the fish will take infant children–one child for one fish. Chernela (1987a:51) suggests that by examining the location of the guardian spirits and the associated taboos, we can begin to see their importance in sustaining the fish population.

Unlike most other researchers, Chernela has speculated about the relationship between taboos and conservation of resources. In one village, five groups had dominion over various areas of the river that could be exploited (Chernela 1987a:51). However, 26 guardian spirits restricted access to these potential fishing locations. One of the five groups (containing 38 people) in the village, for example, controlled fishing on 8.8 km of the Vaupes and 29 of its tributaries–a total of 164.8 km of river front. Fishing was prohibited on 17 of the 29 tributaries (62% of the lengths of all streams). These 17 areas are reported to be spawning areas for one of the most commonly sought after fish in the area. Such taboos serve to conserve fish stocks.

> Since the rising waters are associated with fish spawning runs, the correlation is clear: all potential spawning areas are restricted, either permanently or during spawning seasons. In the long run, the consequences of such proscriptions could prove critical to the maintenance of the fish populations.
>
> Through the belief of spirit guardians, a nutrient-poor river system is managed in such a way that the possibility of irreversible damages to the resident fish populations is greatly reduced.
>
> Traditional Tukanoan human-fish relations are governed by rules of balance and reciprocity. An object taken from one realm is compensated by an object taken from another. Neither man nor fish is the superior party; instead, each is of equal power to destroy. In contrast with Western belief models, for Tukano, boundaries are respected and reciprocity the mode of boundary-crossing. Breaching the rules creates imbalance and retaliation. The numbers of both humans and fish are seen as finite and precarious, locked into a carefully monitored exchange relationship (Chernela 1987a:52).

The Tukano have discovered that by maintaining forests and restricting fishing they can make a potentially sterile system relatively productive. Today they face problems resulting from Western beliefs about productivity in rain forest areas. These beliefs maintain that river margins (*varzeas* in Brazil) are the best–that is, least destructive–areas in tropical rain forests for agriculture. This belief stems from the notion that flooding replenishes the fertility of surrounding areas. In the case of blackwater rivers, however, just the opposite occurs–the flooding, in fact, enriches the river.

Western beliefs are especially destructive to Tukano management belief systems because they contribute to the loss or abandonment of traditional information that young Tukano must learn to maintain the system. Younger Indians who leave the villages and attend boarding schools have little opportunity to learn from their elders. Furthermore, many of the teachers, albeit with good intentions, discourage their students from learning things that they consider backward or superstitious. To show how quickly these knowledge systems can be lost, Chernela reports that in 1979 elders of one village named 26 areas where fishing was restricted or prohibited. By 1985, children from the same village attending a nearby school could name only a few (Chernela 1987a:52).

LOCAL ORGANIZATIONS–A KEY TO CONSERVATION

Local organizations, in contrast to individuals or unorganized communities, are essential to ensure continued respect for indigenous land rights and develop resource management programs by consensus.

These two activities could ensure the survival of the ecosystem upon which the survival of indigenous peoples as groups depends. The cases below describe briefly the important relationship between local-level organizing, land rights and sustainable group management strategies. For more information on this topic see *Cultural Survival Quarterly* 8(4) and other CS publications.

The Inter-Ethnic Development Association (AIDESEP)–Peru

Assistance in strengthening ethnic federations of indigenous peoples is a priority for CS. Strong ethnic federations are perhaps the best mechanism for indigenous societies to create a unified voice with which to demand their rights before national and international agencies–organizations that could easily overwhelm a single community or an isolated tribal group. Ethnic federations also reinforce cultural pride and identity, elements essential for group survival.

The Inter-Ethnic Development Association of the Peruvian Jungle (AIDESEP) is one such ethnic federation. It was formed in 1980 as a private, nonprofit Indian organization to organize and unify the approximately 220,000 tropical forest-dwelling Indians of Peru's eastern lowlands. By 1987, AIDESEP had incorporated, and it now represents over 15 ethnic groups. The organization is currently working to organize and incorporate other groups. As such, AIDESEP now works on behalf of about 60% of Peru's Amazonian Indians.

One of AIDESEP's problems, like all such ethnic federations with limited human and financial resources, is in maintaining a strong presence at both at national level and in communities that are often very isolated and difficult to reach. AIDESEP's directorate of four is responsible for overseeing projects that have community approval. Each local organization has three representatives that maintain regular contact with the directorate and relay decisions, problems and general information from the community level.

To improve its work, AIDESEP has organized itself internally into four departments: (1) Contracts and Projects, (2) Legal Affairs, (3) Scholarships, and (4) Publications. The Department of Contracts and Projects serves as the organization's link with official Peruvian agencies and institutions. This department establishes and maintains formal contacts with universities to enable Indian students (and uncertified Indian teachers) to continue their educations. Recently, AIDESEP has been meeting with the Center for Anthropological Research of the Peruvian Amazon (CIAAP) to develop an appropriate curriculum for training teachers in programs of bilingual/ bicultural education.

The Department of Legal Affairs supports AIDESEP's member organizations directly by helping them obtain legal organizational status (*personería juridicia*) by representing them before government agencies and in the courts and by keeping abreast of the current progress of Indian land titling being undertaken by the Ministry of Agriculture. It has also provided technical assistance to some communities. For example, an Achuar organization recently received training from AIDESEP to allow them to carry out a land demarcation effort. As part of its legal defense work, AIDESEP provides legal advice and advocacy regarding the land and resource rights of member groups. AIDESEP has also handled individual cases. For example, it obtained pardons for Aguaruna and Huambisa leaders who were imprisoned following a land invasion incident sparked by *mestizo* colonists on the Santiago River. Another case resulted in the release of four Amuesha Indians falsely accused of involvement in drug trafficking.

The Scholarships Department supports Indian students who are selected by their respective federations. By 1986, AIDESEP awarded scholarships to more than 35 students from different ethnic groups. These students are seen as an important future resource, helping support efforts for self-determination and economic development of their peoples.

For some years, the Publications Department has produced the newsletter *Voz Indígena* as a means of maintaining contact among the numerous Amazonian Indian groups, sharing news of the progress of their various projects. Periodically, affiliate organizations send representatives to AIDESEP's office in Lima to learn skills in journalism or photography from the *Voz Indígena* staff.

AIDESEP has been extremely successful in developing a program of activities to train, incorporate and represent Amazonian Indian communities and regional ethnic organizations. It provides legal defense and legal advice, supports education of their youth, publishes various journals and books and represents member organizations to government institutions. Most important, in carrying out this work, the leadership of AIDESEP has managed to remain true to its original goals: it has maintained an organization that is truly representative of the expressed needs and wishes of its constituent local-level groups. In recognition of the organization's achievements, AIDESEP (together with the Aguaruna-Huambise and the Coordinadora Amazónica [Amazonian Coordinating Group]) was awarded the prestigious Swedish Right Livelihood Award (commonly known as the Progressive Alternative Nobel Prize) in the fall of 1986. AIDESEP's president, Evaristo Nugkuag, traveled to Stockholm accompanied by José Narciso Jamijoy (secretary of ONIC, the National Indian

Organization of Colombia) and José Uranavi (president of CIDOB, the Indian Federation of Eastern Bolivia) to receive the award on behalf of these organizations. Afterwards, he met with the president of The World Bank to discuss the needs and rights of Indians in the face of large scale international development. Then he returned to Peru's lower Marañón River to monitor AIDESEP's work there–Indian leaders cannot spend too much time away from their roots.

Federation of Indian Organizations of the Napo (FOIN)–Ecuador

During 1987-1988 FOIN (Federation of Indian Organizations of the Napo) shifted its focus from regional organizing of Indians and incorporation of new communities, to land use programs for member communities. CS, along with other international donors, provided funds for the development of a long-range management plan for the organization and the resources claimed by the member communities.

Such work was determined to be at a crucial stage. During the past two years, FOIN has been working to prevent the expansion of agribusiness–largely African palm plantations and palm oil processing plants. They have obtained considerable national and international support for these activities despite pressure from a conservative government anxious to develop the Amazonian region under any conditions. Now FOIN is in the process of developing environmentally sound alternatives to those projects that they have recently been fighting to keep out of their areas. CS, in collaboration with FOIN, has received funds from U.S.A.I.D./Ecuador to develop a land use program for member communities. (Without the presence of such an existing indigenous organization, U.S.A.I.D./Ecuador would not have considered funding such a project.) As a first phase, the FOIN-organized communities have declared a moratorium on all forest clearing until the new land management program is developed.

Organizing and Sustainable Resource Management–Ecuador

CS supports OPIP (Organization of Indian Peoples of Pastaza) in the belief that ethnic/indigenous federations represent one of the most promising strategies for indigenous peoples in their attempts to survive as distinct groups. These federations need to develop sound and sustainable economic programs that satisfy their communities' immediate and long term economic needs as well as the national government's insistence on demonstrable productivity of all arable land. In 1986, OPIP declared its basic goals: land tenure for all member communities, economic self-sufficiency, ecologically sound economic activities and the development of a tightly run indigenous organization to pursue these goals. The group's concern about land titles has become increasingly urgent for two reasons. First, a rapidly expanding road network has brought increasing numbers of colonists, many of whom attempt to take land belonging to Indian communities. Second, Ecuador's president favors the expansion of agribusiness into the Amazon.

Government development activities in the area have centered on direct contact with individual communities in an attempt to break the political power that a larger ethnic federation would represent. OPIP has risen to the challenge by interceding as the unique organization to be dealt with in the area and by sponsoring its own economic programs. For example, OPIP runs its own community stores, eliminating layers of cost formerly added by merchants. OPIP has also established a marketing program for crafts and agricultural products. Its efforts to take control of commercial lumbering on communal lands are also part of its overall efforts to achieve self-sufficiency. OPIP has received additional funding and technical assistance for agricultural and fish farming programs. It is presently undertaking an education program with member communities to inform local residents about various new forms of rational forest resource use. All of OPIP's efforts to use, yet sustain, the resources of the region while generating much needed income for both local residents and the national economy reinforce the land rights of local residents.

Shuar Radio Education Programs–Ecuador

The Shuar, one of the best organized indigenous populations in Latin America, recently received CS support for their bilingual/bicultural radio education program. For some years the Shuar have used radio broadcasts to allow children and adults in remote villages to "attend" school without being forced to move to boarding schools. Children educated through their morning-only school broadcasts have tested higher than the national average, even though testing is only undertaken in Spanish. CS supports the overall thrust of the radio education project, as well as particular radio programs to inform Shuar of traditional resource management practices from other areas, and the Shuar use of radio to reinforce their own management strategies. Such radio programs will help initiate projects among other indigenous groups who live in remote rain forest environments.

CONCLUSIONS

The last two decades have shown that conservation and cultural survival go hand in hand. Thus, conservation is enhanced when strong local organizations and communal tenure systems are supported. Guaranteed land tenure, especially communal land tenure, also encourages conservation. Conservation of resources is most likely to be achieved when bottom-up and top-down strategies are pursued in tandem. In any case the exclusion of local farmers, hunters or communities from conservation, development or land management strategies will have serious consequences in the short term and could well produce even greater problems in the longterm. Conservation is also best achieved; in areas inhabited by indigenous peoples when increasing material needs are acknowledged and innovative sources of generating income are developed. Finally, conservation efforts appear to be enhanced when indigenous peoples develop modified resource management strategies which incorporate both traditional, subsistence activities as well as intensified or new income-generating activities. In such cases, overall dependence on any one activity is reduced.

The most severe threats to the world's fragile environments come from population pressure (usually from the dominant society and only rarely from indigenous peoples), legal restrictions on land rights in general and communal rights in particular, international debt, and the imposition of imported technologies or models of resource use that are inappropriate in fragile areas and create financial dependencies. Excluding populations from their traditional areas has disastrous consequences for the future of that environment, because those peoples are an integral part of the environmental dynamics.

In short, if we want to maintain genetic diversity and healthy bioregions in areas others consider their own, we must pay for it. This does not mean cash payment necessarily, but rather acceptable alternatives for sources of food and income. This task will never be finished. Conservation, like making a living, is a daily activity.

REFERENCES

Chapin, M. 1985. UDIRBI: An Indigenous Project in Environmental Conservation. Pages 31-52 *in* T. Macdonald Jr. (ed.) *Native Peoples and Economic Development*, Cultural Survival, Cambridge, Massachusetts.

Chapin, M. and P. Breslin. 1984. Conservation Kuna-style. *Grassroots Development* 8(2):26-35.

Chernela, J. 1987a. Endangered ideologies: Tukano fishing taboos. *Cultural Survival Quarterly* 11(2):50-52.

________. 1987b. CRIC's (Consejo Regional Indígena del Cauca) Natural Resource Project. Project evaluation (unpubl. report). Cultural Survival, Cambridge, Massachusetts.

________. 1987c. Environmental restoration in SW Colombia. *Cultural Survival Quarterly* 11(4):71-73.

Clay, J. W. 1984. Organizing to survive. *Cultural Survival Quarterly* 8(4):2-5.

________. 1988. *Indigenous Peoples and Tropical Forests: Models of Land Use and Management from Latin America*. Cultural Survival, Cambridge, Massachusetts.

Janzen, D. H. 1986. Regrowing a dry tropical forest. *News and Comment* 14:809.

Leon, L. 1986. Cultural survival projects–1986. *Cultural Survival Quarterly* 10(4):74-80.

Macdonald, T., Jr. 1985. Introduction. Pages 1-4 *in* T. Macdonald (ed.), *Native Peoples and Economic Development: Six Case Studies from Latin America*. Cultural Survival. Cambridge, Massachusetts.

________. 1986. Anticipating colonos and cattle in Ecuador and Colombia. *Cultural Survival Quarterly* 10(2):33-36.

Posey, D. and W. Balée, (eds.). 1989. *Resource Management in Amazonia: Indigenous and Folk Societies*. New York Botanical Garden, Bronx, New York.

Schwartzman, S. 1986. Seringueiros defend the rain forest in Amazonia. *Cultural Survival Quarterly* 19(2):41-43.

13

MAB Biosphere Reserves and Conservation of Traditional Land Use Systems

William P. Gregg, Jr.

INTRODUCTION

The International Network of Biosphere Reserves is the best known project in UNESCO's Man and the Biosphere Program (MAB). MAB's mission is to help provide the knowledge, skills, and attitudes needed for harmonious relationships between man and nature, and, more specifically, for addressing interrelated environmental, land use, and socioeconomic problems. MAB is an important framework for cooperation among the many international agencies and nongovernmental organizations as well as the 114 participating governments which have elected to participate by establishing their own national MAB committees. Nationally, MAB provides an aegis for cooperation in using the results of interdisciplinary research to support management of ecosystems in ways which are ecologically, culturally, economically and institutionally appropriate and sustainable. MAB's benefits depend fundamentally on its ability to improve understanding of natural and human systems and their interactions, to apply this understanding in managing ecosystems, to strengthen research and management capabilities–especially in developing countries, and to break down institutional barriers to cooperative, interdisciplinary approaches (diCastri et al. 1981). In recent years, MAB has focused increasingly on human populations as integral parts of coevolving natural, cultural, and socioeconomic systems–on Man in the Biosphere, both affecting and affected by changes at the local, regional, and global levels (UNESCO 1986, 1987).

Biosphere reserves are the permanent hubs for MAB's efforts to understand natural processes and human use systems in each of the world's biogeographical regions, and for using this knowledge to find ways to conserve biological diversity while improving the well-being of people. Because of their key scientific and educational role, they have been referred to as "landscapes for learning" and, indeed, the objective in each region is to include areas which have the greatest possible potential to generate and share useful knowledge (Gregg 1989). For this reason, the participation of indigenous tribal populations and traditional users--"ecosystem people," in the terminology of Dasmann (1984), who are culturally committed to and dependent upon ecologically sustainable uses--is a cornerstone of many biosphere reserves. Their practical ecological knowledge is essential for developing the scientific and educational purposes of these reserves, just as their involvement and support are essential in planning and demonstrating appropriate methods of sustainable uses.

Together, indigenous tribal populations and traditional peoples are playing an important, active role in the biosphere reserve program. This role will increase as the biosphere reserve concept becomes better known, as more reserves are designated in biologically significant areas where these people are an integral part of ecosystems, and as more financial, logistical, and intellectual resources are directed toward these areas. The potential benefits of biosphere reserves to local people and their ecosystems are only now beginning to be appreciated as practical experience is gained in different parts of the world. However, evidence suggests that the biosphere reserve network, now containing 269 units in 70 countries (Figure 13.1) is becoming an important framework for demonstrating the process of integrating conservation and development, in accord with the objectives of the World Conservation Strategy (International Union for Conservation of Nature and Natural Resources [IUCN] 1980). In the balance of this paper, some of this evidence will be reviewed.

ROLES, OBJECTIVES, AND SPATIAL ORGANIZATION OF BIOSPHERE RESERVES

From the earliest discussions which led to UNESCO's decision to launch MAB in 1971, it was intended that the program have a permanent territorial and logistic base, which would address three fundamental objectives (Batisse 1986):

1. The conservation role: to promote conservation of the world's ecosystems and the genetic diversity they contain.
2. The logistic role: to establish a global information-sharing network for cooperation in carrying out interdisciplinary field research, training, and information exchange.
3. The developmental role: to provide the scientific basis for linking environmental conservation and resource development as a fundamental mission of research and educational activities.

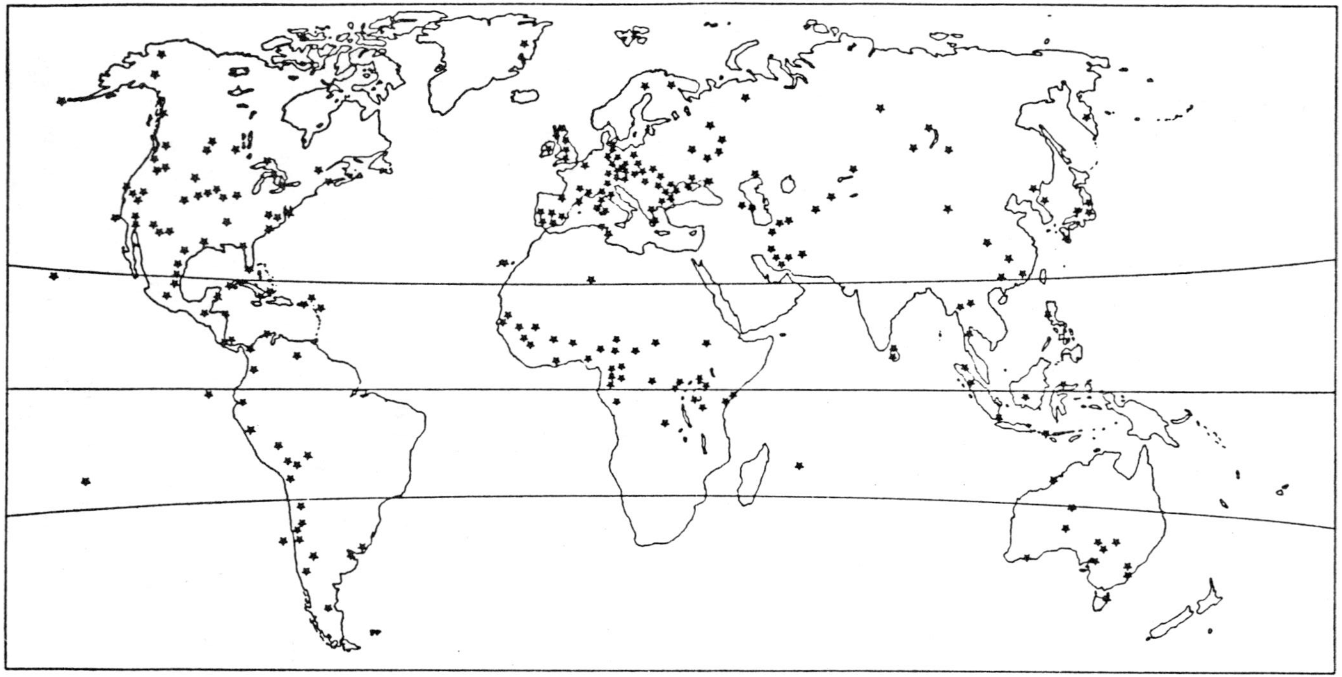

FIGURE 13.1 Map of the International Network of Biosphere Reserves, March 1988. (*Sources*: UNESCO 1984, Metagraphics 1987, and UNESCO Biosphere Reserve list, 1988.)

When the MAB Program was formulated, the establishment of an international network of biosphere reserves became part of MAB Project 8, "Conservation of natural areas and the genetic material they contain." Although UNESCO's operational guidelines addressed the multiple roles (UNESCO 1974), the incorporation of the network within this project resulted initially in disproportionate attention to the conservation role, at the expense of the logistic and development roles. In the years following UNESCO's official designation of the first 57 biosphere reserves in 1976, a large number of national parks and similar strictly protected areas were added to the network. By 1981, there were 208 biosphere reserves in 58 countries. More than 80% were national parks and equivalent reserves. Nevertheless, various early efforts to integrate the several roles were begun in Canada, France, Honduras, México, and elsewhere. Some of these are discussed later because they serve as good examples of the incorporation of local people into biosphere reserve programs.

At MAB's 10th anniversary conference in 1981, the delegates called for a detailed evaluation of the network in light of the experience in different countries. This assessment subsequently took place at the First International Biosphere Reserve Congress which was held in Minsk, Byelorussia, USSR in 1983. On the basis of this evaluation, the "Action Plan for Biosphere Reserves" was approved (UNESCO 1984). The Plan is a unique and strategic document. It sets forth the characteristics and functions of biosphere reserves, establishes nine objectives for the international program, and recommends 35 actions for implementation between 1985 and 1989 by international agencies and organizations and national MAB committees. As part of each biosphere reserve nomination, UNESCO now requires that the administrator(s) of a proposed biosphere reserve sign a declaration of commitment to pursue the general objectives of the "Action Plan" (Table 13.1), to prepare corresponding management guidelines, and to participate in sharing information through the MAB network. Under the Action Plan, the biosphere reserves share a common identity as parts of an international network with a common set of objectives which the administrators are morally, if not legally, committed to pursue.

In a preliminary analysis of a recent UNESCO survey to which 200 of the 269 existing biosphere reserves had responded as of July 1988, 72% of the sites reported that they are pursuing seven or more of the network's objectives, and 47% are pursuing all nine objectives (Gregg and Wargo 1988). Objectives relating to the conservation and logistic roles were more often reported, and more likely to be considered important, than objectives relating to the development role, reflecting the large proportion of strictly protected natural areas

TABLE 13.1
Objectives of the International Network of Biosphere Reserves

The percentage of responding sites and the degree to which they generally support the objective (ranging from "1"—very low to "5"—very high) are indicated in parentheses:

(1) Conserve natural or minimally disturbed ecosystems (98%/4.6).
(2) Provide the legal or administrative basis for resource protection and management (74%/2.7).
(3) Conserve specific genetic resources *in situ* (95%/4.1).
(4) Conduct experimental problem-oriented research on ecosystem management and conservation (91%/3.3).
(5) Conduct long-term environmental monitoring (98%/3.8).
(6) Promote regional planning and integrated rural development (73%/2.0).
(7) Promote local participation in land use and management (74%/2.0).
(8) Promote environmental education (of local people and visitors) and training (92%/3.2).
(9) Generate and share knowledge about conservation and management of the biosphere (72%/2.2).

Source: UNESCO (nomination form) 1988.

and experimental research areas in the network. Traditional land use systems are important considerations in many areas. Conservation of harmonious landscapes resulting from traditional patterns of land use is reported as an objective in 29%, and conservation of traditional agricultural cultivars or domesticated animal stocks in indigenous production systems is reported in 12%. People are living within the boundaries of slightly more than half of the units, with populations ranging from a few individuals to more than 100,000. Human populations living in and around biosphere reserves were characterized wholly or partially as agricultural in 66% of the units, pastoral in 33%, urban/suburban in 32%, and artisanal in 15%. Underrepresented in the network, but more adequately represented in tropical units, are populations described as hunter-gatherers (15% vs. 50%) and nomadic peoples (5% vs. 15%). Research on indigenous populations and traditional uses is being carried out in many reserves, particularly in tropical regions. About 17% are conducting research on traditional land use systems (31% in a tropical subgroup), approximately 17% in cultural anthropology (vs. 23%), and 10% in ethnobiology (vs. 27%). Presence or absence of traditional land use systems in the global network of MAB biosphere reserves is noted in Table 13.2.

Research on sociocultural systems, now occurring in about 27% of the reserves, remains much less emphasized than work on species biology (87%), ecosystem analysis (76%), and other fields of natural science. The results underscore the need to expand the scope of research to include the human populations living in the larger regions which provide the context the existing sites. As shown in Figure 13.2, UNESCO recommends that biosphere reserves be organized into core areas, buffer zones, and transition areas (von Droste zu Hulshoff and Gregg 1985, Gregg and McGean 1985, Batisse 1986, Metagraphics, 1987). These zones describe the spatial distribution of uses and activities in the large landscape area which is typically required to implement fully the multiple roles of the biosphere reserve.

The landscape must contain at least one core area. A core area is a strictly protected natural area containing minimally disturbed examples of the ecosystems characteristic of the biogeographical province (*sensu* Udvardy 1975), including, if possible, outstanding examples of mature or "old growth" communities. Although emphasis is on including a wide range of ecological communities and opportunities for studying natural processes, a biosphere reserve should maintain as much species diversity as possible, including wild relatives of economic species. The core area ideally should be large enough to sustain viable populations of wide-ranging animals, including upper level predators. Nature sanctuaries, scientific reserves, wilderness and primitive areas, and sacred lands within tribal homelands often make suitable core areas. In long-settled regions, such as central Europe, large natural areas may no longer be available, and core areas may consist of small nature reserves within the context of a large traditional landscape. Although most biosphere reserves are located in inland regions, UNESCO is developing guidelines for establishing a global network of coastal and marine biosphere reserves, and the U.S. has already initiated its own program.

A buffer zone includes the area adjacent or surrounding a core area where uses and activities are managed in ways protective of the core area. A buffer zone normally has a legal boundary which often coincides with the boundary of a national park, wildlife refuge, multiple use area, or other administrative areas. Buffer zones may support experimental research, ecosystem rehabilitation, recreation, silviculture and agriculture, scattered settlements, and other uses, and may directly serve conservation objectives by protecting important habitat, agricultural cultivars, and other genetic resources.

The core area(s) and buffer zone(s) of the biosphere reserve are surrounded by the transition area, or zone of cooperation. The zone

TABLE 13.2
Biosphere Reserves, Traditional Land Use Systems and Related Research

Country/Name of Reserve	Biome*	Subsis-tence†	Conserv. Objective**	Current Research‡
Algeria, Tassili	7	YES	b	
Argentina, Costero del Sur	7	YES	b	b
Reserva de Vida Silvestre	7	YES	b	
Australia, Uluru (Ayer's Rock)	7	YES	b	a, b
Bulgaria, Djendema	12	NO	a	
Kamtchia	12	NO	b	
Srebarna	12	NO	b	
Canada, Waterton Lakes	12	YES	b	
Central African Republic Bamingui-Bangoran	4	NO	b	
Chile, Lauca	12	YES	a, b	
China, Wolong	12	YES	a	b
Wuyishan	12	YES	a, b	
Xilin Gol	11	NO	a	
Columbia, Sierra Nevada de Santa Marta	4	YES	a	a, b
Costa Rica, Talamanca-La Amistad	4	YES	a, b	a, b
Cuba, Sierra del Rosario	13	NO		a
Czechoslovakia, Krivotlasko	12	NO	b	
Palava	5	NO	b	
Slovak Karst	5	NO	b	b
Trebon	5	NO	b	
Ecuador, Galapagos	13		a	
Egypt, Omayed	7	NO	b	
France, Cevennes	5	YES	b	a
Gabon, Ipassa-Makokou	1	NO		b
Honduras, Rio Platano	4	NO		a, b
Hungary, Aggtelek	5	NO	a, b	
Hortobagy	5	NO	a, b	a, b
Kiskunsag	5	NO	a, b	
Lake Ferto	5	NO	a, b	
Pilis	5	NO		a
Indonesia, Cibodas	13	YES		b
Gunung Leuser	13	YES	a, b	
Ireland, Killarney	5	NO	b	
Japan, Mount Hakusan	2	NO		a
Mount Odaigahara and Mount Omine	2	NO	b	
Kenya, Mt. Kulal	7	YES	b	b
México, Montes Azules	1	YES	a, b	a, b
Sian Ka'an	1	YES	a, b	a, b

TABLE 13.2 (Cont.)

Country/Name of Reserve	Biome*	Subsis-tence†	Conserv. Objective**	Current Research‡
Peru, Huascaran	12	YES	a	
Manu	1	YES	b	a, b
Poland, Babia Gora	5	NO	b	
Rumania, Rosca-Letea	11	NO	b	
Senegal, Delta du Saloum	4	NO	b	
Spain, Donana	6	YES		a
Mancha Humeda	6	NO	b	
Marismas del Odiel	6	NO	b	b
Montseny	6	NO	a, b	b
Ordesa-Vinamala	6	NO	b	b
Sierras de Cazorla y Segura	6	NO	a, b	
Sierra de Grazalema	6	YES	b	a, b
Sierra Nevada	6	NO		a, b
Sudan, Dinder	7	YES		b
Tanzania, Lake Manyara	4	YES		b
United Kingdom, Caerlaverock	5	NO	a	
Cairnsmore of Fleet	5	NO	a, b	b
Loch Druidibeg	12	NO	a, b	b
Moor House-Upper Teesdale	5	YES		b
Isle of Rhum	12	NO	b	
St. Kilda	12	NO	b	
United States, Aleutian Islands	9	NO	b	
California Coast Ranges	2	YES		b
Carolinian-South Atlantic	5	YES		b
Central Gulf Coastal Plain	5	NO	b	
Central Plains	11	NO		b
Denali	3	YES		b
Everglades	4	YES	b	
Fraser	12	NO	b	b
Glacier	2	NO	b	
Great Smoky Mountains	5	YES	b	a
H. J. Andrews	12	NO		b
Hubbard Brook	5	YES		b
Jornada	7	NO	b	b
Konza Prairie	11	NO	b	
Noatak	9	YES		b
Organ Pipe Cactus	7	YES	a	a, b
San Joaquin	6	NO	b	b
Virgin Islands	13	YES		b
U.S.S.R., Kavkazskiy	12	NO		a
Oka River Valley	5	YES	b	b
Sikhote-Alin	5	YES		a, b
Tsentralnosibirskiy	3	NO		b
Yugoslavia, Velebit Mountain	6	NO	a, b	

TABLE 13.2 (Cont.)

Country/Name of Reserve	Biome*	Subsistence	Conserv. Objective**	Current Research‡
Zaire, Luki	1	YES		a, b
Vallee de la Lufira	4	NO	b	

* (1) = tropical humid forests; (2) = subtropical and temperate rain forests or woodlands; (3) temperate needle-leaf forests or woodlands; (4) = tropical dry or deciduous forests or woodlands; (5) = temperate broad-leaf forests or woodlands and subpolar deciduous thickets; (6) = evergreen sclerophyllous forests, scrubs, or woodlands; (7) warm deserts and semideserts; (8) cold-winter (continental) deserts and semideserts; (9) = tundra communities and barren arctic desert; (10) = tropical grasslands and savannas; (11) temperate grasslands; (12) mixed mountain and highland systems with complex zonation; (13) mixed island systems. (From Udvardy 1975).

† Subsistence activities reported in the transition area.

** (a) = Conserve traditional agricultural cultivars or animal domesticates within indigenous production systems; (b) = Conserve a harmonious landscape resulting from traditional patterns of land use.

‡ (a) = Ethnobiology; (b) = Traditional land use systems.

is a focus for activities involving local people, scientists and other specialists, the administrators of participating administrative areas, and institutions having an interest in using MAB as an aegis for regional cooperation. The transition area supports regionally characteristic uses, and includes, where possible, indigenous tribal populations. It is the principal area for involving local people in improving their own well-being through culturally and ecologically sustainable types and patterns of development, and for implementing educational and training programs which help develop an enthusiastic constituency for the reserve. The transition area is normally open ended, and its dimensions vary in space and time depending on the cooperative activities taking place. It may eventually extend as a functional zone to the limits of a biogeographical or cultural region. The transition area is usually not mapped, although it may contain delineated research and demonstration areas which have elected to join as designated parts of the biosphere reserve.

Increasingly, designation of administrative units as biosphere reserves is being discontinued in favor of naming biosphere reserves

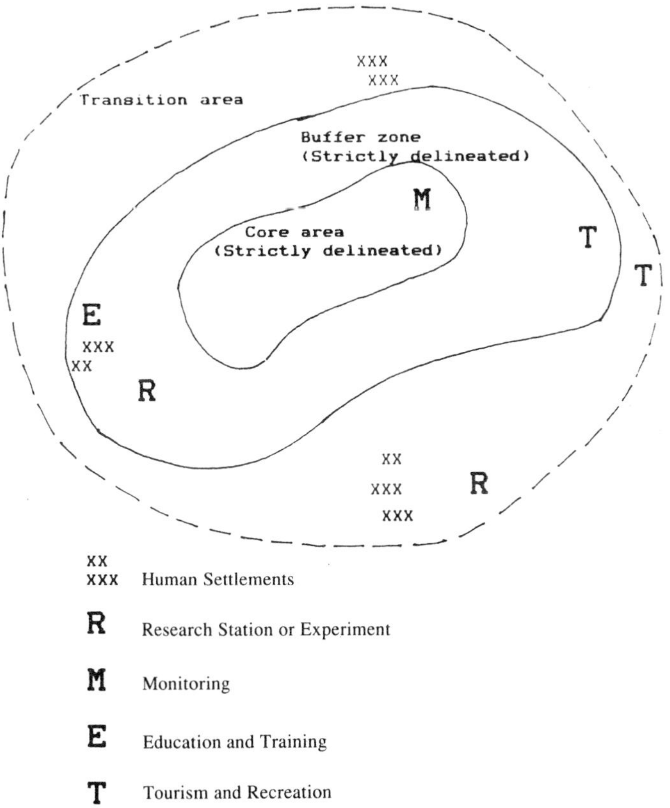

FIGURE 13.2 Schematic zonation of a Biosphere Reserve (Batisse 1986).

after regions which make sense culturally and biologically. If local people personally identify with the region and consider it home, it should be possible to foster pride in the project and enthusiasm for participation. A clear biological identity enhances opportunities for cooperative research and education, and thus the potential regional benefits of the reserve in ecosystem management. Regional names, such as the Pinelands BR (New Jersey), the Hawaiian Islands BR, and Lake Baikal Area BR (USSR), have practical advantage because they allow administrative areas to be added as opportunities arise (Batisse 1986, Gregg 1989).

The zones of the biosphere reserve provide a flexible framework for local and regional cooperation in developing the means of conserving biological diversity in concert with improving the health and well-being of local people whose enthusiastic support will be the key to effective conservation. They facilitate the progressive application of biosphere reserve objectives through a process of voluntary cooperation involving large regions which, in many countries, would be impossible to delineate by law. For administrators of national parks and other conservation areas, a biosphere reserve can offer a bridge to the surrounding region, and an aegis for working with local people and institutions in developing a regional conservation strategy. For local people, it can provide access to education and training, as well as technical and financial resources for appropriate development. For traditional societies, it can provide an incentive for documenting land use practices and cultural history, foster cultural pride, and create a climate in which maintaining traditional agroecosystems and their genetic resources may be possible (Oldfield and Alcorn 1987).

CASE STUDIES

Czechoslovakia's Biosphere Reserves

In Czechoslovakia, the biosphere reserve designation has been applied to four existing protected landscapes. The areas are legally established planning districts in long-settled regions where the original vegetation was almost completely cleared centuries ago. They support harmonious land uses, developed over more than 500 years, which define regional cultures. The judicious use of culturally and ecologically appropriate technologies to improve the productivity of these economic uses, while conserving biological diversity, is a major objective. The the Czechoslovak Academy of Sciences and regional research institutions and universities have participated (Jenik 1987).

In Bohemia, the 70,000 ha Trebon Biosphere Reserve contains an ancient system of shallow artificial impoundments which, prior the development of modern transportation, were a major source of fish in continental Europe. The biosphere reserve is developing the knowledge and skills to manage traditional fisheries, as well as satisfy growing demands for recreation, agriculture and small industries. The core areas effectively protect the small remaining stands of native mixed hardwood forests, and provide benchmarks for research to support regional management of vegetation and wildlife (Jenik and Kvet 1984). The reserve contains some of the last remaining natural stands of mature Bohemian pine, one of the region's most important commercial species. A major public education program has been initiated and a museum facility established in the town of Trebon to communicate the multiple objectives and benefits of the biosphere reserve to the local population. Czechoslovakia's biosphere reserves enjoy considerable governmental support, but developing mechanisms for continuing involvement of local people in setting the research and educational agenda remains an important challenge.

Waterton Lakes Biosphere Reserve, Canada

The Waterton Lakes Biosphere Reserve represents an early effort by a developed country to institutionalize cooperation between a national park and local people through the biosphere reserve program (Lieff 1985, 1989). It was designated by UNESCO in 1979 and consists of the 52,597 ha Waterton Lakes National Park in Alberta, Canada. From the start, cooperation with the local ranching community on resource issues of mutual concern was seen as an important objective. Following a major orientation workshop and a series of meetings involving the government agencies, universities, and the local community, it was agreed to establish a Biosphere Reserve Management Committee, chaired by local ranchers, to serve as a forum for identifying resource issues and the skills for addressing them. Beginning with less controversial issues, the Committee held public seminars in which technical experts provided perspective on problems and potential solutions. An early accomplishment was the development of a cooperative program for monitoring and controlling knapweed (*Centaurea* spp.), a thistle-like exotic species, that is unpalatable to livestock and reduces range productivity. As local suspicions were reduced and confidence in the program improved, attention turned to more controversial issues, such as the management of seasonal movements of elk (*Cervus canadensis nelsonii*) between the park and the adjacent lands, where ranchers were complaining of damage to green crops and haystacks. Depredation was reduced by more effective fencing and scientifically-based revisions for provincial hunting regulations that increased the allowable harvest.

After demonstrating the benefits of science in integrating conservation and range management, the committee moved to establish an interdisciplinary Technical Committee to plan a regional monitoring and research program and communicate accomplishments and recommendations to the Management Committee. The Technical Committee has expanded research on the ecology of elk populations, and initiated experimental range research to further reduce conflicts between elk and livestock. During the past six years, a functional zone of cooperation adjoining the park has been established to implement MAB concepts. An Education Subcommittee is now working on curricula to communicate the MAB concepts in local schools. As testimony to local enthusiasm, the Management Committee arranged for construction of an interpretive highway pull-out overlooking private ranchlands and the park to explain the mutual benefits of the biosphere reserve concept to the park and the local community.

Cevennes Biosphere Reserve, France

The 323,000 ha Cevennes Biosphere Reserve in the hills of southeastern France is a landscape mosaic containing native beech, chestnut and oak-pine forests, and grasslands as well as areas traditionally used for agriculture, grazing, and settlements (Collin 1985, Engel 1985). Species diversity is relatively high, even by Mediterranean standards, and includes more than 1800 species of vascular plants and 50 species of mammals. Perpetuating this diversity involves maintaining the existing pattern of rural land uses in an area which is the contemporary expression of continuous human use since Roman times. This use has not always been compatible. Extensive clearing of beech forests for charcoal, as well as land conversions to agriculture and overgrazing by sheep in the eighteenth and nineteenth centuries caused serious erosion which was abated through extensive plantings of pine (Halffter and Ezcurra 1989). More recently, land conversions for pulpwood production and tourist facilities, financed largely by external interests, caused additional disturbance. Hunting by outsiders became a significant influence on wildlife populations. In recent years, natural succession on unutilized lands that formerly supported indigenous agroforestry has created an increasingly homogeneous landscape. Today, the central limestone plateaus have the lowest population density in France (1.5 people per km^2). The population is declining in part due to deemphasis of traditional grazing and loss of markets for chestnuts, silk, and sheepskin.

The establishment of the Cevennes National Park in 1970 reduced land speculation and helped control recreational construction and use. Hunting is now restricted within the park to local populations. The park has had some success in fostering ecotourism, which focuses on public enjoyment and appreciation of the diversity of the traditional landscape mosaic. However, the park has statutory authority to regulate uses only in a 90,000 ha central zone. In the remaining lands, the park can only advise local municipalities on conservation and land use issues. In practice, park programs have been confined largely to the former zone.

The biosphere reserve includes the central zone of the park, which coincides with the core area and buffer zone, as well as a surrounding zone of cooperation. In this case, the zone of cooperation is actually delineated and, in some places, extends beyond the park's boundary to the limits of the Cevennes ecocultural region. The biosphere reserve complements the park by providing a neutral framework for regional cooperation involving the park, local municipalities, universities, and the local people in planning and implementing research and educational activities to maintain and restore the region's biological heritage while perpetuating traditional culture, and fostering regional pride. Among the reserve's ongoing projects is maintenance of populations of purebred Aubrac cattle, an endangered local domesticate particularly well adapted to the mountain conditions, for use in genetic improvement programs with other cattle breeds.

Central American Biosphere Reserves

In Central America, the most significant opportunities for conserving remaining tropical forests are generally in the homelands of indigenous people, whose cultures are intimately related to these ecosystems (Glick and Betancourt 1983, Houseal and Weber 1989). Since 1980, biosphere reserves including parts of these homelands have been designated in Honduras, Costa Rica, and Panama; additional biosphere reserves are being considered for nomination in Belize, Guatemala, and Nicaragua. In most cases, the designations link indigenous reservations with contiguous national parks, national forests, and other areas to create large biosphere reserves that provide a framework for cooperation among the participating administrative areas. The La Amistad Biosphere Reserve in Costa Rica, which includes nine Indian communities, exemplifies this approach (Torres et al. 1989). Cooperation involving local people is often informal, and focuses on familiarizing the community with biosphere reserve concepts and programs. Several biosphere reserves have management plans that include mechanisms for involving indigenous populations in planning and management.

The proposed Kuna Yala Biosphere Reserve in Panama, which has recently been recommended by the Kuna tribal leaders for nomination, is a special case (Wright et al. 1985, Houseal and Weber 1989). This will be the first biosphere reserve managed exclusively by indigenous people, and illustrates the promising role of biosphere reserves in cultural self-determination. The Kuna Reservation is located in an isolated area of tropical forests extending from the continental divide to the Caribbean Sea. The Kuna, whose permanent settlements are located on small offshore islands and in nearshore villages, have a long-established tradition of fishing and coastal agriculture, as well as selective hunting and gathering in the interior mountains. The catalyst which ultimately led to the biosphere reserve nomination was a government project to construct a road, with support from U.S. A.I.D., through the reservation to link the Pan American Highway with the coast. Recognizing the need to control entry by peasant settlers from the Panama's interior, prevent outside interests from dominating a profitable Kuna-managed tourist trade, and protect their way of life, members of the community attempted to develop a permanent agricultural settlement where the road enters the reservation. Climate and soils in this virgin rainforest area proved extremely unfavorable for farming and cattle-raising, and the project was abandoned. The Kuna's problems were brought to the attention of U.S. A.I.D., which, with Kuna support, subsequently enlisted CATIE (Centro Agronómico Tropical de Investigación y Enseñanza) in Costa Rica to prepare a land suitability analysis for the region. The analysis, which recommended maintaining primary forest as the optimal use, marked the initiation of an intensive cooperative planning effort, which between 1982 and 1987 attracted more than $1 million from national and international sources (Chapin 1987). Through the project, the Kuna have received extensive training in planning and management. Together, the Kuna and their collaborators developed a management plan which stresses phased development of tourism as well as fisheries, agroforestry (reportedly utilizing 48 species of trees and 16 crop plants in various combinations), traditional uses, and a strong emphasis on research and education. Every research project has a Kuna coinvestigator who participates actively in designing and implementing the research. The coordinated pursuit of conservation, logistic, and development roles, and the emphasis on maintaining the Kuna's traditional uses and strengthening their already strong cultural cohesiveness, made the Kuna Homeland a prime candidate for biosphere reserve status. The Kuna's decision to pursue nomination reflected awareness of the intangible benefits of recognizing the unique coevolution of Man and Nature embodied in the relationships between the Kuna and their homeland. It also reflected their awareness of the pragmatic benefits of the international network in sharing information, generating technical and financial assistance, and providing, through international recognition, an additional barrier to outside intervention in Kuna affairs by implicitly recognizing the Kuna's land title, which prohibits landownership by non-Kuna, and their autonomy over planning and management of the ecosystem. Although the Kuna's situation is atypical, it nevertheless illustrates an emerging relationship between biosphere reserves, cultural self-determination, and the conservation of biological resources.

México's Biosphere Reserves

Among tropical countries, México's success in implementing the biosphere reserve concept has been exceptional. Beginning in 1975, the country was among the first to launch its biosphere reserve program. Its architects realized early that conservation objectives would be difficult or impossible to carry out in a developing country if they were undermined by rural people faced with an increasingly difficult subsistence. Involving local people in developing economic alternatives to improve their standard of living was recognized as a principal means for promoting biosphere reserve objectives. Under this approach, community leaders are involved in planning and implementing the reserve's research and demonstration program. Local commitment to conservation objectives develops progressively as local people come to associate practical economic benefits and opportunities with their participation in the biosphere reserve (Halffter 1981, 1985, Halffter and Ezcurra 1989).

México establishes biosphere reserves by executive decree and thus is one of only a few countries to use biosphere reserves as a separate legal category of protected area. The approach has been made possible by the relative weakness of national park and other protected area systems, and the need to build centers to demonstrate ways to integrate conservation and development. México's biosphere reserves have considerable administrative independence. Under the authorizing decree, a Federal or state-affiliated research institution normally has both research and management responsibilities. The capability of these institutions to provide continuity in the biosphere reserve program has been an important factor in the program's success. Finally, the association of the sites as part of a global network has helped gain governmental support within México, as well as assistance from other nations and international organizations.

The Mapimi Biosphere Reserve in the State of Durango, exemplifies the "Mexican modality" (Halffter 1985). The Reserve is located in a remote desert region where marginal livestock grazing and the harvesting of candelilla (*Euphorbia antisyphilitica*) are the traditional economic activities. Here, an important objective was to improve the prospects for survival of the endangered desert tortoise (*Gopherus flavomarginatus*) which was being harvested by local people for its meat. Researchers from the Institute of Ecology enlisted the help of local people in various range improvement and alternative development projects. Although the success of these projects varied, the confidence of the local people was well established by the time the Reserve was designated. As the local people increasingly saw themselves part of an integrated biosphere reserve program, the desert tortoise has become symbolic of the benefits they derive and their hopes for the future. Illegal harvesting has been reduced. In one instance, poachers were apprehended by local people, summarily dispatched from the area, and their inventory of marked tortoises returned to the environment (E. Ezcurra, pers. comm.).

Other Mexican biosphere reserves are being developed under a wide range of ecological, cultural, and socioeconomic conditions. All are based on local participation. The 140,000 ha Sierra de Manantlan Biosphere Reserve in the State of Jalisco is a biologically rich region which contains the critical habitat for the perennial corn (*Zea diploperennis*)–the symbolic catalyst for its establishment. The reserve has less than 1% Federal land, and is being developed as a partnership between researchers at the University of Guadalajara, various government agencies, and local villages. The reserve provides a framework for small-scale rural development projects, and is developing a growing base of local political support needed to counteract the trend toward forest exploitation financed by external interests (Santana et al. 1989). Another example is the Sian Ka'an Biosphere Reserve in a sparsely populated area of the State of Quintana Roo (Yucatan Peninsula), in which conservation of traditional Mayan agriculture and forest utilization practices is a major objective (Tangley 1988).

CONCLUSIONS

The biosphere reserve concept has been widely cited in national and international strategies for maintaining biological diversity and integrating conservation and development (IUCN 1980, Interagency Task Force on Biological Diversity 1985, World Resources Institute and the International Institute for Environment and Development 1986, U.S. Congress 1987). Experience in biosphere reserves figures prominently in the World Bank's efforts to consider indigenous people in integrating conservation and economic development. United Nations agencies (e.g., FAO, UNEP) and international nongovernmental organizations (e.g., IUCN) have joined in implementing the "Action Plan for Biosphere Reserves". However, the potentially important role of indigenous cultures in developing the conservation role of biosphere reserves has yet to be fully realized.

We know that biosphere reserves provide a framework for documenting traditional land use systems and improving their productivity in culturally appropriate ways. We know that biosphere reserves can provide an aegis for demonstrating how to integrate the conservation of significant natural areas with efforts to improve the well-being of indigenous populations living within or adjacent to these areas (Clay 1985, McCaffrey and Landazuri 1987, van Orsdol 1987), to the mutual benefit of both. We know that they can help indigenous societies protect their land use traditions in the face of growing external pressures. Although time has been insufficient for convincing demonstrations, we suspect that they can be a useful means to maintain traditional cultivars and animal domesticates (Oldfield and Alcorn 1987, Halffter and Ezcurra 1989).

It has been suggested that biosphere reserves can help build the ethical basis for a world community based on sustainable partnerships between human societies and Nature (Engel 1989). If this is to occur, indigenous societies may well be in the best position to provide moral leadership through practical example. The following statement by Aurelio Chiari, delegate from the Kuna Nation to a MAB Symposium on Biosphere Reserves at the Fourth World Wilderness Congress in September 1987, suggests this particular potential:

> For the Kuna culture, the land is our mother and all living things that live on her are brothers in such a manner that we must take care of her and live in a harmonious manner on her, because the extinction of one living thing is also the end of another. We preserve the forests because for us they are a pharmacy where our medicines are stored. They are a refrigerator because they keep our meat fresh. They are a store for construction materials because we obtain materials for our houses and boats from them. Some are also sacred and we must defend our religious beliefs.
>
> The concept of a biosphere reserve, even though we don't have this term in our society, has been with us since the existence of our culture. The worry that my people have always had about the strong pressures of deforestation has created the impulse to plan the adequate use of the resources. We have done this through the Kuna Wildlands Project which has prepared a management plan and is now implementing it.
>
> I will be the messenger to the Kuna Congress about the international concern shown here and I will tell them that we are not alone in our efforts.

REFERENCES

Batisse, M. 1986. Developing and focusing the biosphere reserve concept. *Nature and Resources* 22:1-10.

Chapin, M. 1987. Nusagandi: An indigenous project in environmental conservation. (Unpublished manuscript.)

Clay, J.W. 1985. Parks and people. *Cultural Survival Quarterly* 9(1):2-5.

Collin, G. 1985. The Cevennes Biosphere Reserve: Integrating traditional uses and ecosystem conservation. *Parks* 10(3):12-14.

Dasmann, R.F. 1984. The relationship between protected areas and indigenous people. Pages 667-671 *in* J.A. McNeely and K.R. Miller (eds.), *National Parks, Conservation and Development*. Smithsonian Institution, Washington, D.C.

diCastri, F., M. Hadley, and J. Damlamian. 1981. MAB: The Man and the Biosphere Program as an evolving system. *Ambio* 10(2-3):52-57.

Engel, J.R. 1985. Renewing the bond of mankind and nature: Biosphere reserves as sacred space. *Orion* 4(3):52-59.

________. 1989. The symbolic and ethical dimensions of the biosphere reserve concept. Pages 21-32 *in* W.P. Gregg, Jr., S.L. Krugman, and J.D. Wood, Jr. (eds.), *Proceedings of an International Biosphere Reserve Symposium*, 1987 World Wilderness Congress. National Park Service, Atlanta, Georgia.

Glick, D. and J. Betancourt. 1983. The Rio Platano Biosphere Reserve: Unique resource, unique alternative. *Ambio* 12(3-4): 168-175.

Gregg, W.P., Jr. 1989. On wilderness, national parks, and biosphere reserves. Pages 33-40 *in* W.P. Gregg, Jr., S.L. Krugman, and J.D. Wood, Jr. (eds.), *Proceedings of an International Biosphere Reserve Symposium*, 1987 World Wilderness Congress. National Park Service, Atlanta, Georgia.

Gregg, W.P., Jr. and B.A. McGean. 1985. Biosphere reserves: Their history and their promise. *Orion* 4(3):41-51.

Gregg, W.P., Jr. and J.P. Wargo. 1988. The status of biosphere reserves: A preliminary assessment. Report to UNESCO. UNESCO-MAB Secretariat, Paris.

Halffter, G. 1981. The Mapimi Biosphere Reserve: Local participation in conservation and development. *Ambio* 10(2-3):93-96.

_____. 1985. Biosphere reserves: Conservation of nature for man. *Parks* 10(3):15-18.

Halffter, G. and E. Ezcurra. 1989. Evolution of the biosphere reserve concept. Pages 188-206 *in* W.P. Gregg, Jr., S.L. Krugman, and J.D. Wood, Jr. (eds.), *Proceedings of an International Biosphere Reserve Symposium*, 1987 World Wilderness Congress. National Park Service, Atlanta, Georgia.

Houseal, B., C. MacFarland, G. Archibold and A. Chiari. 1985. Indigenous cultures and protected areas in Central America. *Cultural Survival Quarterly* 9(1):10-20.

Houseal, B. and R. Weber. 1989. Biosphere reserves and the conservation of traditional land use systems of indigenous populations in Central America. Pages 234-241 *in* W.P. Gregg, Jr., S.L. Krugman, and J.D. Wood, Jr. (eds.), *Proceedings of an International Biosphere Reserve Symposium*, 1987 World Wilderness Congress. National Park Service, Atlanta, Georgia.

Interagency Task Force on Biological Diversity. 1985. *U.S. Strategy on the Conservation of Biological Diversity: An Interagency Task Force Report to Congress*. U.S. Agency for International Development, Washington, D.C.

International Union for Conservation of Nature and Natural Resources (IUCN). 1980. *World Conservation Strategy: Living Resource Conservation for Sustainable Development*. IUCN/UNEP/WWF, Gland, Switzerland.

Jenik, J. 1987. Selection and role of biosphere reserves in Czechoslovakia. Pages 23-32 *in* Czechoslovak National MAB Committee, *Implementation of the Action Plan for Biosphere Reserves*. Czechoslovak Academy of Sciences, Ceske Budjovice.

Jenik, J. and J. Kvet. 1984. Long-term research in the Trebon Biosphere Reserve, Czechoslovakia. Pages 437-459 *in* F. di Castri et al. (eds.), *Ecology in Practice: Part I: Ecosystem Management*. Tycooly International Publishers, Limited, Dublin, Ireland.

Lieff, B. 1985. Waterton Lakes Biosphere Reserve: Developing a harmonious relationship. *Parks* 10(3):9-11.

_______. 1989. Cast study: Waterton Biosphere Reserve. Pages 134-141 *in* W.P. Gregg, Jr., S.L. Krugman, and J.D. Wood, Jr. (eds.), *Proceedings of the Symposium on Biosphere Reserves*, 1987 World Wilderness Congress. National Park Service, Atlanta, Georgia.

McCaffrey, D. and H. Landazuri. 1987. *Wildlands and Human Needs: A Program Evaluation*. World Wildlife Fund, Washington, D.C.

Metagraphics, Inc. 1987. Biosphere Reserves. (Brochure). Stephenson, Inc. Alexandria, Virginia.

Oldfield, M.L. and J.B. Alcorn. 1987. Conservation of traditional agroecosystems. *BioScience* 37(3):199-208. (See Ch. 3, this volume)

Santana, E.C., R. Guzman M. and E. Jardel P. 1989. The Sierra de Manantlan Biosphere Reserve: The difficult task of becoming a catalyst for regional sustained development. Pages 212-222 *in* W.P. Gregg, Jr., S.L. Krugman, and J.D. Wood, Jr. (eds.), *Proceedings of an International Biosphere Reserve Symposium*, 1987 World Wilderness Congress. National Park Service, Atlanta, Georgia.

Tangley, L. 1988. A new era for biosphere reserves. *BioScience* 38(3):148-155.

Torres, H., D. Masterson and L. Hurtado de Mendoza. 1989. La Amistad Biosphere Reserve: An exercise in environmental planning. Pages 254-263 *in* W.P. Gregg, Jr., S.L. Krugman, and J.D. Wood, Jr. (eds.), *Proceedings of an International Biosphere Reserve Symposium*, 1987 World Wilderness Congress. National Park Service, Atlanta, Georgia.

Udvardy, M.D.F. 1975. A classification of the biogeographical provinces of the world. IUCN Occasional Papers 7:1-44.

UNESCO. 1974. Criteria and guidelines for the choice and establishment of biosphere reserves. MAB Report Series No. 22. UNESCO, Paris.

UNESCO. 1984. Action plan for biosphere reserves. *Nature and Resources* 20(4):1-12.

_____. 1986. International Coordinating Council of the Programme on Man and the Biosphere, Ninth Session, Paris, October 20-25, 1986. MAB Report Series No. 60. UNESCO, Paris.

______. 1987. A practical guide to MAB. UNESCO, Paris.
______. 1988. Biosphere reserve nomination form. UNESCO, Paris.
U.S. Congress, Office of Technology Assessment. 1987. *Technologies for Maintaining Biological Diversity*. Office of Technology Assessment, Washington, D.C.
van Orsdol, K.G. 1987. Buffer zone agroforestry in tropical forest regions. Report to the U.S. Agency for International Development and the USDA Forest Service. Environmental Strategies International, Palo Alto, California.
von Droste zu Hulshoff, B. and W.P. Gregg, Jr. 1985. The international network of biosphere reserves: demonstrating the value of conservation in sustaining society. *Parks* 10(3):2-5.
World Resources Institute and the International Institute for Environment and Development. 1986. *World Resources 1986*. Basic Books, Inc., New York.
Wright, R.M., B. Houseal, and C. de Leon. 1985. Kuna Yala: Indigenous biosphere reserve in the making? *Parks* 10(3):25-28.

14

Linking Rural Development with Biological Conservation: A Development Perspective

Molly Bower Kux

INTRODUCTION

> In Thailand, the principal threat to long-term maintenance of biological diversity and tropical forest resources is agricultural encroachment on already-designated conservation areas by nomadic hill tribe groups and landless lowland Thais. Overcoming this problem is fundamental to the long-term viability of much of Thailand's biological resources and will require concerted efforts in agricultural, rural and economic development, as well as in reforestation and protected area management. For those situations where basic human needs must be met for conservation efforts to succeed, A.I.D. might be able to play a role in both (U.S. Department of State 1986).

This cable excerpt from the Thailand Mission of the U.S. Agency for International Development (A.I.D.) epitomizes the dilemma that faces the international community in coming to terms with global problems related to the loss of living resources and biodiversity. This was only one of more than 80 reactions received in Washington from field offices around the world that provide information and guidance as to how A.I.D.–a development agency that focuses on improving living standards in the Third World–should respond to this relatively new concern. The issues and questions raised by this Thai response are serious. Explicitly or implicitly, they include the following:

- What is the role, form (grant or loan), and amount of foreign assistance that should be allocated for direct conservation and protection of living resources in view of A.I.D.'s primary mandate to deal with problems of poverty? What about other donor responsibilities?
- What balance should be achieved between "pure" conservation programs vs. those that relieve pressure on parks and reserves?

- Should policymakers of developing countries allocate additional human and budgetary resources for biological conservation, particularly if they are faced with mounting debt burdens? Will they be willing to borrow for this purpose?
- If A.I.D., other donors, and developing countries increase their commitments, will the present 5-10 year project period be adequate or will 20-30 year project period be required?
- What economic benefits actually accrue to countries and to local people from investments in parks and protected areas?
- What are the effects of government policies on protected areas that are premised on the separation of humans from wildlife in order to restrict their use of wildlife?
- Can traditional management systems or local values be identified that conserve these resources? Do government policies encourage or discourage them? How can donors best influence policy reforms?
- Is commercial exploitation (legal or illegal) of wild plants and animals adequately addressed through policy and enforcement measures? Is there a need for new approaches that allow sustainable use in some cases?

This paper will try to address these questions through a selective review of the literature, and will summarize promising options for maintaining plant and animal resources that could be or have been incorporated into economic development programs. Finally, it will end with some conclusions and suggestions for future attention.

BACKGROUND

Due to recognition of increasing losses of plant and animal species and the implications of a diminished supply of living resources for future economic development, conservation of biological diversity is receiving increased attention in the United States, other countries, and the international community. Many scientific studies and conferences have outlined the problems and identified general steps that need to be taken [International Union for Conservation of Nature and Natural Resources (IUCN) 1980, National Academy of Sciences (NAS) 1980, U.S. Department of State 1982, McNeely and Miller 1984, World Resources Institute 1985, Wilson 1988a]. In 1983, the U.S. Congress amended the Foreign Assistance Act by adding a new Section 119 on endangered species that authorizes A.I.D. assistance to developing countries to protect and maintain wildlife habitat and develop better management and conservation programs. The amendment also required preparation of a U.S. strategy to conserve biological diversity in developing countries. This was submitted to Congress in 1985 (Interagency Task Force on Biological Diversity 1985). A.I.D., other U.S. government agencies, and international donor organizations are under increasing pressure to fund more activities in this area. In 1986, additional legislation was passed that earmarked FY 1987 funds for specific biological diversity activities and added the new Section 118 on conservation of tropical forests. In addition, the Office of Technology Assessment (OTA), in response to requests from Congress, issued an analysis of technologies to maintain biodiversity (U.S. Congress 1987). The report addressed a broad range of issues such as current rates of biodiversity loss. However, its main purpose was to advise Congress on options for legislative action.

In some developing countries, concerns about the loss of species and ecosystems are shared by policymakers and scientists. This is evidenced by the increase in parks and protected areas in developing countries over the past 14 years. Since 1972, the number of protected areas throughout the world has grown by 47%–with most of this increase in the Third World (Miller 1984). Nine of the ten countries that have set aside over 10% of their land as protected areas are developing countries: Benin, Bhutan, Botswana, Central African Republic, Costa Rica, Rwanda, Senegal, Tanzania and Zimbabwe (Harrison et al. 1984). While these efforts are impressive from an international conservation perspective, creating protected areas is only a first step in solving the problem. Whether these areas do, in practice, conserve biological diversity (even though in most cases this was not a criterion used to establish protected areas), and whether they are properly managed with adequate staff and budget are questions that may well be answered negatively in most developing countries (Machlis and Tichnell 1985). National planners in the Third World are posed with serious dilemmas–how much more land will they be asked to set aside and who will pay? The difficult choices that face politicians and administrators in developing countries were emphasized in 1982 by the incumbent Vice President of Indonesia, Adam Malik:

> In the face of poverty and hunger, how does a government allocate land? What value should be placed on genetic resources to be preserved for tomorrow? What value should be placed on genetic resources for tomorrow in comparison to the value of exploiting natural resources for the alleviation of poverty today? (Malik 1984)

Attempting to shed light on some of these difficult decisions in humid tropical regions, a National Research Council Committee on Selected Biological Problems in the Humid Tropics (NAS 1982) suggested the following mix of technical approaches:

- To retain substantial portions of humid tropical ecosystems intact to serve as: natural sources of genetic materials, parks and reserves, scientific repositories for education and study, and a means to facilitate erosion control and maintenance of air and water quality.
- To use substantial parts of natural ecosystems to produce lumber, pulpwood and plywood, and fuel by emphasizing management practices such as shelterwood forestry, underplanting, selective cutting and limited clear-cutting, and human-assisted reforestation.
- To convert extensive deforested areas and selectively deforest others to increase food production through the application of new agricultural technologies developed in the humid tropics; this option emphasizes the use of fertilizers and promotes land conversion to enhance agricultural use.

This mix of options–protection, sustainable management, conversion and intensified agriculture–makes sense for other biogeographic regions as well. The amount of land devoted to protection will likely depend on factors such as: human population density, availability of underutilized land, options for ecologically oriented national planning, land use options that incorporate economic benefits for local people, and increased support from wealthier nations.

INVESTMENT RATIONALES

There are various rationales for justifying investments in conservation efforts by donor agencies, undergirded by changes in policies (World Bank 1986, A.I.D. 1988a). Among these is the traditional concern for endangered species; but more sophisticated arguments that have a stronger economic slant are coming into use. The following examples of conservation activities are organized under these different rationales. Some are components of much larger programs–an important consideration for agencies that tend to deal with large multi-million dollar loans or fund projects that have many subcomponents. An effort has been made to find information from a range of geographic areas and ecosystems and to include a variety of conservation options supported by donor agencies. The "development/conservation" rationales for these programs include:

- Maintaining ecosystem services to protect economic development investments.
- Addressing the economic aspirations of rural populations that are often the greatest threat to the integrity of protected areas.

- Increasing the potential of nature-based tourism as a development option.
- Protecting specific endangered species, especially those of cultural significance.
- Investigating potential new economic products from the wild.
- Building on indigenous conservation efforts.
- Testing new approaches to sustained yield harvesting.

ECOSYSTEM SERVICES

Several donor-funded projects stand out as examples of development programs that recognize the importance of protecting upland watersheds to maintain the integrity of downstream investments. For example, the Dumonga Watershed Protection Project was started in 1980 in North Sulawesi, Indonesia; it involved the establishment of a forest reserve to help prevent encroachment of shifting cultivators and cash crop farmers on upper watersheds. With a loan of $700,000 from the World Bank, a three-year program was initiated for an extended park area and watershed management to guarantee a sufficient and regular water supply for large, lower elevation irrigation systems (Sumardja et al. 1984).

In Peru, the Central Selva Resource Management Project was started in 1982 in the Palcazu valley of the high jungle, with funds from A.I.D. and the government of Peru totaling $30 million over six years. Its purpose was to test and institutionalize a methodology to promote sustained productivity of the Palcazu watershed, increase productivity on lands suitable for agriculture, and build Peru's institutional capabilities to plan and implement integrated regional development. Covenants included in the loan agreement were that the Government of Peru must legally designate a national park and a protected forest in the watershed and assign technical staff. The project design was derived from an environmental and social assessment of the area that classified land use capability. Subsequent studies concluded that production forestry held the greatest potential for development, that it was not feasible to increase food production via previous plans for resettlement of large numbers of people from outside the valley, and that major attention should be given to managing the area for existing inhabitants–many of whom belonged to native communities (A.I.D. 1982).

The Central Selva project contains ten components, two of which are of greatest interest here: natural forest management for sustained yield, and park and forest reserves. The forest management plan is testing rotational, narrow clearcuts based on new knowledge about gap-phase dynamics and regeneration requirements of tropical forest canopy tree species. It is hoped that this will permit natural regeneration over 30 year cycles. Current estimates are that annual clearcuts of one to two hectares will generate an adequate family income from 80 ha (G.S. Hartshorn, unpubl. data). Park and forest reserve areas were defined not only on the basis of land capability, but also for their potential economic returns for downstream production forestry and agriculture. The project also supported the work of an economic botanist to identify and store (*ex situ* in local gene banks) plant species of potential market value to the Palcazu people (J. Salick, unpubl. data). Problems inherent in testing new and often risky approaches have required that the project be extended several years and that its funding be changed; however, grant funds have been increased in recognition of the research nature of the venture.

Finally, a recently suspended natural resources project started by A.I.D. in Panama had four components related to ecosystem services: watershed management and soil conservation, national parks and wildlands management, natural forest management, and private industrial plantations coupled with farm woodlots. The project rationale was based on the need to: (1) protect economic values of existing agriculture and commercial investments, including the Panama Canal; (2) maintain electricity and water supply to major urban areas, (3) reduce dependence on wood imports, and (4) enhance employment in industrial forestry (A.I.D. 1985).

ECONOMIC ASPIRATIONS OF LOCAL PEOPLE

Human issues are often the most difficult to resolve when designating and maintaining protected areas. First and foremost, these areas are usually "off limits" to local populations. Problems encountered include: resettling people from newly created parks, removing agricultural squatters from existing areas, preventing poaching of commercially valuable wildlife and timber, and overuse of other valuable resources such as water, forage, and other natural products. One or more of these problems exist in most protected areas in developing countries. They are especially difficult to deal with, because management agencies are ill equipped in terms of their staff and budget, and are often viewed with suspicion by local people.

In Africa, and perhaps other regions as well, the situation is exacerbated by the resistance of local people to the perceived imposition of western-style conservation that restricts any human use–a concept that is alien to many subsistance agriculturalists. In his analysis of prospects for protected areas in Africa, Lusigi (1984) proposed rethinking conservation approaches so that they are acceptable to local people as the only feasible long-term solution. His concept centers around the designation of "conservation units" for important or unique ecosystems with three types of land use: national parks, game reserves, and multiple use areas. All of these are to be managed to assure the largest possible involvement and benefits to local people from tourism, controlled game cropping, and continuation of traditional land uses such as pastoralism.

Lusigi's concept is similar to that of UNESCO's Man and the Biosphere (MAB) biosphere reserve program that calls for integration of protected areas with surrounding socioeconomic needs (see also Gregg, Ch. 13, this volume). Ideally, a biosphere reserve should have a protected core surrounded by one or more buffer zones that allow varying intensities of use, e.g., research, tourism, or human settlements (Maldague 1984). The MAB biosphere reserve program has been promoted by UNESCO and participating countries for more than ten years. However, only a few successful examples of putting the total concept into action have been documented. The Mapami Biosphere in México is one of the few examples that appears to have been successfully established. This biosphere reserve is both conserving a representative desert ecosystem and improving the living standards of local people whose main economic activities are gathering candelilla wax, cattle raising, and hunting of wild game (Halffter 1981).

One impediment to the success of the MAB biosphere reserve program may be that it does not generate adequate national level interest or investment because conservation is tied mainly to local needs. The World Wildlife Fund-U.S. (WWF) program on Wildlands and Human Needs could help fill that vacuum in some countries (McCaffrey and Landazuri 1987). Initiated in 1985 with a matching grant from A.I.D., the WWF program represents an evolution in conservation thinking. After more than 25 years of work on international conservation and much success in establishing reserves and promoting protection of endangered wildlife, WWF recognized that the problems of people living in and around protected areas and their traditional use of natural resources had to be addressed. The Wildlands and Human Needs program supports work with local populations to improve their livelihoods and involve them more directly with protected areas from which they have been excluded. Programs have been started in 15 countries in Latin America and Africa in partnership with indigenous nongovernmental organizations (NGOs). This approach is demonstrated by the Sian Ka'an Biosphere Reserve in México, and the Beza Mahafaly and Andohahala Reserves in Madagascar. The goal of this approach is to generate support for reserve protection from local people by improving economic productivity outside the protected area. In Madagascar, assistance has been provided to improve irrigated rice–the favored food crop. In México, these activities include improved management of the spiny lobster–a major source of income, development of alternatives to use of an increasingly rare palm species for trapping the lobsters, and establishment of a demonstration farm for intensive crop production–a possible alternative to extensive agriculture.

A progress evaluation conducted after less than two years of experience focused on suggestions to improve and underscore this new approach (McCaffrey and Landazuri 1987). Strengths identified as contributing to enhancement of the program's credibility were: providing support for existing and new local institutions and innovative programs, tackling issues identified by local communities as important to their survival and well-being, and involving local experts and institutions in the preparation (and hopefully implemention) of reserve area management plans. Identified weaknesses included: lack of social science expertise among field staff, lack of a conservation education program for local people, an annual budgeting process that creates uncertainty and overtaxes small fields staffs, and failure to share experiences across projects. A serious overall constraint identified in all five of the countries visited was national legislation that restricted local communities from using resources within the protected areas. The evaluation recommended continued funding of the program both to allow existing field projects to mature and provide insights that would be valuable for other programs. A proposal for a five-year extension of the A.I.D. matching grant was approved in 1988.

NATURE-BASED TOURISM

Tourism is becoming an increasingly important source of foreign exchange for a number of developing countries. It is near the top of the list in Costa Rica, Kenya, Jamaica, and Thailand. Moreover, tourism is often related in important ways to natural resources–whether they be national parks and spectacular wild animals or magnificent beaches and coral reefs (May 1982, Arbhabhirama et al. 1987, A.I.D. 1988b, D.H. Janzen, unpubl. data). Scientific tourism is seen as one of Costa Rica's major future industries, and is already estimated to generate over $1 million in foreign exchange spent within the country each year (M.S. Kaye, unpubl. data).

In Kenya's Amboseli National Park, compromise between an important national industry and local needs, with benefits to both, was the basis for the strategy that was worked out with difficulty over 15 years (Western 1984). In return for economic benefits (income from accommodation of migrating wildlife on their lands, new jobs, a supply of water for livestock from the park), Maasi landowners agreed to strict protection of a new 480 km^2 park. The national government was, for its part, able to assure that a major source of foreign exchange (i.e., tourism) could continue, but it had to accept a smaller-sized park than originally planned. Nevertheless, payment to the Maasi for migration rights was a practical means of gaining access to the larger area necessary to maintain populations of many large animal species. David Western, the originator of the plan, documented unresolved problems; yet this appears to be one of the more sophisticated approaches for dealing with a complex set of social, technical, and political issues (May 1982).

In underpopulated Belize, the issues are more tractable. A.I.D. is supporting the work of the Belize Audubon Society to develop a management plan for a marine park on the Belize Barrier Reef--a primary attraction of the growing tourist industry. It will permit low impact tourist visitation and fishing in buffer areas--a major source of lobster, conch, and shrimp for the country's commercial fishermen (A.I.D. 1988c). In recognition of the dual need to expand attractions for foreign tourists and protect indigenous species, the A.I.D. Jamaica Mission is planning to support the development of a new national park system (A.I.D. 1988b).

PROTECTING ENDANGERED SPECIES

Considerable moral and financial support for worldwide campaigns to protect endangered species in the tropics have been built by highly publicized and sometimes successful campaigns. Often conducted by expatriate organizations, these efforts have focused on the plight of large, appealing, and sometimes locally valued animals such as gorillas, pandas, seals, elephants, and tigers. The pros and cons of this approach continue to be argued, but it has produced bans on hunting and *in situ* conservation of large areas that may well provide the space that many of these animals require for their survival--and natural habitat for other species of plants and animals and entire ecosystems as well. One example, the Mahaweli Environment Project, funded jointly by the Sri Lankan Government and A.I.D., was designed to protect the habitat of the endangered Asian elephant. Recently, this species has been threatened by construction of a major river basin development project that will include several major dams and irrigation systems and will cover a large portion of the island. Initially spurred by public concern for this national symbolic animal in Sri Lanka, the protection justification also cited the need for watershed protection and conservation of other endangered plants and animals [Tippetts-Abbott-McCarthy-Stratton (TAMS) 1980].

The development of a plan by WWF to help assure the survival of the Royal Bengal Tiger in the Sunderaban forests of Bangladesh is a similar case in which conservation of a highly visible animal was the driving force that justified a protection plan. At stake were extensive mangrove forests and estuaries that have many other ecological and economic benefits (Seidensticker and Hai 1983). A somewhat similar rationale reportedly led to the Air and Tenere National Nature Reserve in the Republic of Niger. Initially, the plan was grounded in the concerns of IUCN and WWF about the disappearance of arid land ungulates such as the addax, scimitar-horned oryx, and several gazelle species (Newby 1984). However as planning progressed, other values were incorporated, such as conservation of plant resources and tourist potential, and permission for local people to remain in the reserve area was granted. According to Newby, one of the plan- ners, issues still to be addressed include culling, traditional wildlife hunting, and involvement of local populations in reserve management.

INDIGENOUS CONSERVATION

Examples of locally driven conservation efforts can be found in Asia and Central America. An ancient form of conservation in India was the "sacred grove"–which continues to be considered inviolate by many local people as recently as the 1970s. By that time, most forests surrounding the groves had been cleared for agriculture, and the groves had become one of the few remaining sources of forest products (medicines, fuelwood, timber, water for cattle) as well as potential sources of germplasm resources for commercial timber and fruit orchard species. As local taboos were weakening, it was recommended that the government establish the groves as a network of preservation plots for maintaining biological diversity (Gadgil and Vartak 1976).

In the past ten years, conservation of biological diversity in India has become increasingly a people's movement. Examples include the public debate over construction of a dam in Kerala State that threatened a large rain forest in the Silent Valley–now declared a national park, and the Chipko movement in the western Himalayas– an organization of local people that opposed commercial logging and initiated measures to conserve soil, regulate grazing, and reforest steep mountain slopes. Opportunities now exist to merge local conservation efforts with establishment of a national network of biosphere reserves, two of which are near the Chipko region. Because traditions of nature conservation include taboos on the use of alpine pasture, the MAB concept might work if it means that the Chipko people will not be displaced or prevented from collecting much needed fuel from parts of the area (Gadgil 1984).

In Panama, the Kuna Indians, who control their own territory (the Comarca of San Blas), are developing a 20 km^2 forest park and wildlife refuge for scientific and nature tourism. It will protect their natural resources by buffering their territory from incursion by non-Kuna farmers and cattle ranchers and their cultural and political identity as well. The selected area on their border is classified as humid tropical forest and is thus ill-suited for agriculture and colonization. To establish their control over the area, the Kuna, after failure of an agricultural production scheme, eagerly accepted the idea of a nature reserve–a concept that already existed in their culture in the form of botanical parks protected by spirits where no farming is allowed and certain trees are protected (Breslin and Chapin 1984, Chapin 1985).

NEW ECONOMIC PRODUCTS

With increasing global dependence on a very small number of high-yielding varieties of major food crops for which relatively good *ex situ* networks for collection and storage of germplasm resources exist, concern has grown about the lack of attention to conservation of the so-called minor crops, landraces, wild crop relatives and other wild, useful plants (Prescott-Allen and Prescott-Allen 1983). Many of these contain valuable pest resistance attributes, have potential for either local subsistence or commercial development, and may be well suited for use on marginal lands. Supported by A.I.D., staff from the New York Botanical Garden and Missouri Botanical Garden are collaborating with Ecuadorian scientists to assess the economic potential of newly identified plants as sources of medicines, commercial timber, and nutritious foods. In India, a component of the new A.I.D. Plant Genetic Resources Project will map the distribution of wild relatives of economically important crop species and collect and assess them for potential value in crop improvement (A.I.D. 1988c).

SUSTAINED YIELD MANAGEMENT

Game ranching, culling, and hunting of wild animals continue to be proposed as sustainable uses of African savannas, particularly in areas infested by the tsetse fly (Coe 1980). Examples of successful private game ranching operations are reported to exist in Zimbabwe, South Africa and southern West Africa. In Kenya, a combined wildlife and livestock operation has been tested for a number of years and appears to be ecologically and financially sound (May 1982, D.

Hopcroft, unpubl. data). According to the owner, David Hopcraft, wildlife is more profitable than cattle (i.e., produces more protein per acre), and its productivity can be maintained better in times of drought due to more sustainable use of plant resources. He also reports the interest of the Tanzanian government in a wildlife cropping system in cooperation with tribal areas that would provide an additional source of income for local people (D. Hopcraft, pers. comm.). Problems that need to be addressed in wildlife production schemes include: legal constraints (because many countries prohibit outright the hunting and killing of indigenous, endangered wild animals), economic competition, establishment of markets, disease control and transmission, and suspicions of many conservation groups that these activities hinder rather than help their efforts.

Other methods have been reported for sustainable use of both domesticated and wild animal species, ranging from little-known Asian species (e.g., banteng, kouprey) to water buffalos, crocodiles and butterflies (NAS 1983). An example of sustained yield management of primates for medical research purposes has been reported for tropical humid forests in South America (World Resources Institute 1985) and others have been proposed (Arambulo et al. 1988). This is an area where more innovative programs are needed.

METHODS OF PLANNING AND ANALYSIS

The options reviewed above are representative samples of a range of opportunities to address conservation concerns. But how have development practitioners justified including them in their programs? To facilitate progress on this front, improved methods of economic analysis are being devised, and better planning techniques are being used more broadly. Most present forms of economic analysis are inadequate to address the real costs of natural resources depletion (Warford 1986, Panayotou 1987). Investments in these areas are therefore difficult for developing country governments to justify on the basis of short-term financial returns, particularly in times of severe budget deficits and low levels of economic growth. More and better work is needed by economists and ecologists alike to remedy the deficiencies at both theoretical and practical levels to address macroeconomic planning at the national level and more appropriate cost-benefit analyses at the project level. More accurate analyses of the value of renewable natural resources that are becoming nonrenewable, very expensive, or impossible to replace should be supported. Revisions in 1987 to the A.I.D. manual on economic cost-benefit analysis represent a step in this direction by incorporating environmental and natural resource concerns. Work is also beginning on macroeconomic policies (Tarrant et al. 1987, Panayotou 1988, Repetto 1988), but it is still unclear what attention these new approaches will receive from governments and donors.

Existing approaches can be used for national, regional, and local planning. Techniques for land capability analysis for areas still to be converted, useful methods for analyzing existing production patterns on lands already in use, and environmental impact assessments can be applied to better assure sustainability and equity, relieve pressure on natural systems, and contribute to maintenance of biological diversity. They can be categorized as political or technical–the former supplying the groundwork for national policy, and the latter for field applications. At the political level, national conservation strategies (NCS), country environmental natural resource profiles or "state-of-the-environment" reports have been, or are being, prepared in a number of countries (World Resources Institute 1990). They are comparatively recent phenomena, and it is still too early to evaluate their impact. In general, they address conservation of biological diversity and/or endangered plant and animal species or communities, and provide a framework for national priorities. Environmental profiles have provided the basis for A.I.D. investments in natural resources management in a number of countries.

The format and content of national conservation strategies (NCS') are more specific in this regard if, as seems to be the trend, they follow conceptually from the World Conservation Strategy and the outline for NCSs recommended by IUCN (IUCN 1984). The NCS process results in the incorporation of conservation goals and priorities in national development plans, along with the allocation of financial and human resources required for implementation. With support from a number of bilateral donors including A.I.D., NCSs have been completed in Zambia, Nepal, and Botswana; and others are currently being prepared.

Technical approaches widely applied in Latin America at the local level include the Ecological Land Use Capability Classification System developed by Tosi at the Tropical Science Center in Costa Rica (J.A. Tosi, unpubl. data) and the integrated regional planning methods of the Organization of American States (OAS) (OAS 1984). The former, based on the World Life Zone System of Ecological Classification devised by L.R. Holdridge, involves mapping land capability at five different use levels that can be applied from the national level down to local farm size land units. The five types of use are: clean tillage, pasture, permanent ground cover crops, production forestry, and absolute land protection. This classification system is particularly useful in unsettled "agricultural frontier" areas (Tosi 1985). In comparison, the OAS methodology combines diagnosis of economic problems in a given geographical region and its physical potential (based on the Holdridge system) followed by formulation of a package of sectoral projects designed to address the problems. It generally includes proposals for the establishment of protected areas within the study region. However, since the OAS Office of Regional Planning provides only technical assistance, it depends on governments or donor organizations to provide investment funds for pieces of the package. Protected areas may or may not receive the same priority as roads, schools, or other development activities.

In Asia, where land has been under intensive agriculture for centuries longer than in Latin America, a different technical approach has evolved, in part because impressive food crop production increases may have peaked, and attention is now being redirected to agricultural systems for marginally productive areas, i.e., dry, hilly uplands, tidal swamps, grasslands, and forests. Agroecosystems analysis, a form of systems analysis, has grown out of the work of Conway and others in a number of countries (Thailand, Indonesia, the Philippines, Bangladesh, and Pakistan) with support from the Ford Foundation and other national and international institutions (Rambo and Sajise 1985, Conway 1986). The basic concept of this type of analysis is multidisciplinary investigations of specific agricultural systems drawing on the knowledge of ecology, biology, and cultural systems to define research and development priorities for improving local agriculture based on principles of sustainability, equitability, productivity, and stability. The limitation of this approach may be its perceived isolation from broader ecosystem evaluation, but its strengths in defining key research and development questions and resolving environmental conflicts are considerable. Its role vis-a-vis biological diversity lies with its potential to identify more effective and ecologically sustainable food production systems for use by small farmers on marginal lands–a problem that circumscribes the globe in its impacts on living resources. The potential for combining the Tosi and Conway techniques needs further investigation, especially for applying them in different geographic settings and more widely across continents.

Another planning approach that has evolved largely in the United States and is now being tested for developing countries, is integrated coastal zone management. Spurred by the passage of the Coastal Zone Management Act of 1972, a number of states have prepared integrated plans. This technology employs environmental impact assessment as a tool, and includes identification of critical areas for protection. Conflict resolution among development and protection interests is one aspect of the process, along with state and local responsibility for decisionmaking and implementation. Developing countries with important commercial or artisanal fisheries, coastal agriculture and forestry, or attractive beach resources for tourism (especially those vulnerable to natural hazards) have been identified as those that could derive particular benefits from the adaptation of integrated approaches to their specific institutional and natural environments (Sorensen et al. 1984). A.I.D. is providing technical assistance to the ASEAN (Association of South East Asian Nations) countries on a regional basis and to Ecuador, Sri Lanka and Thailand individually in an effort to apply this methodology. One of the important aspects of these programs will be the identification and protection of critical coastal habitats such as estuaries, mangrove forests, and coral reefs that need to be maintained for both biological and economic reasons.

These types of planning methodologies provide an overall framework that can incorporate conservation concerns, but more specific data are needed to define critical ecosystems and species in need of protection. The U.S. Congress has required A.I.D. to address conservation needs in each of its country strategies. To carry out this directive, overseas missions are conducting background assessments on tropical forestry and biological diversity conservation needs. This information is shared with other donors.

ESTABLISHMENT OF PRIORITIES

Many environmental groups and scientists contend that existing policies and programs are inadequate. At one end of the spectrum, there are arguments that taxonomic studies are needed to identify the totality of living organizations in the world wherever they are yet to be discovered, described, and named (Wilson 1988b). Although the scientific rationale for such extensive work is sound, the reality is that national priorities for baseline taxonomic and inventory work will be determined by a combination of forces–political, scientific and economic–and, based on the interplay of these, the resources made available.

In the final analysis, local priorities and decisions may be the most important ruling factor in maintaining living resources. If actual control at the local level is nonexistent, so too is responsibility. One can detect a certain pattern of resource deterioration that emerges from various studies, workshops, and field observations in many developing countries. It is characterized by strong central governments where most decisions about land use are made in capital cities by sectorally oriented line ministries that do not consult with each other or with local governments or local people; the result has been conflict and confusion over land and water use.

In an assessment of the state of the watersheds above the Mahaweli irrigation projects in Sri Lanka, it was found that 25 agencies were operating in the upper catchments and adjacent areas, but no single agency had overall responsibility (TAMS 1980). In the Philippines, the mayor of Buhi City cited a similar problem: lack of control over the municipality's immediate and surrounding land and water resources which caused serious social and physical problems (Conway and Sajise 1986). OAS considered the identification of a central point of geographic responsibility for regional planning as a key to ultimate success (OAS 1984).

CONCLUSIONS

Some preliminary answers can be given to the issues and questions raised earlier. Development assistance agencies are broadening their perspective on the role of conservation of living resources in economic development. This is in part due to the pressure applied by public interest environmental groups (Rich 1985), but also to internal recognition. Without external interest, however it is doubtful that continued progress will occur. The World Bank spent three years compiling a technical background paper and operational policy on wildlands. Its approval coincided with mounting public pressure for improved environmental performance. That policy is very specific in laying out those activities that should normally contain wildlands components. They include projects involving agriculture and livestock, transportation, power generation, and industrial development. Similarly, A.I.D.'s present strategy places greatest emphasis on incorporating conservation components into mainstream rural and agricultural development activities. These policy applications need to be monitored.

The level of funding that will be available for conservation is still unclear, but for A.I.D., at least, it should be possible to increase such investments considerably through greater use of debt for nature swaps and local currencies generated from sales of agricultural commodities provided by the United States to some developing countries (M.B. Kux, unpubl. data). An amendment to the Agricultural Trade and Development Act in 1987 authorized the use of P.L. 480 funds for activities such as the purchase of land for protected areas, inventories of tropical forests, education and training, and restoration of degraded ecosystems (U.S. Senate 1987). Some of these have already been financed with local currencies, but the potential is unrealized.

The willingness of developing countries to borrow for conservation is being tested, and there is skepticism among development planners as to whether it will occur on a broader scale. Grant funds are a relatively scarce commodity; some donors have none at all. Even if countries increase their borrowing somewhat, where will increased support come from on the scale required to address the problem? There are some glimmers of hope on the horizon. A study of innovative mechanisms to fund conservation is being conducted by the World Resources Institute with support from the United Nations Development Program, Canadian CIDA, the MacArthur Foundation and A.I.D. It will look into debt swaps, debt forgiveness, blocked currencies, etc. Another potential source of funds is the private foundation community in the United States. In 1987, 15 or so of these institutions along with A.I.D. formed a Consultative Group on Biological Diversity with the expressed purpose of addressing conservation needs in the tropics. As discussed earlier, better economic analyses of the actual importance of tropical forests, wetlands, grazing lands, and wildlife are needed as justification for these investments and the benefits of rehabilitating degraded lands to relieve pressure on undisturbed ares. The latter is particularly important in the drought-affected areas of Africa and in countries such as Haiti where population pressure is severe. Finally, the appropriate duration of projects that deal with natural resources needs to be rethought. Typical 3-6 year time frames are unrealistic given the severity of the problems, the time required for returns on investments, and the need to test new technologies.

Very little attention and funding are being invested in research on management of tropical forests, unlike that alloted to plantation forestry and agroforestry. Studies are needed on primary and secondary forests and wetlands in all geographic regions of the tropics. Greater opportunities and increased responsibility should be given to local people to manage natural resources including wildlife for either subsistence or additional income. Changes in policy and legislation on tenure and ownership, as is now beginning to happen in the Philippines on public forest lands (Lynch and Talbott 1988) will be required in many countries. This could reduce environmental degradation and illegal exploitation, but analytical work will be required for the increasing focus on buffer zones to determine whether they can be designed in a manner that maintains the integrity of protected areas over time.

The key elements in the success of future conservation efforts abroad include: enlightened national and international economic development policies, decentralized responsibility to encourage building on proven traditional resource management techniques, and maintenance of existing (at the very least) living conditions for economically disadvantaged people. If advances are not made in these areas, investments in more conventional conservation approaches in the developing countries of the tropics may themselves be unsustainable.

REFERENCES

Agency for International Development (A.I.D.). 1982. Central Selva Resource Management Project. Project Paper No. 527-0240. A.I.D., Washington, D.C.

______. 1985. Natural Resources Management Project No. 525-0248. Project identification document. A.I.D. Panama Mission, Panama.

_______. 1988a. Environment and Natural Resources. A.I.D. policy paper. Bureau for Program and Policy Coordination, A.I.D., Washington, D.C.

_______. 1988b. Protected Areas Resources Conservation Project No. 532-0148. Project identification document. A.I.D. Jamaica Mission, Jamaica.

_______. 1988c. *Progress in Conserving Tropical Forests and Biological Diversity in Developing Countries*. 1987 Annual Report to Congress on Implementation of Sections 118 and 119 of the Foreign Assistance Act (as amended). A.I.D., Washington, D.C.

Arambulo, F.V., J. Moro, F. Encarnacion and C. Malaga. 1988. *Management of Neotropical Primates in Their Natural Habitat as a Renewable Natural Resource*. Pan American Health Organization, Washington, D.C.

Arbhabhirama, A., D. Phantumvanit, J. Elkington and P. Ingkasuwan, eds. 1987. *Thailand Natural Resources Profile*. Thailand Development Research Institute, Bangkok, Thailand.

Breslin, P. and M. Chapin. 1984. Conservation Kuna-style. *Grassroots Development* 8(2):26-35.

Chapin, M. 1985. Udirbi: An indigenous project in environmental conservation. Pages 39-53 *in* T. MacDonald (ed.), *Native Peoples and Economic Development*. Cultural Survival Occasional Paper No. 16. Cultural Survival, Inc., Cambridge, Massachusetts.

Coe, M. 1980. African wildlife resources. Pages 273-302 *in* M.E. Soule and B.A. Wilcox (eds.), *Conservation Biology: An Evolutionary-Ecological Perspective*. Sinauer, Sunderland, Massachusetts.

Conway, G.R. 1986. *Agroecosystem Analysis for Research and Development*. Winrock International, Bangkok, Thailand.

Conway, G.R. and P.E. Sajise. 1986. *Buhi Agroecosystems Analysis*. Program on Environmental Science and Management (PESAM), University of the Philippines, Los Baños, Philippines.

Gadgil, M. 1984. Conserving biological diversity: The Indian experience. Pages 485-491 *in* F. DiCastri, F.W.G. Baker and M. Hadley (eds.), *Ecology in Practice*. Tycooly International Publishing Ltd., Dublin, Ireland.

Gadgil, M. and V.D. Vartak. 1976. The sacred groves of the western Ghats in India. *Economic Botany* 30:152-160.

Halffter, G. 1981. Local participation in conservation and development. *Ambio* 10:93-96.

Harrison, J., K. Miller and J. McNeely. 1984. The world coverage of protected areas: Development goals and environmental needs. Pages 24-38 *in* J.A. McNeely and K.R. Miller (eds.), *National Parks, Conservation and Development: The Role of Protected Areas in Sustaining Society*. Smithsonian Institution Press, Washington, D.C.

Interagency Task Force on Biological Diversity. 1985. *U.S. Strategy on the Conservation of Biological Diversity: An Interagency Task Force Report to Congress*. A.I.D., Washington, D.C.

International Union for Conservation of Nature and Natural Resources (IUCN). 1980. *World Conservation Strategy: Living Resource Conservation for Sustainable Development*. IUCN, Gland, Switzerland.

________. 1984. *National Conservation Strategies, A Framework for Sustainable Development*. Gland, Switzerland.

Lusigi, W.J. 1984. Future directions for the Afrotropical realm. Pages 136-146 *in* J.A. McNeely and K.R. Miller (eds.), *National Parks, Conservation and Development: The Role of Protected Areas in Sustaining Society*. Smithsonian Institution Press, Washington, D.C.

Lynch, O.J. and K. Talbott. 1988. Legal responses to the philippine deforestation crisis. *New York University Jornal of International Law and Policies*, Vol. 20, (3):679-713.

Machlis, G.E. and D.L. Tichnell. 1985. *The State of the World's Parks*. Westview Press, Boulder, Colorado.

Maldague, M. 1984. The biosphere reserve concept: Its implementation and its potential as a tool for integrated development. Pages 376-401 *in* F. DiCastri, F.W.G. Baker and M. Hadley (eds.), *Ecology in Practice*. Tycooly International Publishing Ltd., Dublin, Ireland.

Malik, A. 1984. Protected areas and political reality. Pages 10-11 *in* J.A. McNeely and K.R. Miller (eds.), *National Parks, Conservation and Development: The Role of Protected Areas in Sustaining Society*. Smithsonian Institution Press, Washington, D.C.

May, C.D. 1982. Preservation for profit. *The New York Times Magazine*, Sept. 12, 1982.

McCaffrey, D. and H. Landazuri. 1987. *Wildlands and Human Needs: A Program Evaluation*. World Wildlife Fund, Washington, D.C.

McNeely, J.A. and K.R. Miller, eds. 1984. *National Parks, Conservation and Development: The Role of Protected Areas in Sustaining Society*. Smithsonian Institution Press, Washington, D.C.

Miller, K.R. 1984. The natural protected areas of the world. Pages 20-23 *in* J.A. McNeely and K.R. Miller (eds.), *National Parks,*

Conservation and Development: The Role of Protected Areas in Sustaining Society. Smithsonian Institution Press, Washington, D.C.

National Academy of Sciences (NAS). 1980. *Research Priorities in Tropical Biology*. National Academy Press, Washington, D.C.

_______. 1982. *Ecological Aspects of Development in the Humid Tropics*. National Academy Press, Washington, D.C.

_______. 1983. *Little-Known Asian Animals With a Promising Economic Future*. National Academy Press, Washington, D.C.

Newby, J. 1984. The role of protected areas in saving the Sahel. Pages 130-135 *in* J.A. McNeely and K.R. Miller (eds.), *National Parks, Conservation and Development: The Role of Protected Areas in Sustaining Society*. Smithsonian Institution Press, Washington, D.C.

Organization of American States (OAS). 1984. *Integrated Regional Planning: Guidelines and Case Studies from OAS Experience*. OAS, Washington, D.C.

Panayotou, T. 1987. Economics, environment, and development. Development Discussion Paper No. 259. Harvard Institute for International Development, Cambridge, Massachusetts.

_______. 1988. Management of natural resources for sustainable development: Market failures, policy distortions and policy options. Report prepared for U.S.A.I.D./Thailand, Bangkok, Thailand.

Prescott-Allen, R. and C. Prescott-Allen. 1983. *Genes from the Wild*. International Institute for Environment and Development, London.

Rambo, A.T. and P.E. Sajise. 1985. Developing a regional network for interdisciplinary research on rural ecology: The Southeast Asian University Agroecosystem Network (SUAN) experience. *Environmental Professional* 7:289-295.

Repetto, R. 1988. *The Forest for the Trees? Government Policies and the Misuse of Forest Resources*. World Resources Institute, Washington, D.C.

Rich, B.M. 1985. Multilateral development banks: Their role in destroying the global environment. *The Ecologist* 15(1/2):56-68.

Seidensticker, J. and M.A. Hai. 1983. *The Sundarbans Wildlife Management Plan: Conservation in the Bangladesh Coastal Zone*. IUCN, Gland, Switzerland.

Sorensen, J.C., S.T. McCreary and M.J. Hershman. 1984. *Institutional Arrangements for Management of Coastal Resources*. Research Planning Institute, Columbia, South Carolina.

Sumardja, E.A., Tarmudii and J. Wind. 1984. Nature conservation and rice production in the Dumonga area, North Sulawesi, Indonesia. Pages 224-227 *in* J.A. McNeely and K.R. Miller (eds.),

National Parks, Conservation and Development: The Role of Protected Areas in Sustaining Society. Smithsonian Institution Press, Washington, D.C.

Tarrant, J., E. Barbier, R.J. Greenberg, M.L. Higgins, S.L. Lintner, C. Mackie, L. Murphy and H. Van Veldhuisen. 1987. *Natural Resources and Environmental Management in Indonesia: An Overview*. A.I.D., Jakarta, Indonesia.

Tippetts-Abbott-McCarthy-Stratton (TAMS). 1980. *Environmental Assessment, Accelerated Mahaweli Development Program*. Main Report, Vol. 1. TAMS, New York.

Tosi, J.A. 1985. Possibilities for integration of ecological land use capability classification and agroecosystem analysis techniques. International Institute for Environment and Development, Washington, D.C.

U.S. Congress, Office of Technology Assessment (OTA). 1987. *Technologies to Maintain Biological Diversity*. U.S. Government Printing Office, Washington, D.C.

U.S. Department of State. 1982. *Proceedings of the U.S. Strategy Conference on Biological Diversity*. U.S. Department of State, Washington, D.C.

_______. 1986. Comments on biological diversity action plan. Bangkok Cable No. 23611, 16 May, Washington, D.C.

U.S. Senate, Committee on Agriculture, Nutrition, and Forestry. 1987. Report to accompany bill S659, 100th Congress. Report 100-33. U.S. Government Printing Office, Washington, D.C.

Warford, J.J. 1986. Natural resource management and economic development. Paper presented at IUCN Conference on Conservation and Development on Implementing the World Conservation Strategy. Ottawa, Canada.

Western, D. 1984. Amboseli national park: Human values and the conservation of a savanna ecosystem. Pages 93-100 *in* J.A. McNeely and K.R. Miller (eds.), *National Parks, Conservation and Development: The Role of Protected Areas in Sustaining Society*. Smithsonian Institution Press, Washington, D.C.

Wilson, E.O., ed. 1988a. *Biodiversity*. National Academy Press, Washington, D.C.

Wilson, E.O. 1988b. The current state of biological diversity. Pages 3-18 *in* E.O. Wilson (ed.), *Biodiversity*. National Academy Press, Washington, D.C.

World Bank. 1986. The World Bank's Operational Policy on Wildlands: Their Protection and Management in Economic Development. The World Bank, Washington, D.C.

World Resources Institute. 1985. *Tropical Forests: A Call for Action*. World Resources Institute, Washington, D.C.

World Resources Institute, Center for International Development and Environment. 1990. Country Environmental Studies, An annotated biolography of Environmental and Natural Resources Profiles and Assessments. W.R.I. Washington, D.C.

Epilogue

Ethics, Economies, and Conservation

Janis B. Alcorn

INTRODUCTION

> It is worth dying for, if we can save the forest for future generations. (Ban Huey Kaw farmer, Thailand, quoted in Anonymous 1989)

> This is our community forest that was just put inside the new national park. No one consulted us! We protected this forest before the roads were put in. We set up a roadblock on the new road to stop the illegal logging. We caught the district police chief and arrested him for logging. We warned him not to come again. (Tambon leader, Lampoon District, Thailand, 1990)

> I don't understand why Protected Areas aren't working in the tropics. They work so well here [in the U.S.]. (Consultant on biodiversity issues, World Bank 1991)

Together the essays in this book make the point that two strategic actions are absolutely necessary to achieve global conservation action. The first strategic action is to empower the local, traditional conservation ethic wherever it still exists. The second strategic action is to construct a new, modern conservation ethic appropriate for the emerging global capitalist economy. Planners and decisionmakers have not been aggressively pursuing either of these two strategic actions, even though large investments are being directed toward conservation efforts. Instead, they have been developing and investing in mirages–strategies rendered incomplete or useless because of three incorrect assumptions about biological conservation and ecodevelopment.

First, planners mistakenly view their task as working to encourage a transition from archaic consumption of nature to modern conservation of nature. They interpret the sights of peasants burning or otherwise consuming the forest as evidence of achaic consumption, for example. They visualize modern conservation as the National Parks and Protected Areas Systems currently functioning in the First World. They mistakenly assume local people must be "trained" and "educated" to change their behavior. As one prominent United Nations Development Programme (UNDP) professional recently put it, "If we can manage the people, we can manage the forest."

The real transition in the tropics over the past several centuries, however, has been from archaic conservation under local economies to modern consumption under the global capitalist economy. Archaic conservation limited extraction according to local rules that prevented destruction of nature, because nature was the pre-eminent resource base. In contrast, capital and encourage a transition from "archaic consumption" to "modern conservation" of nature. For example, they interpret sights of peasants burning or otherwise consuming forests or natural areas as evidence of archaic consumption, whereas they visualize modern conservation in terms of the national parks and Protected Areas Systems (PASs) currently functioning in the First World. However, the reality in the tropics over the past several centuries has actually involved a transition from "archaic conservation" under local economies to "modern consumption" under what can be most appropriately characterized as the emerging global capitalist economy. Archaic conservation operates in subsistance economices by placing limits on the extraction of natural resources; local rules therefore tend to prevent destruction of nature because raw material inputs derived from nature comprised the pre-eminent resource base. In contrast, capital and labor comprise the pre-eminent resource base in capitalist economies, and nature is transformed into capital for investment and commodities for consumption. Modern conservation, derived from First World models, is incompatible with modern consumption patterns in Third World nations, because Third World consumption patterns are driven by the combined appetites of First World and Third World consumers who stimulate extraction of natural resources from Third World rural areas. Archaic conservation, not modern conservation, is responsible for the continued presence of most of the world's biodiversity, because the transition from archaic conservation to modern consumption is still incomplete in most developing countries. I believe careful study would confirm my field-based observation that the majority of the Earth's biodiversity is not held in patrolled wildland reserves, but in landscapes managed by local people. Most of the world's biodiversity is on lands and waters used by traditional peoples marginal to the global economy. This condition is not likely to persist through the 21st century, however.

Second, conservation planners have assumed that First World park systems are functioning well in capitalist nations and therefore are a goal appropriate to Third World countries. Conservation planners make two mistakes here. They reason that First World parks offer proof that capitalism and the market economy create a solid framework for making the (misperceived) developmental transition from archaic consumption to modern conservation. They are, however, forgetting that First World businesses are drawing on Third World natural resources to meet First World demands, and, therefore, not meeting those demands by extracting the natural resources in their own nations' park and reserve systems. It is also an open question how well conservation is working in the First World. Analyses of the impact of the U.S. Endangered Species Act of 1973 and the politics of its implementation by U.S. government agencies (e.g., Tobin 1990), and reports of poaching in U.S. national parks (e.g., nationally televised documentaries about local residents' poaching of bears in Smoky Mountain National Park) demythologize the idealized Protected Area System (PAS). In fact, there is perhaps greater opposition to conservation among the general populace in the U.S. than in the tropics. For example, the general populace is often suspicious that declaration of critical habitat that includes their land will affect their economic gain from selling that land. Tobin (1991:178) cites an example where the U.S. Fish and Wildlife Service withdrew a proposal to designate a critical habitat for the Chihuahua chub fish, because "they received so many threats from people claiming they would deliberately destroy the chub if its habitat was formally designated." Yet, on the other hand, there is also evidence that democratic participation by concerned citizens and NGOs can have a positive impact on conservation action. For example, in response to popular opposition to federal agency decisions regarding conservation, federal judges have intervened in the case of U.S. old growth forests (Washington Post 1991a, 1991b) and the definition of critical habitat for the monk seal, one of the world's most endangered animals, in Hawaii (Tobin 1990). Unfortunately, many international consultants are apparently unaware that the U.S. PAS shares many of the same problems found in the tropics–a misperception typified by the third introductory quote, a comment made by a long term consultant to the World Bank during conversation with me.

The third faulty assumption made by conservation planners is that governments exist to use and conserve national assets (including a nation's endowment of natural assets) for the public good. They assume that Third World failure to conserve biodiversity lies in governments' failure to recognize the economic value of biodiversity. Therefore, they reason that the appropriate strategic action to achieve conservation is to raise governments' consciousness about the true economic and commercial values of biodiversity while delimiting Protected Areas Systems for governments to protect. In line with this reasoning, many First World conservationists have worked to assign economic values to tropical forests based on everything from medicinal plants to ecological services. Grassroots Third World NGOs and grassroots conservation activists do not make this assumption, however. Instead, they design their programs and campaigns to force changes in governments that have historically represented the interests of the elites and exploited public lands for their private gain. Economic values do not influence decision-makers who are more interested in the immediate financial values of converting and exploiting biodiversity in response to market opportunities. This is a particularly sensitive topic that few conservation planners are willing to discuss in public fora.

Modern conservation planners' real task is to encourage a transition from fading archaic conservation and growing modern consumption to modern conservation (as yet unrealized) appropriate to large rural populations, regional economic relationships, local economies, and local, subcultural variations of global culture. As part of that task, conservation planners should develop strategies and plans that monitor and enforce government accountability for conservation action.

Many have prophesized that a quarter to half of the world's biodiversity will be gone by 2050 unless trends are reversed. Today's struggle to halt the loss of biodiversity will only be successful if 21st century "modern" culture balances marketplace guidance with an ethical commitment to stewardship. If current trends continue, by 2050, the global capitalist economy and modern consumer culture will have penetrated to the remotest areas and displaced archaic conservation mores. Unless modern conservation efforts forge a vigorous partnership with archaic conservation traditions in the periphery of the global economy, and unless a modern conservation tradition emerges to guide the inner circle behavior of capitalist elites, the doomsday prophesies will be fulfilled.

PAPER PARKS AND ANGRY VILLAGES

The first two quotations that introduce this epilogue, like the official cable from Thailand quoted by Kux (see Ch. 14, this volume), remind us that the future of biodiversity is tightly linked to the futures of rural people and governments. Unlike the words of the official cable, however, the words of these Thai people startle us. They are spoken by rural people angry about the destruction of biodiversity. In the first case, a village's community forest was usurped and bulldozed by a powerful government official to develop a personal resort home by means of a state-sanctioned forest improvement scheme. When village residents attempted to fight the take-over, one villager was assassinated. A coalition of villagers, nearby urban conservationists, and university students successfully reclaimed the forest, but villagers continued to worry about the future of their forest.

The second quote refers to a case where villages' rice fields, traditional irrigation systems, and community forests were incorporated within the boundaries of a new national park. Only homes were left outside the border. These villagers were not consulted, nor were they formally notified of the demarcation. They fear that state management of their watershed forests will result in deforestation, because they have fought state-sanctioned logging operations in the past. Villages have united to fight this new threat--a national park that may well become a "paper park," known on paper documents and maps, but exploited by influential politicians on the ground under the protection of the state. As the tambon leader told me, "We can protect the forest, but we need the state behind us. Without the help of the state, it cannot be done. We want to follow the Middle Way. We need cooperation between the village and the government." A recent appraisal of Thai conservation action (Lohman 1991) asks, "If corporations and the state, backed by international agencies, are the principal destroyers of biological diversity in Thailand, who is most effectively defending it?" His analysis provides the answer--"nearly everywhere it is the rural villagers who are playing the key roles."

The Thai situation is not unique. It is representative of what is happening in every region high in biodiversity. Globally, rural people who depend on biodiversity value it, defend it, and fight for their rights to manage it. Globally, failing "paper parks" constitute most of the world's PASs, because the powerful interests who control the state convert biodiversity into personal capital. The introduction of the PAS concept did not include transfer of the concept of stewardship with the title "protected area." In the tropics, "Protected Areas" are de facto a new category under which the state builds on a tradition of "State Forests" and "Public Lands" managed to function as mints run for the private financial benefit of elites--not as endowments treasured and protected for the public good (e.g., Hurst 1990).

The third introductory quote, made by a biodiversity consultant to the World Bank, illustrates that, even when faced with the failure of PASs and evidence that rural people defend their resource base, most modern conservationists cling to the belief that only the state has the legitimate right to protect biodiversity. They take action based on a misbegotten belief that dialogue on conservation plans begins and ends at the level of state agencies. Although they may include meetings to get national NGOs "on board," conservation planners rarely hold meetings with communities affected by their plans until after decisions are reached. The U.S. system of protected areas remains the basic model that is being transplanted–that is, global, regional, and national strategies to conserve biodiversity rely heavily on public sector administration of PASs.

Money is being invested in establishing PASs, PAS management plans, and biodiversity action plans as if plans guarantee results. But without monitored implementation, plans lead nowhere. Many planners now recognize that there are no short-term profits to be gained from investing in biodiversity protection, and thus that First World money must purchase Third World conservation. They seem to have forgotten, however, that some other indicator must be substituted for profit, which usually functions as a bottom line indicator of success or failure. Without a monitoring/accounting system to substitute for profit measurement, the impacts of conservation investment decisions are not being measured, and there is no feedback system to trigger decisions that do lead to successful conservation. Because funds must be raised for conservation, success stories are promoted so that investors will feel that they have purchased, or "bought into," something of proven value. Conservationists are reluctant to air failures for fear that investment monies will dry up because prospective buyers (donors) may feel that they are throwing away their money. Reports from the field, however, indicate that tropical PASs are having very limited impact and that few are likely to survive for another century–much less thousands of years into the future.

The problems of modern conservation run much deeper than failing PASs. Even perfect PASs will be rendered ineffective by global climate changes. Another major investment channel, ex situ gene banking, has been criticized for: 1) inability to save the evolutionary processes necessary to produce resistance and adaptation to changing climates, 2) dependence on sustained funding flows, and 3) inability to conserve more than a small sample of the genetic diversity of a limited number of species. The papers in Part Two of this volume summarize the ex situ issues. The overwhelming majority of biodiversity cannot be conserved in PASs and gene banks.

For biodiversity protection to be successful, in situ action must be simultaneously encouraged on two fronts: 1) natural ecosystems, and 2) human-dominated ecosystems (Oldfield 1984, Alcorn et al. 1989).

Natural ecosystems include lands little disturbed by their human inhabitants. A small subset of these lands is now demarcated in PASs. Human-dominated ecosystems range from suburban backyards and coastal port waters to ranches, farms, and commercial plantation forests. Human-dominated ecosystems have received little attention by modern conservationists, but biodiversity contributes to the ecological and economic sustainability of production systems. Traditional, local conservation systems have operated as an integral parts of the production systems employed in the general landscape. Successful modern in situ conservation action in the total landscape will require a serious rethinking of modern approaches–not only to address the biological parameters of biodiversity conservation, but also social, political and economic parameters.

The essays in this book challenge conservationists to rethink their stance, to listen to the people who live with biodiversity in their neighborhood. These essays do not offer "the answer," but they lay out the perspectives that must be brought together in the search for answers. They challenge conservationists to sort out notions that are "romantic," "naive," and "paternalistic," and to take a hard look at what has had real on-the-ground impact as opposed to what sounds promising at the planning or early implementation stage. They are examples of words spoken by the voices engaged in a growing dialogue about the future of cultural and biological diversity. In this epilogue, I place these essays in the context of that larger, ongoing dialogue. That evolving dialogue, and these essays, make it clear that traditional culture-bearers should not be viewed as useful libraries of traditional information, but rather as essential participants in biodiversity conservation during the transition to a fully integrated, global capitalist economy.

The time is right for a paradigm shift in the way conservationists and governments approach biodiversity conservation from global and national levels. The appropriate paradigm shift will occur only if traditional peoples actively participate in developing a new conservation ethic linked to the emerging capitalist economy but grounded in local, traditional understanding of the ecological underpinnings of economic enterprises.

CONSERVATION IN SUBSISTENCE AND CAPITALIST ECONOMIES

Today, two conservation systems–traditional and 20th century modern–exist in areas with high concentrations of biodiversity. Traditional conservation systems maintain biodiversity for subsistence and survival purposes. On the other hand, modern conservation systems depend on state commitment and expenditure to protect biodiversity from subsistence use and commercial exploitation. Neither system, however, is functioning at full capacity. The 20th century system is not functioning well, because it has not been invested with strong state commitment. At the same time, the traditional system is slowly disappearing as the shift to modern culture is completed. Both systems are being undermined by commercial exploitation of natural resources for short-term profit and the subsistence needs of a burgeoning global human population.

The shift from traditional to modern culture is inextricably linked with a shift from traditional subsistence economies to a global capitalist economy. Cultural transition is one aspect of economic transition. Globally, modern culture is linked to capitalist economy, whereas traditional cultures are linked to subsistence economies. While few people today depend directly on nature for their entire subsistence, the majority of the world's rural people still live at subsistence levels. They meet their basic needs by combining cash-generating activities with direct harvest of goods from nature. Many could not survive without goods harvested from nature; in effect, these people survive because they receive subsidies from nature.

Young people are the next generation of decision-makers who should both carry forward traditional conservation systems and participate in the evolution of modern conservation systems. However, they are being enculturated as modern culture-bearers and growing up ignorant of the long term services provide by nature, the intrinsic and extrinsic values of biodiversity, or the traditional rules which guaranteed their long term maintenance. The words of Pehuenche parents in Chile (quoted in Quarto 1990) are typical of those being said around the world–"The old people have their traditional . . . ways, but the young people are losing it. The only hope we have is the piÑon (trees) to keep our connections to the past. . . . The old people, they have all their feelings in their blood, and they won't change . . . but the young people will change." While culturally-coded resistance to outsiders has protected natural resources and buffered minority ethnic groups from incorporation into intruding, dominant cultures for centuries (e.g., see Toledo, Ch. 8, this volume), the attraction of consumer goods has captured the imaginations and conquered the cultural defenses of many resistant cultures during the past decade. Changes in the culture of the world's younger generation do not bode well for the continuation of traditional conservation systems in areas where the transition out of subsistence economies is, and may continue to be, incomplete.

The key questions for conservation during the 21st century may well be:

* How can we reinforce existing local level incentives and empower local people who currently preserve biodiversity on the basis of subsistence ethics?
* How can we, at the same time, ensure that the transition from traditional to modern culture (and subsistence to capitalist economy) will include a transition in the ethics and the means of biodiversity preservation?

Conservation in Subsistence Economies

Biodiversity has persisted up to the present time partly because of limited human population size and limited human technological abilities. At the dawn of the modern capitalist era, however, biodiversity was still rich in areas that have supported high population densities for centuries. One of the major reasons lies in the moral underpinnings of subsistence economies (Scott, 1976) which "provide a perspective from which the typical peasant views the inevitable claims made upon his resources by fellow villagers, landowners, or officials. It implies, above all, that such claims are evaluated less in terms of their absolute level than in terms of how they complicate or ease his problem of staying above the subsistence crisis level" (Scott 1976:29). The subsistence economy ethic also provides a perspective to measure available options of natural resource use. In subsistence economies, traditional culture includes a respect for nature and the natural processes that provide, from a modern point of view, the factory from which subsistence flows. Systems of resource use evolved to prevent destruction of the resource base and protect subsistence over the long term (e.g, see Toledo Ch. 8 this volume, Nabhan et al. Ch. 7 this volume, Alcorn 1989b, Quarto 1990, Lynch and Alcorn 1991). Traditional religions and livelihood strategies are based on the premise that nature is providing economies with a renewable subsidy (e.g., see Gliessman, Ch. 10, this volume). While labor is not invested in producing those things provided by nature (e.g., firewood is not planted and tended if it can be harvested free from the forest), natural processes are protected as resources (e.g., the forest's regenerative processes are not destroyed by firewood harvesting) (Alcorn 1989b, Geist 1988). In subsistence economies, ethical evaluation of human actions includes a moral commitment to the human right to meet basic needs (subsistence). The strength of the moral commitment to a community's subsistence resources should not be underestimated. Rebellions and revolutions have arisen when subsistence resources, including forests, have been alienated from peasants by nation states and the elites they serve. Uprisings have proven numerous times that peasants will die to protect biodiversity important to their subsistence (see the first introductory quote for present-day demonstration of the ethical commitment to community subsistence rights).

Traditional systems worked well, because they were based on strong ethical commitments. In addition, populations dependent on harvesting natural products were not so large that choices had to between death or conservative use. Furthermore, these systems worked because they did not require implementation through a bureaucratic system liable to corruption, underfunding, or poor, long-distance planning by those with little stake in success. Traditional systems were implemented by people who lived in the immediate vicinity of biodiversity, and who had the power to make and enforce management decisions about the natural resources in their neighborhood. Traditional systems are still in place in many areas high in biodiversity. The biologically-rich rainforests of the world support millions of traditional forest dwellers who have maintained their rainforest environment (Lynch 1990). Biologically diverse, traditional agroforestry systems that incorporate high levels of biodiversity are found throughout tribal regions of Southeast Asia, South America, Africa, South Pacific, and South Asia. Traditional farming systems tend to maintain a more diverse landscape than modern agricultural systems, because secondary vegetation and mature forest patches are maintained as necessary elements of the farm community's livelihood base (c.f., Oldfield and Alcorn, Ch. 3, this volume). Traditional conservation systems still operate in the world's deserts as well (c.f., Nabhan et al., Ch. 7, this volume). On the marine side, traditional South Pacific sea tenure regimes continue to guarantee conservative management of marine resources in some areas (e.g., see Hames, Ch. 9, this volume, Bailey and Zerner 1991, Ruddle and Johannes 1985, Baines 1989, 1991, Cordell 1991).

Under traditional cultural regimes functioning in subsistence economies, decisions about land and resource use are not solely based on financial analyses, but also on a sophisticated understanding of the economic and option values of ecological services (e.g., Alcorn 1989a, Randall 1986). Many rural people still give existence value to forest and species for which there are no known uses. They also base their decisions on their vision of what constitutes "the good life." The aesthetic beauty of birds, butterflies and flowers is appreciated, and intellectual satisfaction is derived from understanding the lifeways of animals. These factors enter into decision making. The freedom of rural people to make choices about how they use their land, however, is being severely constrained by the lack of economic opportunities for their growing population. Rural people will only continue their traditional sustainable resource use systems if they choose to do so, and if the people at the "top" (i.e., the elites in power) recognize the importance of enabling these land/water use systems to continue. In this volume, Clay, in Ch. 12, and Kux, in Ch. 14, provide specific examples of efforts to empower local resource managers.

Conservation in the Global Capitalist Economy

Conservation action in the global capitalist economy is carried out at private, public, national and international levels–the latter level being largely First World conservation action directed toward Third World countries. Capitalists, however, are also responsible for the global depletion of biodiversity, because nature has been treated like any other asset (Daly and Cobb 1989). Capitalism entails investment in enterprises that generate greater capital. If there are better opportunities for investing capital, assets can be converted into new capital that can be invested elsewhere (e.g., Clark 1973). "Mining" of biological resources for sale as commodities and conversion of critical habitat to establish factories and other enterprises (e.g., see Oldfield 1984:281) are done for short-term profits. The market rewards the capitalist who makes quick profits that can be invested in current opportunities. Capitalists who choose to conserve must bear the "opportunity cost" of passing up such options. Private sector businesses have been responsive to market incentives to convert natural resource assets into private capital, particularly when it can be done at little or no cost. Profits generated from biodiversity/nature in the Third World are largely spent in the First World and contribute primarily to the growth of First World capital. Ecological impacts of short-term profit taking are not incorporated into the cost-benefit analysis used in commercial ventures; instead they are "externalities" born by someone outside the profit loop. The myopic vision of commercial enterprises (Panayotou 1990) has contributed to the perception that development based on stimulating capitalistic endeavors is at odds with long term environmental management.

In First World capitalist societies, such as that of the United States of America (U.S.A.), private conservation action is supported by non-governmental organizations (NGOs), often with donations from large corporations which earn their profits in the global capitalist economy. Conservation action includes: 1) public awareness and advocacy campaigns to promote legislative or policy changes; 2) private participation in government sponsored programs that provide incentives for conservation; and 3) NGO purchase of privately-held wildlands. Almost all wilderness is administered by the public sector, and virtually no one in First World society derives their subsistence directly from nature. First World NGOs have become increasingly concerned about biodiversity conservation in the Third World, and are attempting to work with Third World governments, NGOs, and local communities to secure biodiversity. First World philanthropical foundations have also begun to channel funds into conservation in the Third World (Abramovits, 1989). And corporations which extract natural resources from tropical nations are becoming interested in conservation in order to improve their public image (Menagh 1991). These patterns offer options for future private conservation action.

The public sector in the capitalist First World is likewise active in conservation action. Nationally, public sector institutions enforce laws which are theoretically designed to protect the "public good," including biodiversity. While these institutions are imperfect, they have achieved reasonable success to date. Internationally, First World public sector resources are channeled through aid agencies, international agencies and multilateral development banks to Third World governments and, to a limited extent, to non-governmental organizations (NGOs).

The public sector also formally controls most of the natural assets in Third World countries rich in biodiversity. While one might assume that this has been done to ensure the wise conservation of a public good for public benefit, history has shown otherwise. The distinction between the state and private enterprise is blurred. Parastatal (quasi-government) forest-based industries, for example, are often controlled by close friends of those who control the government. They receive special rights to logging contracts, guaranteed low-priced wood inputs, and other subsidies from the state. In some cases, state government officials openly use their positions as assets to generate income for themselves. They take funds from state-sponsored programs and from private citizens who apply to receive benefits from those programs. While some might label these activities as "graft" and "corruption," they are accepted norms in many countries.

The state has failed in its role to protect biodiversity for several reasons. First, the motivation of the state has generally been to channel profits to a select group of individuals (generally those that are in, or supportive of, the ruling group) involved in the global capitalist economy. Informal systems exist whereby governments garner political support by awarding free access to these "public" resources to a limited set of businesses. Secondly, states are often weak and are forced to keep their coffers open to those that might otherwise move to overthrow the status quo. Thirdly, there is no political constituency (the rural people have little or no political power) or the political will to protect biodiversity. Thus, state institutions charged with natural resources management and conservation are underfunded and understaffed. Fourth, lack of due process and fair judicial systems mean that laws protecting biodiversity are not applied equally to all citizens. Fifth, the state fails to empower local people to protect biodiversity or consult with them about management decisions. Sixth, governments have failed to gather data on the livelihoods of rural people and the subsidies from nature upon which they depend. Seventh, governments have failed to recognize standing natural forest as a legitimate use of land, often because it does not generate desired tax revenues.

Administration and protection of protected areas in biodiversity rich countries is generally poor. Low budgets, inadequate staff, poor enforcement of regulations, and low political will at "the top" have hamstrung those bureaucratic units dedicated to wildlife conservation and protected area management. In some cases, those employed by the government to enforce regulations against logging and wildlife poaching are threatened or bribed by men in the employ of government officials who are involved in logging and other businesses that extract profits from "protected" lands.

What is the role of rural people in conservation in the global capitalist economy? Rural people who live in or near Third World protected areas generally feel that "reserves" are reserved for the profit of those in the government. The state blocks rural people from traditional conservation action, because conservation is defined as protection from people, a role to be performed only by the state. This situation places stress on traditionally conservative resource use patterns. Faced with the sight of others grabbing short-term profits from nature, local people sometimes choose to compete and convert nature to capital for themselves (illegally) before someone else (sanctioned by the state) does it. In effect, the state turns communal property into open access areas. There are many documented cases of the community strife that arises when some members want to take short-term gains from natural areas and others want to manage for long-term sustainability (e.g., VIKSAT/CEE 1990; Groenfeldt et al. 1990).

The story of deforestation in India (summarized from Alcorn and Molnar 1990, n.d.; Guha 1983) is illustrative of the situation in most developing countries, although this history is now being played out much more rapidly in the remaining pockets of high biodiversity. Although Indian communities fought for rights to make decisions about forest management, colonial and state governments usurped their rights in the name of modern management and conservation. Prior to incorporation into the global capitalist economy, India's forests were used by large populations of people and livestock for thousands of years. Historical records demonstrate that forests were under sufficient pressure that they were protected by local rulers over 2,000 years ago. Indications are that in pre-British times, authoritarian rules regarding forest use and maintenance were established and enforced through tribal authorities, local elites, or princes. These included taxes, land use zoning, fines, and labor used toward maintaining the resource base as well as village forest protection committees and forest guards. However, during the 18th and early 19th century, the British East India Company ignored local rules for forest conservation and management and clearcut princely teak plantations and sacred groves. Beginning in the mid 19th century, the British government denied people the right to continue to maintain their forest assets in species mixtures useful to the local economy, and progressively reduced their rights to forest products for subsistence or sale. In a long term drive beginning in 1865, the state passed a series of Forest Acts and Policies that slowly shifted from granting limited community "rights" over forest products (i.e., the state usurped local authorities' rights to grant rights) to complete state control of forest products and forest management. The state consolidated its control over Indian forests and mined them for commodities and capital. After Independence, the Indian state assumed the British role and completed the deforestation process to generate capital for the political elites. The state forest department was a tool, and a scapegoat, in this process. After forests were mined into wastelands while under state control–wastelands from which the state can extract little revenue, the Indian state began to accept the longstanding demands for community participation in forest management.

The legal structure of the capitalist economy grants rural people few rights to economic benefits generated from their traditional knowledge, traditional crop genetic resources, or their conservation of biota from which herbal medicines or pharmaceuticals are developed. Third World voices are increasingly demanding royalties or payments for use of these resources (e.g., Kunming Declaration 1990). But the rules under which capitalist societies currently share income from such resources are drawn up under the assumption that such resources must give the holder a competitive advantage that the holder wants to keep for himself (not share) in order for that holder to claim patent or trade secret rights, for example. The special contracts that capitalist drug companies and research institutes are beginning to draw up to pay royalities or establish trust funds are examples of what may be viewed as concessions given by the First World players to Third World players in the global economy. While the conservation impacts of these payments are as yet unknown, they are examples of ways that the First World can pay for maintaining biodiversity resources in the Third World.

FACING THE TRANSITIONAL CHASM: OPTIONS FOR TAKING BIODIVERSITY TO 2100 AND BEYOND

We are facing a global transition, and our predictions of what lies on the other side are imperfect. The global capitalist economy is reaching into the most remote areas, and the completed transition to modern culture is imminent. While capitalists trumpet the coming of a New World Order based on free enterprise, the highly-publicized Brundtland Commission Report on "our common future" has sounded the alarm that a chasm has been sighted in the path of transition–a chasm that threatens to cause a "system shift" (i.e., accurate forecasts cannot be made using models based on past experience) (World Commission on Environment and Development 1987). Pundits reading signs in the sky forecast "the end of nature" during the transition (e.g., McKibben 1989). Unless special effort is made to construct a bridge to accomodate biodiversity during the transition, a significant portion of Earth's biodiversity will be left behind as dry museum specimens and archaeologically-interesting detritus of Earth's earlier production system lost in the chasm of 2050, leaving humans to face the consequences.

The moral conservation imperative of subsistence economies and traditional cultures is fading. Financial considerations drive decisions in the global capitalist economy. Local subsistence economies often included some capital accumulation and trading of commodities, but after the transition, rural residents are unable to meet their subsistence needs without income from trade and employment as labor. At the same time, without alternate employment opportunities, these new bearers of modern culture, who lacking ethical commitment to biodiversity conservation, will consume or sell the biodiversity around them and undermine their own insurance for future subsistence goods.

The loss of biodiversity will become a global crisis for the rural poor. For these people, the loss of biodiversity translates into loss of food, construction materials, medicine, fuel, and material inputs essential for their survival. The Third World population living below the poverty line is growing, and there are strong indications that millions of people will be unable to meet their basic needs through participation in the capitalist economy (World Bank 1990). The poor are becoming enculturated to rely on unreliable, state-supported subsidies (over which they have little control) instead of continuing to rely on managing their immediate natural environment for famine relief and health services. As they devour their survival resource base instead of preserving it to meet their future needs, they join other modern culture bearers–moving away from conservation.

In order for biodiversity to be retained, all actors with access to biodiversity–from rural poor to international corporations–must be kept as players at the conservation table. They must all be made stakeholders in long term conservation. This premise must underlie any bridge constructed to take biodiversity across the chasm before us. Building that bridge will require dynamic and responsible action addressing many elements:

- financial altruism and ethical commitments,
- public awareness and political will,
- empowerment of traditional conservation systems,
- experimentation with protected area systems (PASs),
- indicators and monitoring systems,
- policy and legislative reforms,
- user-group stewardship,
- guarantees of basic human needs,
- guarantees of tenure rights and due process, and
- sustained dialogue between all stakeholders.

Financial Altruism and Ethical Commitments

Perhaps most critical for conservation success will be firing and sustaining the altruistic impulse of private business–as stated by E.O.Wilson (see Forward, this volume). For it is private enterprise that must begin to internalize environmental externalities and pay the costs that someone will eventually have to pay. Assertions that ecotourism, park-generated revenues, microenterprises, and new industries based on biodiversity extraction will save biodiversity must be critically reexamined. On-the-ground observations generally do not support these assertions, and in fact indicate that these profit-making activities are accelerating the loss of biodiversity. Some examples include the commoditization of ivory, sandalwood, and many medicinal species, and coastal degradation accompanying tourist complexes. Capitalist markets do not reward those who practice conservation for the benefit of future generations (c.f., Oldfield 1989). Only cultural ethics, societal rewards, and economic incentives (that require someone to pay an attendant economic/financial cost) will be able to compete with the market for decision-making guidance.

Altruistic action by the "movers and shakers" of the global capitalist economy will be absolutely essential. In an advertising supplement entitled "The Environment: A Corporate Challenge" in Newsweek's June 17, 1991, issue, England's Prince Philip writes: "The connection between the management of industrial and commercial enterprises and the conservation of nature needs to be clearly understood. Companies are entitled to maximize profits . . . but not without regard for the consequences of their operations. Just as they take responsibility of the health and safety of their employees, or for the quality of their products . . . , managers must accept moral and practical responsibility for limiting damage to the natural environment." He goes on to state, "I am glad to say that recent years have seen the development of a remarkable partnership between business and the conservation organizations to achieve more effective protection of the natural environment. . . . I do not think it is too much to hope that every honest and responsible manager should have sufficient understanding of the facts of the natural environment to enable him to manage his company with a proper regard for the long-term well-being of the whole living world."

The ethical dimensions of decisions to conserve, or not conserve, biodiversity are being explored as part of the International Union for the Conservation of Nature and Natural Resources's (IUCN) effort to develop a new World Conservation Strategy–"Caring for the World: A Strategy for Sustainability" (Engel 1991). A central premise of the "Caring for the World" strategy is that "only a new worldview and morality can change the basic relation of people to the earth" (Engel 1991:9). The IUCN Ethics, Culture, and Conservation Working Group is working from the assumption that "(i)t is the challenge of the next decade to find and create ethics of biodiversity adequate to the 21st century" (Engel 1991:3). The Biodiversity Ethics Task Force of that Working Group has proposed that a new component, a "Biodiversity Ethics Project," be added to the Biodiversity Strategy Programme under which World Resources Institute, IUCN, and United Nations Environment Programme are developing the Biodiversity Strategy and Action Plan-BSAP–for presentation at the United Nations Conference on Environment and Development (UNCED–a.k.a. "Brazil 92"–the follow-on to the Stockholm conference in 1972, predicted to be the largest international conference ever held). This would add experts in ethics and religion to the BSAP team. Ideally, ethical commitments in the hearts and minds of corporate and government decision-makers will create the political will necessary to conserve biodiversity. These commitments should lead to the production of appropriate advertisements and subtle media messages to reinforce a modern conservation ethic in mainstream modern culture, just as modern consumerism is reinforced by the media today.

Public Awareness and Political Will

Building a bridge will require public awareness of biodiversity issues, and the political will to halt accelerating losses of biodiversity. The example of INBio in Costa Rica (Gámez 1991) holds out great promise as a model of what can be achieved through public awareness and political will. INBio is conceived as though Costa Rica were a corporation holding biodiversity and human resource assets and looking for investors. According to its director, Rodrigo Gámez, it is based on "the premise that biodiversity must offer Costa Ricans intellectual and economic rewards in order for Costa Ricans to conserve biodiversity" (?? Gámez, pers. comm.). Intellectual rewards come from the excitement of scientific study of nature, the same activities that promote public awareness and interest in biodiversity issues. INBio promotes the intellectual rewards of nature study at the levels of youth groups, community groups, parataxonomists, schools, and universities. Economic rewards will come from grants, royalties, and fees paid by First World research institutions, private industry, and governments. INBio is just beginning, and it is too early to judge its success. Its builders recognize that they have a difficult task before them. Involvement of indigenous people is in incipient stages, but there are indications that traditional economic and intellectual valuation of biodiversity may be incorporated into the INBio model. Deforestation continues, however, and INBio is in a race to reverse current trends.

Empowerment of Traditional Systems and Experiments with PASs

The bridge will require a commitment to building on the national heritage of traditional conservation systems, and a commitment to experimenting with modifications of PASs. PASs are commonly described as the "bullets and barbed wire system," because they are often implemented by a paramilitary park protection force. PAS success, under current design, rests heavily on the capacity and commitment of local guards to stop poaching, encroachment, and other "illegal" activities inside protected areas. Alternative approaches to protection of biodiversity are being explored, however. The World Wildlife Fund/U.S.'s global "Wildlands and Human Needs" program, the Woodlands Mountain Institute's Nepal and Tibet program in "Conservation Areas," and Wildlife Conservation International's efforts to develop "negotiated protected areas" in

Papua New Guinea are all examples of conservation organizations' efforts to work with local people to create appropriate incentives for conservation while meeting locally-perceived needs around protected areas.

Some new approaches attempt to strengthen traditional conservation systems. For example, Worldwide Fund for Nature (WWF-International) is documenting the unwritten environmental management policies of communities in Marovo Lagoon in the Western Solomon Islands, because "agents of economic development–be they government officers, investors, or development assistance agencies–are often unaware that such laws exist" (Baines 1991). In Western Samoa, private donors (including a pharmaceutical firm), the Swedish Society for the Conservation of Nature, Swedish International Development Agency (SIDA), biological researchers, villagers and Samoan chiefs have negotiated covenants to create marine and rainforest reserves on lands/sea still retained under customary tenure, regulation, and use (Cox and Elmqvist 1991). The covenants were negotiated under traditional Samoan law. Even the U.S. National Park Service is taking steps to accomodate indigenous customary tenure. For example, in 1988, a new National Park in American Samoa was established by means of a long term lease agreement between the U.S. National Park Service and villages who own the land under customary tenure (Cox and Elmqvist 1991). Villagers may continue to harvest medicinal plants, develop small gardens, and forage from reefs using traditional techniques. The Park Service is forbidden to allow construction of major tourist facilities or roads into the forest. An advisory committee including local chiefs will collaborate in developing this park, which was intitiated by Samoan petition to the U.S.Congress. Such developments are exciting, but protected areas created under indigenous people's terms are still rare among PASs worldwide.

Despite the promise of the biosphere reserve formula, local people are rarely involved in management decisions in biosphere reserves (see Gregg, Ch. 13, this volume). The O'odham (or Papago) peoples have long occupied lands that now fall into the Organ Pipe Cactus National Monument in Arizona, which was declared a biosphere reserve in 1978, and lands in the Sierra El Pinacate Protected Zone in Sonora, Mexico, which has been nominated for biosphere reserve status (Flores et al. 1990). The possibility of joining the two in a transnational border biosphere reserve has been actively discussed. In 1988, the Tohono O'odham Nation developed a position paper (Flores et al. 1990:26) which stated, "The Pinacate region is recognized as the aboriginal homeland and the actual place of creation of the Hia Ced O'odham, people who are recognized citizens of the Tohono O'odham Nation.. . . (T)he designation of the Pinacates as an international biosphere is of paramount interest to this Nation. Such a designation should provide (further protection) for the delicate and exquisite ecology . . . and is therefore supported by the Tohono O'odham Nation." But the statement goes on to note the the O'odham have concerns that their "human rights have been violated, . . . or at best not even considered, by the framers of such an internationl biosphere designation." It remains to be seen whether today's negotiations between the U.S., Mexico, and the O'odham Nation will lead to a path-breaking experiment in biosphere reserve management.

Indicators and Monitoring Systems

Another major element necessary to maintain biodiversity during modernization is the development of appropriate national-level indicators for use by decision-makers. Existing conservation data centers and global monitoring centers focus on species and habitats. But the language of Latin names and ecological terms, used by these data and monitoring centers, is unintelligible to decision-makers. Lists of Latin names not appeal to aesthetics, ethics, or economics. These systems have failed to monitor biodiversity through indicators important to either the global capitalist economy or domestic economies. The role local subsistence economies play in supporting the global capitalist economy has not been appreciated. There have been numerous calls to re-examine the pre-analytic bias of cost-benefit analyses used by economists to program development (e.g., Oldfield 1984, Daly and Cobb 1989). Repetto et al. (1989) and others have focused attention on reforming national income accounts to take into account the costs of environmental degradation for future commercial ventures. Attempts are also being made to incorporate intergenerational equity into decision-making formulae. But the impact of environmental degradation on present-day subsistence economies and the subsequent impact on economic growth have been ignored.

Agarwal's (1985) proposal that India should develop a Gross Nature Product indicator to measure biomass production is a major step toward integrating a concern for subsistence and biodiversity into national economic indicators. This measure would be an indicator for biomass production levels–and biodiversity levels–both of which are critical for survival and economic prosperity in rural areas. But economists have not yet heeded his suggestion. In the meantime, the costs of destruction of the subsistence base are undermining national economic growth. The Center for Science and Environment in India has concluded that loss of biodiversity in India is affecting nearly three-fourths of the rural population. But they cannot quantify the cost of that degradation, because the data base does not exist. No one is keeping records on the economic contributions of the periphery to the center, or the contributions of nature necessary for calculating a Gross Nature Product. What does it matter to the state that people can harvest needed items directly from nature? What would a Gross Nature Product indicator monitor? It monitors the services nature provides to the global capitalist economy. Nature frees capital that might have to be used for social security or unemployment insurance and makes it available for investment. It supports reproduction of the labor force and subsidizes a partially employed workforce. Globally ,loss of these services is having a negative impact on the capitalist economy.

In order to build a bridge to the future, nations should monitor traditional systems, and enact and enforce to support them, or create substitutes when they are threatened by modernization. For example, in Morocco, the traditional "aqudal" system mandates that forage species must not be destroyed during fuelwood collection on threat of a stiff fine (Geist 1988). Centuries of traditional management of argan forests in the Souss area of Morocco have resulted in semi-natural woodlands rich in indigenous species (Mellado 1989). But pressures to make quick profits by selling extracted products, e.g., firewood, have undermined traditional incentives to sustain the biological resource base. The cash economy has also altered traditional systems in other ways. For example, traditional rules and feedback mechanisms that regulated herd size in Morocco are being broken as urban rich and expatriates living in Europe invest in herds that run (cost-free) on their old community's rangelands. This occurs because investment in free-ranging livestock is viewed as safer than savings accounts in banks (Duvall, 1988), and communities have not instituted regulations requiring investments in community rangeland conservation. At the same time, donor attempts to improve rangelands have often failed, because they have ignored the social diversity and diverse tenure and management systems of herdspeople (Artz and Jamtgaard, 1986).

Modern, high-input agricultural systems with short productive lives have led to the abandonment of traditional agricultural systems, loss of biodiversity, and degradation of lands that were farmed sustainably for thousands of years. For example, Clarke (1991) has offered the term "agro-deforestation" to cover the ongoing loss of biologically rich, traditional agroforestry systems (e.g., Clarke 1991, Alcorn 1990, Peluso and Padoch 1991, Thannam 1988). Unsustainable monocropping of cash crops, plantations and pastures for short term gains are replacing traditional land use systems that have maintained biodiversity in densely inhabited landscapes. Islands of natural vegetation are dropping out of the traditional mosaics of managed forests and cash crops, and trees are being removed from agricultural fields to make way for mechanization of agriculture.

Traditional conservation can be maintained during the transition to a capitalist economy. For example, in Thailand, farmers are taking the initiative in defining ways that biodiversity can be conserved within a capitalist system (Lynch and Alcorn 1991). Rural colonists who had been converting forests into farmland for cash crops have made community-based decisions to conserve specific tracts of forest and even to allow cash crop fields to revert to forest, because they recognized that the diminution of forest had had a negative impact on their livelihood system. Communities have drawn up modern legal documents to formalize traditional rules and penalties regarding forest conservation. Mountain farmers ("hill tribes") have collaborated with university teams to analyze the ecological limits of their land use system under population growth pressures, and have consequently altered the spatial distribution of fields and crops to conserve water, soil, and natural forest. In the struggle to preserve forests, Thai farmers are joined by Buddhist monks (Figure E.1) who hope to prevent the destruction of community forests by government-sanctioned concessionaires who plan to plant eucalyptus plantations. The robes being tied on the trees in Figure E.1 warn any person who fells a tree that he will be guilty of an act equivalent to that of murdering a monk. This is but one example of how religion can support grassroots action to conserve nature during modernization.

Aquatic resource management systems are also placed in jeopardy by modernization. For example, in the Peruvian Amazon, fish stocks, once regulated by local patrons, have been declining because of overfishing to meet increasing market demands from urban Iquitos. In response, communities are taking the initiative to formalize community tenure to lakes. They are instituting regulations to prevent over-fishing, including taxes on commercial fishing (Coomes 1991). Yet, such community-based measures were taken centuries ago in Indonesia only to be undermined today by local government's movements to capture revenues for local elites. "The centralization of (comunally defined) rights over valuable resources excludes local fishers from control over systems of regulation which they themselves invented" (Bailey and Zerner 1991:7). In the Maluku Islands, this change in control has led to documented overexploitation of trochus (Bailey and Zerner 1991). On the basis of analysis of the situation in Philippines and Indonesia, Bailey and Zerner (1991:1) have urged

FIGURE E.1 Buddhist monks wrap saffron robes around tree trunks to assist Thai farmers in defending their community forest rights. (Photo: Project for Ecological Recovery, Bangkok, Thailand.)

that "government should resist the temptation to extend bureaucratic control over fisheries resources but rather should work in partnership with local management systems where they exist, and foster their extension into new areas wherever possible."

In situ conservation of traditional crop varieties can also be promoted during the development process. In many areas of the world, traditional farmers are still maintaining their traditional crop genetic heritage. Rice farmers in Bangladesh, for example, maintain ten thousand traditional rice varieties in complex systems complemented by production from over one hundred species in intensively managed dooryard orchards. Yet, outside donor agencies that pour billions into Bangladesh pay little attention to farmers, or ask their advice. Farmers are being pressured by incentives and disincentives to change their ways to those recommended by outsiders who are ignorant of local ecology or the true value of their crop genetic resources. They are haranged and disparaged for not modernizing and for holding back Bangladesh's development. The in situ conservation work being done, with no outsider-generated compensation, by farmers globally should instead be supported (see Nabhan 1989, Wilkes, Ch. 5, this volume). NGO-supported seed-savers' networks are beginning to encourage and support farmers' ongoing in situ conservation work in Africa, Southeast Asia, and Northern Mexico.

Policy and Legislative Reforms

Policy and legislative reforms will be necessary to protect forests by offering incentives for commercial wood production to replace harvesting from natural forests. Forests will also have to be recognized as part of farming systems. Farmers have long understood the ecological function of maintaining forest as part of their livelihood system. Outsiders have failed, however, to recognize that forests are part of the larger farming system, and believe that forests are unused, or are simply "frontier zones" for farmers. Farmers are now moving to demand that outsiders explictly recognize traditional tenurial rights to forests that are part of local farming systems and are demanding that the state support their rights to protect and manage their own forests. In Thailand, for example, farmers are putting up signs to tell outsiders that forests near their farms belong to the community of farmers (Lynch and Alcorn 1991). These grassroots efforts to conserve local biodiversity should be encouraged and supported.

User-Group Stewardship

User-group stewardship provides a model for action. Forest protection committees and community regulation of forests are common in many countries. Empowerment of such existing entities is preferred over creating new authorities. The role of the government should be to support the authority of these community bodies to take action against outsiders, poachers, and other trespassers who demand access to resources. For example, five states in India (West Bengal, Gujarat, Himachal Pradesh, Orissa, and Haryana) have set up local, joint management committees that have successfully protected forest areas in return for usufruct and management rights. In 1990, the Ministry of Environment and Forests endorsed their efforts and encouraged other Indian states to experiment with similar models. Changes in policy that allowed user-group stewardship have contributed to successful reforestation in Nepal (Fox 1991).

In many areas, forests alienated from communities by the state have been destroyed, and unproductive wastelands remain in their place. Grassroots groups are now concerned with "rebuilding nature" in these areas by planting trees. The vision of a nationally eminent Indian NGO–the Center for Science and Environment (CSE)–is that "each rural settlement must have its own clearly and legally defined environment to protect, improve, care for and use" (CSE 1990). These new community forests contain relatively few species, a faint shadow of what was once maintained in community forests. The Indian experience foreshadows what may happen on a global scale if action is not taken soon to empower local communities and legally recognize their traditional rights to manage their own resources.

Guarantees of Basic Human Needs

Ultimately, the battle to save biodiversity will be won or lost based on whether the global economy can meet the basic needs of all Earth's citizens. This requires addressing human needs–both physical and legal–in concert with establishing nature reserves.

Guarantees of Tenure and Due Process

Lack of clear tenure rights to resources and lack of due process when resources are alienated are also undermining natural resource conservation. This situation encourages management decisions aimed at maximizing short-term profits. This affects conservation in the general landscape as well as in protected areas. Millions of people are dependent upon biological resources on lands declared to be "public." In many cases, communities have lived on and used these lands for generations. Alienation of the land for use by those associated with the governing elite has reduced the effectiveness of traditional conservation systems. Government sanctioned destructive logging in community watershed forests where villagers are themselves fined for cutting any trees, for example, has repeatedly triggered public protests in many countries. The establishment–in terms of national laws or international agreements–of clear legal rights to natural resources held under traditional tenure systems will help to build the bridge for biodiversity as we enter the 21st century.

Promotion of due process is part of the larger human rights agenda which has an impact on biodiversity conservation. Lack of due process for those charged with crimes (e.g., hunting or cutting firewood in parks or reserves) hampers compliance with regulations. If due process is not available, people use informal systems to bribe their way out of the hands of local authorities which have seized them. Without due process and a fair and impartial system of justice, no one respects the regulations established to protect the "protected area"–neither government representatives nor citizens.

But the due process issue is about more than simply the enforcement of regulations. As Peter Brosius (1991) has so eloquently stated, "The issue of deforestation . . . is not simply economic, but it is laden with value and is profoundly moral as well." People at "the bottom" all over the world are talking about their ideas for managing their local environment in an ecologically sustainable manner, but their voices are not being heard by the top. People are willing to take on personal responsibility to restore natural resources or to manage biodiversity. They have seen the consequences of degradation of the natural resource base. They understand how and why they will suffer from further loss of biodiversity. But their action is often blocked by national or state level government policies and laws that deny them rights to take action. Denial of rights to take action is, in the end, an issue of legitimacy and respect. As Brosius (1991) argues: "The entire dialogue between Penan and the (Sarawak) government is about legitimacy.. . . Penan view the commoditization of the landscape as indicating an utter disregard for their humanity, a lack of respect.. . . Penan feel they have been treated not just with disrespect, but violated." (This sense of violation has in part been created by destruction of Penan grave sites by bulldozers and felling of trees traditionally incorporated into their *molong* tenure system.) He further observes, "The Penan defense of their land is cast in profoundly moral terms, with reference to the same types of self-evident moral truths that bind us all to our various societies.

They are trying to engage and participate in the larger moral community that is Malaysia.. . . It is this which has led to acts of resistance." Brosius's observations about the particular case of the Penan hunter-gather society's struggle to save their forests in Malaysia could apply to the ethical dimensions of almost any marginal group's struggle to maintain their rights to manage forests and lands anywhere in the world.

Sustained Dialogue Between All Stakeholders

The indigenous knowledge, intelligence and commitment of local people is not being used by the state; it is being ignored as though these people were invisible. At the same time, the ethical dimensions traditionally reflected in indigenous dialogue about biodiversity are also reflected in the words spoken during biodiversity discussions among modern policy-makers and biologists. But instead of ethical commitment to "Mother Earth," modern speakers' words reveal a different set of ethics. For example, internationally respected biologists, hired as consultants to World Bank and national governments to assist in biodiversity conservation plans, make value judgements when they make recommendations. Such biologists often appear to feel that biodiversity was created for them, for their study and pleasure. One such biologist recently stated in an international conference on people and forests, "Many of you have been recommending that shifting cultivators be allowed to remain in reserves. How would you like to have a shifting cultivator in your rock garden?" Such a statement implies that reserves are like rock gardens maintained by property owners for aesthetic enjoyment (i.e., reserves are for the enjoyment of biologists). At the same conference, another eminent botanist asserted that rattan collectors have no right to be in reserves, and that the biodiversity in reserves does not belong to the rattan collectors, or to the country in which the reserve is found. Instead, he said, the reserve belongs to the world. He went on to acknowledge that, "the beneficiaries of conservation should pay rent, but the world isn't used to the idea yet." Longterm biodiversity conservation will only be achieved when the voices of the "bottom" and the ethical vocabulary of traditional conservation become part of the dialogue that develops working solutions which meet the needs of all stakeholders–not just the elites.

Solutions to the global biodiversity crisis will require long term commitment over many human generations and centuries. But the conservation systems we attempt to put into place must have built-in flexibility to evolve under changing economic and cultural conditions, as well as the changing climatic conditions–a topic that is so much under discussion today. Future humans will nurture these systems if they have a moral commitment to both biodiversity and their fellow human beings.

CONCLUSIONS

As Raymond Dasmann emphasized in Chapter 1, it will take humility, patience, and time to link conservation with development. It will also take a tireless commitment to evaluate what works and what doesn't work--an evaluation that must give equal weight to the assessments of local people and outside experts. It will require continuing dialogue between rural poor and urban rich who share a desire to conserve biological diversity for future generations. Perhaps most importantly, it will require nurturing an ethical commitment to biodiversity conservation and sustainable development--not just sustained economic growth as measured by GNP and GDP.

No crystal ball or computer model can predict what lies ahead. No biodiversity futures are being traded in the futures exchange. Few spot markets exist to monitor changes in biological diversity due to economic transactions (Oldfield 1984). Two likely candidate futures demand two different adaptations for biodiversity conservation. If we succeed in halting the population growth that threatens to overload Earth's resource base, we must at the same time succeed in establishing the means and ethical motivation to conserve biodiversity as part of the economic development process linked with capitalism. If, on the other hand, population growth overwhelms the carrying capacity of Earth under the current "growth economy," or if the global capitalist system is dissolved through resource wars, we must have have succeeded in securing the archaic conservation ethic that will motivate people to defend biodiversity as they did for thousands of years prior to the global penetration of capitalism. In either case, only an ethical commitment will secure the future of biodiversity.

We humans of the First World will never be able to save tropical biodiversity ourselves through direct action, but we can have an impact through three actions at home. We can alter the global, modern culture that shapes actions in the Third World by publicly and privately urging modern capitalists, wherever they operate, assume an ethical responsibility toward the Earth. Legislation, public awareness campaigns, and boycotts are among the tools to achieve this end. Second, we can insist that "development" assistance of the bilateral and multilateral development agencies and banks (such as A.I.D. and World Bank) must include financial assistance to strengthen an ethical commitment to biodiversity conservation. Third, we can strive within our jobs, from our positions within private sector business, international conservation organizations, or as individual concerned citizens, to ensure that modern conservation planning is based on real, long term local participation that includes the voices of indigenous peoples who live in nations rich in biodiversity. Indigenous people are frequently not even recognized as citizens of the countries where they live, and their human rights are often ignored. Action taken by outsiders is frequently the critical catalyst whereby indigenous people are offered opportunities to participate in their own nation's life, or to contribute their cultural heritage toward modern biodiversity conservation in the countries where they reside.

The challenge is clear. The need for action is urgent, but action being taken under the current conservation paradigm is not achieving success. The bridge across the transitional chasm to the year 2100 awaits construction. And conflicting priorities hamper cooperation. It's time to construct a solid bridge—from the shared interests and strengths of the rural poor, the ruling classes, the Third World and the First World. Success will require forging substantive, cooperative links in the tension-filled vacuum between the interests of the governing elites and the interests of the public good. The hour is late. The chasm lies before us. We have no time to waste!

REFERENCES

Abramovitz, J.N. 1989. *A Survey of U.S.-Based Efforts to Research and Conserve Biological Diversity in Developing Countries.* World Resources Institute, Washington D.C.

Agarwal, A. 1985. Human-nature interactions in a Third World country. Fifth Annual World Conservation Lecture, WWF/UK, London.

Alcorn, J.B. 1989a. An economic analysis of Huastec Mayan forest management. Pages 182-206 *in* J. Browder (ed.), *Fragile Lands of Latin America: Strategies for Sustainable Development.* Westview Press, Boulder, Colorado.

Alcorn, J.B. 1989b. Process as resource: The agricultural ideology of Bora and Huastec resource management and its implications for research. Pages 63-77 *in* D.A. Posey and W. Balee (eds.), *Natural Resource Management by Indigenous and Folk Societies in Amazonia.* New York Botanical Garden, Advances in Economic Botany Series, Bronx, New York.

Alcorn, J.B. 1990. Indigenous agroforestry systems in the Latin American tropics. Pages 195-210 *in* M. Altieri and S. Hecht (eds.), *Agroecology and Small Farm Development,* CRC Press, Boca Raton, Florida.

Alcorn, J.B., C. Jensen, S. Pfeifer, C. Potter, and M. Symington. 1989. Working paper on: "The Biodiversity Conservation Strategy of the U.S. Agency for International Development". Prepared for the AID Working Group on Biological Diversity. Mimeo. AID, Washington, D.C.

Alcorn, J.B. and A. Molnar. 1990. Deforestation and forest-human relationships: What can we learn from India? Paper presented at the "Tropical Forest Ecology, the Changing Human Niche, and Deforestation," Symposium. American Anthropological Association, New Orleans, Louisiana.

Alcorn, J.B. and A. Molnar. 1991. In Press. Deforestation and forest-human relationships in India. *In* L. Sponsel, R. Bailey, and T. Headland (eds.), *Human Adaptation to Tropical Forests and Deforestation.*

Anonymous. 1989. "100 villagers take over forest plot of MP's wife." Bangkok Post July 19, 1989.

Artz, N.E. and K. Jamtgaard. 1986. The use of cluster analysis in distinguishing land-use patterns in a traditional Moroccan pastoral system. *In* E.E. Whitehead, C.F. Hutchinson, B.N. Timmerman, and R.G. Varady (eds.), *Arid Lands: Today & Tomorrow.* Westview Press, Boulder, Colorado.

Bailey, C. and C. Zerner. 1991. Role of traditional fisheries resource management systems for sustainable resource utilization. Report presented at Forum Perikanan II, Perikanan Dalam Pembangunan Jangka Panjang Tahap II: Tantangan dan Peluang, June 18-21, Sukabumi, West Java.

Baines, G.B.K. 1989. Traditional resource management in the Melanesian South Pacific: A development dilemma. *In* F. Berkes (ed.), *Common Property Resources: Ecology and Communty-Based Sustainable Development,* Belhaven Press, London.

Baines, G.B.K. 1991. Asserting traditional rights: Community conservation in Solomon Islands. *Cultural Survival Quarterly* 15(2):49-52.

Center for Science and Environment (CSE). 1990. Towards Green Villages. CSE, New Delhi, India.

Clark, C.W. 1973. Project maximization and the extinction of animal species. *Journal of Political Economy* 81(4):950-961.

Clarke, W.C. 1991. Agro-deforestation in Melanesia. *Cultural Survival Quarterly* 15(2):45-48.

Coomes, O.T. 1991. Rain forest extraction, agroforestry, and biodiversity loss: An environmental history from the Northeastern Peruvian Amazon. Paper presented at the Latin American Studies Association Annual Meetings, Washington, D.C.

Cordell, J. 1991. Negotiating sea rights. *Cultural Survival Quarterly* 15(2):5-10.

Cox, P.A. and T. Elmqvist. 1991. In Press. Indigenous control of rain forest preserves: An alternative strategy for rain forest preservation. Ambio.

Daly, H.E. and J.B. Cobb, Jr. 1989. *For the Common Good: Redirecting the Economy Toward Community, the Environment, and a Sustainable Future.* Beacon Press, Boston, Massachusetts.

Duvall, L. 1988. The status of biological resources in Morocco. USAID/Morocco, Rabat.

Engel, J.N. 1991. The role of ethics, culture, and religion in conserving biodiversity. Paper presented at the Pacific Science Congress, May 26-June 2, Honolulu, Hawaii.

Flores, M., F. Valentine, and G.P. Nabhan. 1990. Managing cultural resources in Sonoran Desert Biosphere Reserves. *Cultural Survival Quarterly* 14(4):26-30.

Fox, J. 1991. Firewood, fodder, and grazing: Collection of non-timber forest products in a Nepal village between 1980 and 1990. Paper presented at the Pacific Science Congress, May 26-June 2, Honolulu, Hawaii.

Gamez, R. 1991. Costa Rica's national Biodiversity Institute: Biodiversity working sustainably for society. *Diversity* 7:79-81.

Geist, A. 1988. Kitchen "ranges": Household energy systems in rural southern Morocco. USAID/Morocco, Rabat.

Groenfeldt, D., J. Alcorn, S. Berwick, D. Flickinger, and M. Hatziolos. 1990. Opportunities for eco-development in buffer zones. Report to USAID/India. World Wildlife Fund, Washington, D.C.

Guha, R. 1983. Forestry in British and post-British India. *Economic and Political Weekly,* October 29, 1983, pp. 1882-1896; and November 5-12, 1983, pp. 1940-1947.

Hurst, P. 1990. *Rainforest Politics.* Zed Books, Ltd., Atlantic Highlands, N.J., and London.

Lohman, L. 1991. In Press. Who defends biological diversity? Conservation strategies and the case of Thailand. *The Ecologist.*

Lynch, O.J. 1990. Whither the people: Demographic, tenurial and agricultural aspects of the Tropical Forestry Action Plan. World Resources Institute, Washington, D.C.

Lynch, O.J. and J. Alcorn. 1991. *Empowering Local Forest Managers: Towards More Effective Recognition of Tenurial Rights, Claims, and Management Capacities of People Occupying "Public" Forest Reserves, i.e., Paa Sangin, in Thailand.* World Resources Institute, Washington, D.C.

McKibben, B. 1989. *The End of Nature.* Doubleday, New York.

Mellado, J. 1989. S.O.S. Souss: Argan forest destruction in Morocco. *Oryx* 23:87-93.

Menagh, M. 1991. The greening of management. Page 4 of advertising supplement, The Environment: A Corporate Challenge. *Newsweek* CXVII(24).

Nabhan, G.P. 1989. *Enduring Seeds.* North Point Press, San Francisco.

Oldfield, M.L. 1984. *The Value of Conserving Genetic Resources,* U.S. Department of the Interior, Washington, D.C.

Panayotou, T. 1990. *The Economics of Environmental Degradation: Problems, Causes, and Responses.* Development Discussion Paper No. 335. Harvard Institute for International Development, Boston, Massachusetts.

Peluso, N.L. and C. Padoch. 1991. Changing resource rights in managed forests of West Kalimantan. Paper presented at the "Interactions Between People and Forests in Kalimantan" Conference, June 21-23, New York Botanical Garden, Bronx.

Quarto, A. 1990. The Pehuenche and the monkey-puzzle tree. *Cultural Survival Quarterly* 14(4):21-25.

Randall, A. 1986. Human preferences, economics, and the preservation of species. Pages 79-109 *in* B.G. Norton (ed.), *The Preservation of Species: The Value of Biological Diversity.* Princeton University Press, Princeton, New Jersey.

Repetto, R., W. Magrath, M. Wells, C. Beer, and F. Rossini. 1989. Wasting assets: Natural resources in the national income accounts. World Resources Institute, Washingon, D.C.

Ruddle, K. and R.E. Johannes, eds. 1985. The Traditional Knowledge and Management of Coastal Systems in Asia and the Pacific. UNESCO Regional Office for Science and Technology for Southeast Asia, Jakarta, Indonesia.

Scott, J.C. 1976. *The moral economy of the peasant: Rebellion and subsistence in Southeast Asia.* Yale University Press, New Haven, Connecticut.

Thaman, R. 1988. Fijian agroforestry. *In* J. Overton (ed.), *Rural Fiji,* University of the South Pacific, Suva, Fiji.

Tobin, R. 1990. *The Expendable Future: U.S. Politics and the Protection of Biological Diversity.* Duke University Press, Durham, North Carolina.

VIKSAT/CEE. 1990. Buffer zone restoration in India: Issues and approaches. Nehru Foundation for Development, Ahmedabad, India.

Washington Post. 1991a. The owl and the law. Excerpts from the opinion of U.S. District Judge William L. Dwyer in Seattle. The Washington Post Section A, page 19, June 25, 1991.

Washington Post. 1991b. Owls, trees, and loggers. Editorial Page. The Washington Post Section A, page 18, June 25, 1991.

World Bank. 1990. *Poverty: World Development Report, 1990.* Oxford University Press, Washington, D.C.

World Commission on Environment and Development. 1987. *Our Common Future.* Oxford, New York.